Ocean Engineering & Oceanography

Volume 13

More information about this series at http://www.springer.com/series/10524

Binbin Pan · Weicheng Cui

Multidisciplinary Design Optimization and Its Application in Deep Manned Submersible Design

浙江科学技术出版社
ZHEJIANG SCIENCE AND TECHNOLOGY PUBLISHING HOUSE

Springer

Binbin Pan
Shanghai Ocean University
Shanghai, China

Weicheng Cui
Marine High-tech Park
Shanghai, China

School of Engineering
Westlake University
Hangzhou, Zhejiang Province, China

ISSN 2194-6396 ISSN 2194-640X (electronic)
Ocean Engineering & Oceanography
ISBN 978-981-15-6454-3 ISBN 978-981-15-6455-0 (eBook)
https://doi.org/10.1007/978-981-15-6455-0

Jointly published with Zhejiang Science and Technology Publishing House
The print edition is not for sale in China (Mainland). Customers from China (Mainland) please order the print book from: Zhejiang Science and Technology Publishing House.

This Springer imprint is published by the registered company Springer Nature Singapore Pte Ltd.
The registered company address is: 152 Beach Road, #21-01/04 Gateway East, Singapore 189721, Singapore

Preface

With the rapid expansion of the population on the earth, the supply of land resources is approaching its limit. Countries have shifted their focus of economic development to the ocean. This is because the vast ocean, which occupies more than 2/3 of the total area of the earth surface, is rich in marine chemical resources, seabed mineral resources, marine power resources, marine biological resources, etc. Human beings will step into the marine economic era in the twenty-first century. Ocean development will form a number of emerging industries, such as marine oil and gas industry, marine chemical industry, deep-sea mining industry, etc. In order to develop and utilize the ocean better, it is necessary to build the deep-sea equipment system in time to meet the needs of deep-sea mineral resources investigation, deep-sea comprehensive investigation and research, deep-sea scientific research, marine resources exploration, and the maintenance of marine rights and interests. Deep-sea equipment system can include all kinds of submersibles for exploration and operation, for example, Human Occupied Vehicle (HOV), Remotely Operated Vehicle (ROV), Autonomous Underwater Vehicle (AUV), and Autonomous Remotely Operated Vehicle (ARV) or Hybrid Remotely Operated Vehicle (HROV); the surface support mothership carrying the submersibles for operation; the general or special deep-sea operation tools for underwater exploration and operation; the submarine observation station for long-term observation of the marine environment; the deep-sea laboratory for scientific research under the real deep-sea environment conditions; the large deep-sea workstation for deep-sea environment observation, scientific test, deep-sea resource development, etc. Among them, at least one HOV, ROV, AUV, and related deep-sea operation tools are the most basic equipment of deep-sea exploration system. This is because these three types of submersibles have their own characteristics and complementary functions. In some cases, they need to work together and rescue each other in case of accidents. If there is only a single submersible, its use is limited in function and there is an "operational risk", which will greatly reduce the efficiency of the submersible. Since 1980s, China began to carry out research on unmanned submersibles and manned submersibles. By 2012, "Jiaolong" manned submersible, which has the deepest submergence depth of operational manned submersibles in the world, was successfully developed, and finally entered the international club of developed countries in deep submersible technology. During the "12th Five-Year Plan", the Ministry of Science and Technology approved and supported three economical and practical 4500 submersibles. The development of these medium-depth ROV, AUV, and HOV aims to digest and absorb the introduced technology in the development process of Jiaolong and consolidates the technical foundation.

During the 13th Five-Year Plan, the Ministry of Science and Technology will fully support the development of 11000 m deep-sea manned and unmanned submersibles, making the development of deep-sea equipment technology enter a golden age. In order to further speed up the development of China's manned deep diving technology, I established the first Hadal Science and Technology Research Center in China's universities since March 2013 under the strong support of the leadership of Shanghai Ocean University. In November 2014, this center was approved as "Shanghai Engineering Research Center of hadal science and Technology (Preparatory)". We recruited a technical team to develop a movable laboratory for Hadal Science and Technology which is composed of three 11000 m landers, a 11000 m hybrid unmanned submersible (ARV), a 11000 m manned submersible (HOV), and a mothership of 4,800-ton displacement. At the same time, some ocean scientists are recruited to specialize in hadal sciences, hoping to climb the peak of manned deep diving technology and make positive contributions to fill the gap of hadal sciences in China. We work closely with Shanghai Rainbowfish Ocean Technology Co., Ltd. to adopt the new mode of "private funds + state support" and name the whole project "Rainbowfish challenging the Challenger Deep". With the supports from Shanghai local government and all walks of the social society, the whole project is progressing smoothly.

The hadal science and technology mobile laboratory concept of the "Rainbow Fish hadal Challenge" program

As an important deep-sea scientific research and operation equipment, the deep-sea manned submersible integrates the most advanced deep-sea technology in materials, mechanics, machinery, control, acoustics, optics, electricity, and other disciplines. In order to coordinate the balance between multiple disciplines and subsystems in the design stage of the submersible, Multidisciplinary Design Optimization (MDO) theory is needed to deal with the interdisciplinary interaction and coupling. Realizing the importance of multidisciplinary design optimization theory, I have been guiding graduate students to carry out theoretical and application researches since 2002. A large number of theoretical and applied researches on multidisciplinary design optimization have been carried out in the design of Jiaolong deep manned submersible and Pan Binbin's doctoral dissertation has systematically combed and integrated the research results of my research group for more than 10 years. On the basis of Dr. Pan Binbin's thesis, the book has made a lot of modifications to the systematization and readability. The publication of this book has been greatly supported by Zhejiang Science and Technology Press. They have made great efforts. We sincerely thank them for their help! I hope that the publication of this book can play a supporting role in the coming golden development period of deep-sea equipment technology.

Shanghai, China
July 2017

Prof. Weicheng Cui
Chair Professor, Westlake University
and Adjunct Professor of Shanghai Ocean University

Contents

Symbols and Acronyms

AP	Approximation, 近似
AP-RBMDO	Approximation-based RBMDO, 基于近似解耦的 RBMDO
AU	Aleatory Uncertainty, 偶然不确定性
CMC	Crude Monte Carlo, 原始蒙特卡洛法
CO	Collaborative Optimization, 协同优化
CO-RBMDO	Collaborative-Optimization-based RBMDO, 基于协同优化解耦的 RBMDO
CPT	Classical Probability Theory, 经典概率论
DLRBDO	Double Loop RBDO, 双循环 RBDO 算法
EU	Epistemic Uncertainty, 认知不确定性
FORM	First-Order Reliability Method, 一次二阶矩法
GPT	Generalized Probability Theory, 广义概率论
IBDO	Interval-Based Design Optimization, 基于区间分析的设计优化
IPT	Interval Probability Theory, 区间概率论
IS	Importance Sampling, 重要性抽样
MDF	Multidisciplinary Feasible Method, 多学科可行算法
MDF-RBMDO	基于多学科可行算法的 RBMDO
MDO	Multidisciplinary Design Optimization, 多学科设计优化
NU	Numerical Uncertainty, 数值不确定性
PBDO	Possibility-Based Design Optimization, 基于可能性的设计优化
PMA	Performance Measure Approach, 性能估计法
RA	Reliability Analysis, 可靠度分析
RBD	Reliability-Based Design, 基于可靠性的设计
RBDO	Reliability-Based Design Optimization, 基于可靠性的设计优化
RBMDO	Reliability-Based Multidisciplinary Design Optimization, 基于可靠性的多学科设计优化
RIA	Reliability Index Approach, 可靠度指标法
RSM	Response Surface Method, 响应面法
SFSORA	Safety-Factor-Based SORA, 基于安全系数的序列优化可靠度分析法

SLRBDO	Single Loop RBDO, 单循环 RBDO 算法
SORA	Sequential Optimization and Reliability Analysis, 优化可靠度分析交替进行法
SORM	Second-Order Reliability Method, 二次二阶矩法
UBD	Uncertainty-Based Design, 基于不确定性的设计
UBDO	Uncertainty-Based Design Optimization, 基于不确定性的设计优化
UUHT	Unifying Uncertainty Handling Theory, 统一不确定性信息描述理论
WJF	Welding Joint Factor, 焊接系数

Chapter 1
Introduction

1.1 Difficulty of the Design of Large Complex Engineering Systems

This book is mainly aiming to demonstrate the application of multidisciplinary design optimization theory in the design of large complex engineering systems like deep sea human occupied vehicle (HOV). As an important phase of the lifecycle of an engineering project, engineering design means the activity to provide the design document and drawing base on solid technical calculation, evaluation and estimation for the implementation of the engineering project. Engineering design is an important process to conduct the overall plan and intention of implementation of the project, it also can be considered as the way to turn science/technology into actual productivity. That means during the engineering design, the balance of technology and cost has to be determined, thus engineering design will heavily affect the total cost of the engineering project. As described, design is essentially the process to create the idea and plan of product with human scientific knowledge and technical skill, and it is involved in almost all human activities. The cost of the engineering design itself usually is just a small proportion of the total cost of the product (8–15%), but it has a decisive influence on the advancement and competitive power of the product, it usually also determines up to 70–80% of the manufacturing cost and the following service cost of the product. Therefore, engineering design is known as the pillar of modern industrial civilization, the core part of industrial innovation, and the flagship of modern social productivity. The engineering design level and ability represents the industrial innovation and competitive power of a country or a region. A suitable design will make the cost estimation and control of an engineering project much easier.

The first feature of large complicated engineering system is its large scale that a great amount of people are involved. For better management and higher efficiency,

B. Pan and W. Cui, *Multidisciplinary Design Optimization and Its Application in Deep Manned Submersible Design*, Ocean Engineering & Oceanography 13,
https://doi.org/10.1007/978-981-15-6455-0_1

the large complex engineering system is usually divided into several subsystems and subsystems can be divided into sub-subsystems…such division continues until the engineering design are divided into very specific design jobs that can be finished by one designer.When there comes a engineering project, the subsystem division of the project has to be carried out base on the specific engineering target and the ability of current engineering design team, rather than simply copy the division experience of other teams. Therefore, the first problem at front of the chief designer of large complex engineering system is how to accomplish a subsystem division that matches the specific engineering problem and fully considered the characteristics of the design team. After the subsystem division, the second problem comes to the chief designer is to check the coupling interfaces of different subsystems, sub-subsystem etc., and all interface of subsystems and sub-subsystems etc. has to be complete and no repeat. But it is very hard to make sure that all coupling interfaces of subsystems have been considered without a systematic algorithm. Once one or more coupling of subsystems or sub-subsystem etc. are improperly neglected, the whole design may have to rework when it is found that the coupling is unneglectable. This problem will cost many repeat design work and lower the design efficiency, it usually also lead to a low level design.

In order to accomplish a suitable subsystems division of a complex engineering system and effectively manage the coupling between subsystems (sub-subsystems etc.), Liu Zhengyuan et al. learn the design experience of the United States Air Force that gradually decomposes the system from the top to the bottom, and propose an engineering system design algorithm call "four-element algorithm" during the development of the "8A4 underwater vehicle" in 1989. This method has been further improved during the development of the "Jiaolong" manned submersible and other underwater engineering projects (Cui et al. 2008). After the successful application of this algorithm in many engineering projects, designers have realized that it is a powerful tool to decompose large complex engineering systems, clarify the composition of the system, and straighten up the relationships of subsystems. In the four-element algorithm, the design of the large complex engineering system is first divided into subsystems designs according to the work and discipline division of the design team, and the subsystem design is further subdivided into sub-subsystems designs. And during the division process, the responsibility of each subsystem and sub-subsystem is bound to a specific person. That is, the division of system, subsystems and sub-subsystems is related to the people in the design team directly, the subsystem will not even exist if there is no one taking the responsibility. For example, in the design of the "Jiaolong" HOV, the system chief designer takes the technical responsibility of the design of the whole system, the subsystem chief designers take the responsibility of the corresponding subsystem, the component designers take the responsibility of the corresponding component. After all subsystems, sub-subsystems and the corresponding responsibility are determined during the system division, the designers at different positions can start their design jobs, but as described, subsystems and sub-subsystems are not independent most of the time, there usually are many relations or coupling between different subsystems(sub-subsystems) like design constrains or requirements from other subsystems (sub-subsystems). The four-element algorithm then summarizes the design of the subsystems and sub-subsystems into four elements: input, output, support, and constraints.

(1) Input: It is the requirement of the overall system and other subsystems (sub-subsystems), It is the design precondition of current subsystem (sub-subsystem).
(2) Output: It is the design result of current subsystem (sub-subsystem). It must meet the requirements of the input and the limitations of the supports and constraints.
(3) Support: Requirements for the system and other subsystems(sub-subsystem) to accomplish the design of current subsystem(sub-subsystem).
(4) Constraints: The limitation from the system and other subsystems(sub-subsystems) that need to be considered during the design of current subsystem(sub-subsystem).

The steps for the preparation and implementation of the four element algorithm is as follows:

(1) The system chief designer divides the large complex engineering system into subsystems according to the work and discipline division of the design team and cooperating organization, and determined the responsibility of each subsystem(bound to specific person).
(2) The four elements of the respective subsystems are preliminarily prepared by the subsystem chief designer, and the subsystems that can be further decomposed are divided into multi sub-subsystems according to the requirements of the development task and the characteristics of the design team. During the further division, the responsibility of each sub-subsystem needed to be related to the corresponding designer. After that, the four elements of each sub-subsystem and components will be prepared by the responsible designers.
(3) The system chief designers then coordinates the four elements of each subsystem (sub-subsystem), and makes sure that the interfaces of four elements of all divided subsystems (sub-subsystems) become a closed loop, i.e. clarify the relationship of different subsystems (sub-subsystems).
(4) In the design process, if the subsystem (subsystem) designer proposes new problems related to other subsystems(sub-subsystems) that needed to be solved, Adjustment of the four element table will be made accordingly, and the Adjustment of the four-element table will be evaluated and implement by the system chief designer.

The four-element system method has the following two important characteristics:

(1) Closed-loop and internal dynamic balance. The division of the system is Closed-loop, and subsystems (sub-subsystems) may be added and merged at different design stages. but the four elements table of the entire system must be closed-loop at any time. That is, the output of each subsystem (sub-subsystems) must meet the requirements and limitations of overall system and other subsystems (sub-subsystems). At the same time, the support of each subsystem (sub-subsystems) must also be supplied by the overall system and other subsystems (sub-subsystems). Besides that, the four elements table will be changed

according to the change of the system division, it means the four elements table keeps a dynamic balance.

(2) Non-uniqueness. The four elements table of a system is not unique. Different design schemes will have different support requirements, for example, some input can be treated as the output of the subsystem, and some constraints can be treated as input to the subsystem. Therefore, the input, output, support, and constraints of the subsystem (sub-subsystem) are related to the design jobs and experience of the designers. Although the four elements tables of the same engineering project is not unique, a determined four element table still need to be proposed to clarify the system composition, rationalize the relationship between the systems, and lay the foundation for planning arrangements and project node control.

The second feature of the engineering design of a large complex engineering system is that there are many disciplines involved, which exceeds the knowledge that a single person can master and requires the cooperation of many designers who are professional at different disciplines. Hence, a relatively feasible design process is needed, that is, to a particular engineering system designing problem, a working sequence in which different designer with professional skills can carry out their own design work without conflict and help each other need to be proposed. Take the traditional ship design as example, the design procedure follows the design spiral of: main dimension design–hull shape design–engine power design–space design–tonnage design–draught design- general layout–endurance design–seakeeping design–structure design- cost estimation… Designers sequentially design subsystems along the design spiral, and coordination among subsystems is based on the data of the subsystems that have been completed in previous design cycle, existing accumulated data, and designers' experience. When the performance or cost of the design cannot meet the design requirement, the designer needs to perform new design cycles along the design spiral for better performance and cost. Obviously, the major disadvantages of design spiral algorithm is it's long design cycle and low efficiency, and the coupling or conflict between different subsystems is determined by the designer basing on their experience, intuition, limited analysis and testing, the designer himself cannot be sure whether the balance of subsystems is good or bad. Thus design spiral method usually lead to a design that just meets the requirements rather than an optimum design.

How much system improvement can be achieved by the proper coordination and balance of subsystems? Chen (2001) answered this question vividly with an example of an aircraft design. In the mid-1960s, the Soviet MiG-25 fighter set 8 new flight speed world records, 9 flight altitude world records and 6 climbing time world records, and in the 4th Middle East War of 1971, the MiG-25 aircraft overshadows the Western Advanced "F-4 ghost" fighter aircraft with "Sparrow-1" air-to-air missiles for such a long time that lot's of Western military experts have speculated with uneasiness that such aircraft must have adopted some kind of epoch-making new technology. On September 6th, 1976, the Soviet pilot, Viktor Belenko, defected to the Hokkaido Airport in Japan on the mysterious MiG-25. Then MiG-25 was disassembled and every part was examined by Japanese and the USA military and aircraft expert. However, Western military experts found that there was no breakthrough technology in the MiG-25. Some of the aircraft's parts and designs are even backward, but the normal parts and traditional design was evenly combined into an effective man-machine system, and brought out surprisingly combat power. Therefore, there are comments that: "…Steel structures have made the Western expert surprise, and the backward electron tube technology has enabled the Western to deeply understand the lag of electronic technology of Soviet… But the Western still marveled at the superiority of the system integration technology of Soviet (Chen 2001). This example vividly illustrates the importance of subsystem coordination. Some aircraft designers already were aware of this since the early 1970s. Ship designers have also recognized that the optimal design always depended on the balance between subsystems rather than the advanced technology since the 1980s.

Design methods has been developed through three stages. The most primitive engineering design method is the serial design method like the design spiral. In the serial design method, all subsystems are sorted into a sequence and the design work are carried out serially, it means only one individual subsystem is designed at any stage of the design and the design of the next subsystem is performed after the design of current subsystem is finished. This serial process of serial design not only ignores the influence of the subsequent subsystem on the upstream subsystem, but also makes the design of the subsequent subsystem has to wait until the design of the upstream subsystem complete, And multiple rounds of complete serial design cycles are required to reach a design solution that meets the requirements. All these disadvantages greatly prolong the whole engineering design process. With the development of technologies of computer aided design and digital virtual design, people have proposed a design method called concurrent engineering (CE), the core principle of CE is to decompose the system design task into subsystems design that can be executed independently at the same time, so that each subsystem can be designed in parallel rather than in serial, which significantly improves design efficiency and shortens design process. Since then, engineers have been looking for better ways to solve the coordination of subsystems. Multidisciplinary Design Optimization (MDO) is one of such design algorithm that is gradually developed on that demand. It enables engineers to find an optimal design base on the full use of current available technologies and resources and the reasonable coordination of all subsystems.

It can be found that the idea of the four-element algorithm developed in the field of underwater engineering and applied to the design of "Jiaolong" HOV can be treated as the simple version MDO algorithm. The "Jiaolong" HOV project can also be considered as a demonstrated application of MDO theory in underwater engineering.

1.2 Basis of MDO Theory

MDO integrates most of the techniques of traditional optimization theory and proposes new concepts to handle the coupling of multiple disciplines. Therefore, the optimization related basic concepts of MDO are similar to traditional optimization theory, and the parallel execution idea of MDO also extend the Foundation of traditional optimization theory. The basic concepts of MDO will be introduced in this section, then the rise and development history of MDO will be briefly reviewed.

1.2.1 Basic Concept

1.2.1.1 MDO

Sobieszczanski-Sobieski J. is considered to be the founder of MDO theory. He defines MDO as: multidisciplinary design optimization is a design algorithm that considers the interdisciplinary interactions within the system (Sobieszczanski-Sobieski 1995). In MDO, the design of a discipline (subsystem) affects not only itself, but also has an important impact on the performance of the overall system.

The Multidisciplinary Optimization Branch (MDOB) of the National Aeronautics and Space Administration's Langley Research Center defines MDO as: multidisciplinary Design optimization is a method that aims to design complex engineering systems and their subsystems and explore the collaborative mechanisms of subsystems.

1.2.1.2 Discipline

In multidisciplinary design optimization, the concept of discipline refers to the relative independent design modules or subsystems. Note that the relative independence here does not mean that there is no connection between disciplines, there is usually exchange of design parameters between disciplines(subsystems) in MDO, which is called coupling or interdisciplinary intersection of disciplines. In Fig. 1.1, x_1 and x_2 are the input variables (design variables) of discipline 1 and discipline 2 respectively, and z_1 and z_2 are the output variables of the two disciplines (also called the state

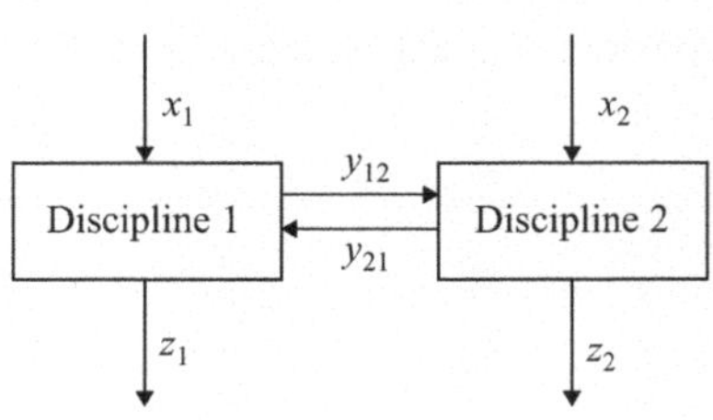

Fig. 1.1 Coupling of disciplines

variables of the subsystems), and y_{12}, y_{21} are the coupling parameters or coupling variables between discipline 1 and discipline 2. It can be found that the concept of discipline in multidisciplinary design optimization is not exactly the same as the traditional concept of discipline, disciplines of MDO is more close to the design of subsystems.

1.2.1.3 Subsystem Analysis

Subsystem analysis refers to the analysis and calculation process of the subsystem to obtain the output parameters according to the given input parameters. Take Fig. 1.1 as example, subsystem analysis is the process that discipline 1 calculates z_1 and y_{12} according to the input variables x_1 and y_{21}, and the process that discipline 2calculates z_2 and y_{21} according to the input variables x_2 and y_{12}. Compared to the subsystem analysis in traditional optimization design, the input parameters of the MDO subsystem analysis of the the i'th subsystem includes both the design variables x_i of the subsystem itself and the coupling variables y_{ji} from other subsystems (j represents all subsystems that pass coupling parameters to the i'th subsystem), and the output parameters of the subsystems analysis of the i'th subsystem includes not only the output variables z_i but also the coulpling variables y_{ij} for other subsystems.

1.2.1.4 System Analysis

System analysis refers to the process of calculating and analyzing the output parameters of the overall system according to the given input parameters. The system analysis usually includes subsystem analysis of all subsystems, and needs multiple equilibrium iterations to achieve the balance of the coupling subsystems. As shown in Fig. 1.1, the system consisting of two sub-disciplines (subsystems), and the system analysis is the process to calculates the output parameter of z_1 and z_2 based on the input variables x_1 and x_2, but as there is coupling between two sub-disciplines, several internal iterations of subsystem analysis need to be run to achieve the balance of the two sub-disciplines, that is, to determine the appropriate values for coupling parameter, y_{12} and y_{21}.

1.2.1.5 Uncertainty

By observing natural phenomena, humans analyze and summarize the laws of nature to establish scientific knowledge. However, natural phenomena often occur outside of our known knowledge. These objective accidental phenomena often cannot be explained or described with existing sciences or knowledge, and it usually shows random oscillations and fluctuations activities. Some of these unexpected activities cannot be explained now because of the lack of knowledge, and some because the characteristics of the research objects themselves may contain properties that cannot be accurately described.

1.2.1.6 Aleatory Uncertainty

Accidental uncertainty refers to the objective existed and unreducible uncertainty. Accidental uncertainty describes the characteristics of the irregular changes (random changes) inherent in the actual system or environment, so it is often referred to as inherent uncertainty.

1.2.1.7 Epistemic Uncertainty

Epistemic uncertainty refers to subjective uncertainty caused by lack of knowledge. Epistemic uncertainty is caused by lack of knowledge or incomplete information, so it is also called subjective uncertainty.

1.2.1.8 State Function

In the design stage, the performance of the engineering system can often be expressed as a function of the input variables. This function is called state function. Usually, to meet the performance requirement of the system, the value of the state function value is needed to be limited within a given range.

1.2.1.9 Reliability

The ability (or probability) of an engineering system or product to accomplish specified function under specified conditions and within a specified time is called reliability; correspondingly, the probability that an engineering system or product cannot perform a specified function is called the probability of failure. The process of calculating the reliability or failure probability of the system is called reliability analysis, and the basis of reliability analysis is uncertainty theory.

1.2.1.10 Definition Related to Traditional Optimization

As described before, Multidisciplinary design optimization includes traditional optimization theory, so the concept of objective functions, constraints, design variables, etc. are basically consistent with the definitions in traditional optimization theory.

1.2.2 Brief History of MDO Research

Multidisciplinary design optimization originated in the field of aircraft design. The popular aircraft design method in the aviation industry in the 1970s was still the subsystem serial design method, in which all subsystems were first sequenced in design, and then each subsystem was designed serially in order. When designing a subsystem, the connection between other subsystems and this subsystem is temporarily ignored. The design of all next subsystems cannot be carried out until the design of this subsystem is finished. If it is found that the output of the system design or a subsystem design cannot meet the requirement in this design cycle, design parameters will be adjusted and new design cycles with adjusted design parameters are needed…many design cycle will be needed to find an acceptable design. There are obvious drawbacks to this design process: first, there is actually coupling between current subsystem and other subsystems, but serial design method artificially cuts off the connection between current subsystem and other subsystems when designing current subsystem, and makes the assumption that each subsystem are independent from all other subsystems. But in real engineering problems, when the design of the subsystem changes, it will lead to the change of other subsystems, and the changes of other subsystems will also cause the change of the performance of this subsystem. Secondly, this design method is carried out serially, meaning that the design of all subsystems must wait for the upstream subsystem design to complete, and this serial character wastes a lot of time on waiting.

In the 1980s, some aircraft designers realized that aircraft design should be based on a Collaborative Engineering perspective, and taking into account all aspects of the aircraft's entire life span in the design stage (Schrage et al. 1991), leading by Sobieszczanski Sobieski J. and Kroo I, a group of aviation scientists and engineers proposing several methods for the analysis and design optimization of complex systems. Sobieszczanski-Sobieski (1982) proposed a method for dealing with optimization problems using linear decomposition, which already included one of the core ideas of the multidisciplinary design optimization method: decomposition. These ideas and methods for analyzing and designing complex systems keep developing gradually, and finally formed the multidisciplinary design optimization theory (Xiang and Weiji 2003).

Sobieszczanski-SobieskiJ. officially proposed that multidisciplinary design optimization is a method suitable for engineering system design, especially for complex engineering systems in 1993 (Sobieszczanski-Sobieski 1995). After being proposed, MDO method has been highly valued by the academic and engineering circles. AIAA established the Multidisciplinary Design Optimization Technical Committee (MDOTC), and the white papers of this technical committee in 1991 (Schrage et al. 1991) and 1998 (Giesing and Barthelemy 1998) summarizes the research status and application status of MDO, discussed the requirements of actual engineer problems to MDO theory and MDO practical technology and pointed out the research direction of multidisciplinary design optimization according to these requirements. AIAA also published the MDO album in Journal of Aircraft (Zhang et al. 2006).

NASA's MDOB has established a problem set for testing MDO methods called MDO Test Suite (Padula et al. 1996), the MDO Test Suite already becomes an important criterion for testing and validating new multidisciplinary design optimization methods, and helps the commercial MDO software. AIAA/UASF/NASA/ISSMO also corporate together to organize the Multidisciplinary Analysis and Optimization (MA&O) seminar every two years.

Driven by these academic institutions and organizations, the multidisciplinary design optimization theory developed rapidly in the 1990s, and proposed Multidisciplinary Feasible Method (MDF), Individual Disciplinary Feasible Method (IDF), Successive Approximate Optimization (SAO), Bi-Level Integrated Synthesis (BLISS), Concurrent Subspace Optimization (CSSO), Collaborative Optimization (CO), Analytical Target Cascading (ATC) and other MDO methods. By the beginning of the 21st century, the theory and main methods of multidisciplinary design optimization have been basically mature, and theoretical research has entered a period of relatively gradual development. Today's MDO theory research is more concentrated in the improvement and adjustment of existing MDO methods for the MDO application in specific engineering peoblems. Popular multidisciplinary design optimization methods and their improvements will be introduced in this book.

The implementation of multidisciplinary design optimization requires the support of a large amount of computer algorithm and technology. As the applications of MDO increase, the market demand for multidisciplinary design optimization program and software also grows. Multidisciplinary design optimization commercial software has been gradually developed and reached the level of actual engineering applications, such as iSIGHT (Engineous), ModelCenter (PhoenixInt.), modeFRONTIER (Esteco) and other general-purpose multidisciplinary design optimization platforms. Many specialized multidisciplinary design optimization software has also been rapidly developed. For example, Chen and other people (2006) established the Satellite Integrated Design Environment (SIDE) platform for satellite design; Iqbal (2009) developed an Excel-based integrated system that can be used in the conceptual design stage of aircraft by integrating high-level CAD and CAE software such as CATIA, FLUENT, ANSYS, and SURFCAM for high-precision design. The booming development of these software platforms also drives the multidisciplinary design optimization to spread out all over the world from its birthplace—United States. The acquisition of iSIGHT by Dassault Aviation in France reflects the Emphasis of the European aviation community to multidisciplinary design optimization techniques.

Multidisciplinary design optimization theory is developed for practical engineering problems and its development is inseparable from the testing and promotion of engineering applications. Therefore, the application research of multidisciplinary design optimization is as important as the theoretical research. In its origin—the aviation industry, multidisciplinary design optimization now has become an important design tool: Boeing applied lots of MDO methods and algorithms in its F18, F117 fighters, B2 bombers (Young et al. 1998; Wakayama and Kroo 1998) and military helicopters (Tarzanin and Young 1998), and the application depth and breadth of MDO is increasing continuously. GE has used iSIGHT to study the collaborative optimization framework of turbine engines, an approximation model has been developed and optimization base on the approximation model has been implemented (Röhl et al. 1998; Golovidov et al. 1998). Lockheed Martin also used a lot of multidisciplinary design optimization technique in the conceptual design, aeroelastic design, and overall design of the modified F16 fighter and the advanced F22 fighters (Carty 2002), (Love 1998; Radovcich and Layton 1998); Rafique (2009) uses genetic algorithms to perform MDO analysis on multi-stage rocket concept designs including four subsystems of propulsion, aerodynamics, flight control and weight; Gündüz (2010) integrates commercial CAD, CAE Software, Matlab, and the private program of Georgia Tech by ModelCenter to determine most of the parameters of the helicopter design in the conceptual design stage, in order to reduce the design changes and costs of future testing, manufacture,and maintenance.

It can be found from the existing literature that the current development and application of multidisciplinary design optimization is mainly carried out in aero field. A variety of multidisciplinary design optimization theories and applied research are also being carried out at universities: Professor Kroo of Stanford University has proposed the famous CO algorithm (Kroo et al. 1994). Virginia Institute of Technology has established the Advanced Aircraft Multidisciplinary Analysis and Design Center to develop research and teaching work of MDO, Multidisciplinary design optimization has been used to study the truss-supported wing type supersonic aircraft with wing end mounting engine, it has been found that compared to traditional design method, the optimal design obtained by MDO has better overall performance: it reduces the total takeoff weight, fuel weight, and operation space weight and increase the cruising radius and the number of miles traveled per unit of fuel.

The Space Systems Analysis Center of the college of Aerospace Engineering of Georgia Institute of Technology conducted research on concurrent engineering, approximation models, and multidisciplinary design optimization methods that consider uncertainty. In the preliminary design stage of the helicopter, subsystems of stability, maneuverability, engines selection, propulsion performance, conveyor, weight and balance, aerodynamics, structural analysis, noise, economy, overall performance were integrated in the MDO model and studied (Khalid and Schrage 2006), the school's Aviation Systems Design Center (SpaceSystemsDesignLab) adopt the idea of multidisciplinary design optimization in the design of the Branching Trajectories system, and the idea of MDO was introduced to handle the coupling between the spacecraft and the aircraft (Ledsinger and Olds 1998).

The University of Buffalo developed the multidisciplinary design optimization method simulator CASCADE to compare the effectiveness of different multidisciplinary design optimization methods (Hulme and Bloebaum 1997, 2000); the University of Florida studied the application of MDO in missile design, Rotterdam University used parallel subspace method to design the aircraft brakes, and investigated methods to improve the efficiency of multidisciplinary optimization sensitivity analysis and optimization, they also studied the influence of uncertain parameters on the design result of multidisciplinary design optimization; the University of Ontario has developed a distributed multidisciplinary design optimization framework based on web technologies (Wang et al. 2003); the University of Toronto has developed a multidisciplinary design optimization framework based on the Python language and comprehensively compared the efficiency of various multidisciplinary design optimization methods through multidisciplinary design optimization test problem sets. (Tedford 2007); the University of Southampton in the UK established the Center for Computational Engineering and Design to study multidisciplinary design optimization techniques.

In addition to being widely used in the aviation industry of developed countries, multidisciplinary design optimization is spreading out globally, Korea (Yi et al. 2007), Japan (Kazuhisa et al. 2005), Germany (Hönlinger et al. 1998) and other countries are also developing their own multidisciplinary design optimization methods and software. MDO are also extending from the aerospace field into other areas, such as the underwater weapon (Yukish et al. 2001), trimaran design (Hefazi et al. 2005), submarine (Shingler et al. 2005), thermal protection system (Sun and Zhang 2006), space structural locks (Zhang et al. 2008), automatic reconfigurable systems (Ferguson 2008), ship concept design (Hart 2010), etc.

Since Hua Luogeng's article "program evaluation and review technique" was written into textbook of primary school, the concept of overall planning and optimization has already been deeply rooted in the hearts of Chinese people.. Chinese scholars have also made outstanding achievements in the study of traditional optimization theory. For example, Dalian University of Technology has developed the international leading structural optimization software MCADS system (Gu and Cheng 1995; Wang and Cheng 2003). Since the rise of multidisciplinary design optimization in the United States, researchers in China have recognized the importance of multidisciplinary design optimization and start the theory and applied research on multidisciplinary design optimization. Nanjing University of Aeronautics and Astronautics (NUAA) started the research of MDO methods and software since the 1990s (Yu 1999; Yu et al. 2004; Liu and Yao 2007), and explored MDO application in the aeroelasticity design of aircraft wings (Liu et al. 2007). Beijing University of Aeronautics and Astronautics (BUAA) rebuilt the system model by design matrix analysis to reduce the number of feedback coupling parameters, which reduces the complexity of the optimization model and may even turn the coupling model into a pure feedforward model (Liao and Wang 2007), and began to study the impact of uncertainty on the results of multidisciplinary design optimization (Han and Deng 2007); National Defense University of Science and Technology carried out a number of studies on MDO and its application in missile design and explored various MDO technologies like distributed collaborative algorithms (Chen 2003; Chen et al. 2001), experimental design and response surface model (Luo et al. 2003), sensitivity analysis (Yan et al. 2005) and the application of these technologies in the design of missiles.

References

Carty A (2002) An approach to multidisciplinary design, analysis & optimization for rapid conceptual design. AIAA 2002-5438

Chen W, Optimized combination creates strong fighting capacity. PLA Daily Newspaper, 20010314, P 12. (陈卫. 科学组合出战斗力. 解放军报, 2001年03月14日, 第12版.)

Chen Q (2003) Research on multi-disciplinary design optimization method for distributed coevolution of aircraft. Doctoral thesis of National University of Defense Technology, advisor Dai Jinhai. (陈琪锋. (2003). 飞行器分布式协同进化多学科设计优化方法研究. 国防科技大学博士论文, 指导老师戴金海.)

Chen Q, Li X, DAI J (2001).Cooperative coevolutionary MDO algorithm for missile overall parameter optimization design. J National University of Defense Technol, 2001(05). (陈琪锋,李晓斌,戴金海. (2001). 导弹总体参数优化设计的合作协同进化MDO算法.国防科技大学学报,2001,(05).)

Cui W, Liu Z, Xu Q (2008) Four elements method for Large complex engineering system design. China Ship Build 49(2):1–12. (崔维成, 刘正元, 徐芑南. 大型复杂工程系统设计的四要素法[J]. 中国造船, 2008, 49(2): 1-12.)

Ferguson Scott M (2008) Design of autonomous reconfigurable systems for use in extreme operating environments. Dissertations & Theses - Gradworks

Giesing JP, Barthelemy JM (1998) A summary of industry MDO applications and needs. In: An AIAA white paper of AIAA technical committee, 7'th AIAA/USAF/NASA/ISSMO symposium on multidisciplinary anslysis and optimization

Golovidov O. FLEXIBLE IMPLEMENTATION OF APPROXIMATION CONCEPTS IN AN MDO FRAMEWORK. AIAA 98-4959, 1998

Gu Y, Cheng G (1995) Development and application of computer-aided structural optimization design software McAds. Chinese J Comput Mech 12(003):304–307. (程耿东, 顾元宪. 我国机械优化研究与应用的综述和展望[J]. 机械强度, 1995, 017(002):68–74.顾元宪, & 程耿东. (1995). 计算机辅助结构优化设计软件mcads的开发与应用. 计算力学学报, 12(003), 304–307.)

Han M, Deng J (2007) Uncertainty modeling for multi-disciplinary design optimization. J Beihang University, January 2007, 33(1). (韩明红, 邓家禔. (2007). 多学科设计优化中的不确定性建模. 北京航空航天大学学报, 2007年1月, 第33卷第1期.)

Hart CG (2010) Multidisciplinary design optimization of complex engineering systems for cost assessment under uncertainty. PhD thesis of Naval Architecture and Marine Engineering in The University of Michigan

Hefazi H, Schmitz A and Shide R (2005). Automated multidisciplinary design optimization method for multi-hull vessels. CCDoTT Final Project Report

Hulme KF, Bloebaum CL (1997) Development of a multidisciplinary design optimization test simulator. Structural optimization, 14(2-3):129–137

Hulme KF, Bloebaum CL (2000) A simulation-based comparison of multidisciplinary design optimization solution strategies using CASCADE. Structural and Multidisciplinary Optimization, 19(1):17–35

Iqbal LU (2009) Multidisciplinary Design and Optimization (MDO) methodology for the aircraft conceptual design. Ph.D. thesis of Purdue University

Kazuhisa C, Shinkyu J, Shigeru O and Hiroyuki M (2005) Data mining for multidisciplinary design space of regional-Jet Wing. 0-7803-9363-5/05/$20.00 ©2005 IEEE, pp 2333–2340

Khalid AS, Schrage D (2006) Development and implementation of rotorcraft preliminary design methodology using multidisciplinary design optimization. A Ph.D. Dissertation of Georgia Institute of Technology, December 2006

Kroo I, Altus S, Braun R et al. (1994) Multidisciplinary optimization methods for aircraft preliminary design[R]. AIAA-94-4325

Ledsinger LA, Olds JR (1998) Multidisciplinary design optimization techniques for branching trajectories. AIAA 98–4713

Li X, Li W (2003) Collaborative optimization method based on hypersphere approximation subspace and its application. J Northwestern Polytechnical University, 21(4):461–464. (李响; 李为吉. 基于超球近似子空间的协同优化方法及应用研究. 西北工业大学学报, 2003, 21(4):461–464)

Liao X, Wang Z (2007) Design process modeling and reconstruction techniques in multidisciplinary design optimization. Aeronaut Manufact Technol, 3, 2007. (廖馨, 王振华. (2007). 多学科设计优化中的设计过程建模及重构技术. 航空制造技术, 2007年第3期.)

Liu K, Yao W, YU X (2007). Aerodynamic multidisciplinary design optimization of wing structures using low freedom collaborative optimization. Acta Aerophenica Sinica, Sept 2007, Vol. 28, No. 5. (刘克龙, 姚卫星, 余雄庆. (2007). 运用低自由度协同优化的机翼结构气动多学科设计优化. 航空学报, 2007年9月,第28卷第5期.)

Love MH (1998) Multidisciplinary design practices from the F-16 Agile Falcon. AIAA 98-4704

Luo S, Luo W, Wang Z (2003) Multidisciplinary design optimization of hypersonic cruise aircraft based on test design and response surface approximation. Missile and Space Delivery Technology, 2003, 6, Total 266. (罗世彬, 罗文彩, 王振国. (2003). 基于试验设计和响应面近似的高超声速巡航飞行器多学科设计优化. 导弹与航天运载技术, 2003 年第6 期, 总第266 .)

(刘克龙, 姚卫星. (2007). 多学科设计优化的低自由度协同优化方法. 南京航空航天大学学报, 2007年6月, 第39卷第3期.)

Padula SL, Alexandrov N, Green LL (1996) MDO test suite at NASA Langley Research Center. AIAA Paper 96-4028

Radovcich N and Layton D (1998) THE F-22 Stuctural/Aeroelastic design process with MDO examples. AIAA

Rafique AF, He LS, Zeeshan Q, Kamran A, Nisar K, Wang XW (2009) Integrated system design of air launched small space launch vehicle using genetic algorithm. In: 45th AIAA/ASME/SAE/ASEE joint propulsion conference and exhibit. Denver, Colorado. AIAA 2009-5506

Schrage D, Beltracchi T, Berke L, Dodd A, Niedling L, Sobieszczanski-Sobieski J (1991) Current state of the art on Multidisciplinary Design Optimization (MDO). In: An AIAA white paper of AIAA Technical Committee. ISBN 1-56347-021-7

Shingler K, Goff D, Shrewsbury D, et al. (2005). Design report littoral Warfare Submarine (SSLW). Report of advanced tactics littoral alternative submarine ocean engineering design project AOE4065/4066

Sobieszczanski-Sobieski J (1982) A linear decomposition method for large optimization problems. Blueprint for development

Sobieszczanski-Sobieski J (1995) Multidisciplinary design optimization: an emerging new engineering discipline[M]. Springer, Netherlands

Sun JL, Zhang G, Vlahopoulos N, Hong SB (2006) Multi-disciplinary design optimization under uncertainty for thermal protection system applications. In: 11th AIAA/ISSMO multidisciplinary analysis and optimization conference. Portsmouth, Virginia, AIAA 2006-7002

Tarzanin F, Young DK (1998) Boeing rotorcraft experience with rotor design and optimization. AIAA 98-4733

Tedford N (2007). Comparison of MDO architectures within a universal framework. A master thesis of the University of Toronto

Wakayama S, Kroo I (1998) The challenge and promise of blended-wing-body optimization. AIAA 98-4736

Wang J, Cheng G (2003) Topological optimization design of continuum structures under multi-condition stress constraints. Mech strength. (王健,程耿东. (2003). 多工况应力约束下连续体结构拓扑优化设计. 机械强度, 2003年1期.)

Wang YD, Shen W, Ghenniwa H (2003) WebBlow—a Web-agent-based multidisciplinary design optimization environment. Computers in Industry 52:17–28

Yan L, Chen X, Wang Z (2005) Sensitivity analysis method for aircraft multidisciplinary design optimization. Aeronaut Comput Technol, March, 2005, 35:1. (颜力, 陈小前, 王振国. (2005). 飞行器多学科设计优化中的灵敏度分析方法研究. 航空计算技术, 2005年3月, 第35卷第1期 .)

Yi SI, Shin JK, Park GJ (2007) Comparison of MDO methods with mathematical examples. An REVIEW ARTICLE of Struct Multidisc Optim at Springer-Verlag. DOI https://doi.org/10.1007/s00158-007-0150-2

Young JA, Anderson RD, Yurkovich RN (1998) A description of the F/A-18E/F design and design process. AIAA 98-4701

Yu X (1999) New aircraft design technology – multidisciplinary design optimization. Aeronaut Sci Technol. 1999. 1. (余雄庆. (1999). 飞机设计新技术——多学科设计优化. 航空科学技术, 1999.1.)

Yu X, Yao W, Xue F, MO X, Liu K, HUANG A (2004) Discussion on multidisciplinary design optimization computing framework. March, 2004, vol. 23, No. 3.(余雄庆, 姚卫星, 薛飞, 穆雪峰, 刘克龙, 黄爱凤. (2004). 关于多学科设计优化计算框架的探讨. 2004年3月, 第23卷第3期.)

Yukish M, Bennett L and Kurtz P (2001). Simulation-Based design for undersea weapon. iMAST Quarterly, No.2, pp 3-6

Zhang K, Li W, Wei H (2006) Collaborative method for optimal allocation of design indicators. Mach Sci Technol, 25(7):797–801

Zhang J, Su D, LIU Y (2008) Reliability analysis and optimization of mechanical products [M]. Beijing: Electronic Industry Press, 2008.(张建国, 苏多, 刘英卫. 机械产品可靠性分析与优化[M]. 北京: 电子工业出版社, 2008.)

Chapter 2
Multidisciplinary Design Optimization Theory

Sobieski, the founder of multidisciplinary design optimization, believes that there are six key technologies for multidisciplinary design optimization: mathematical modeling, design-oriented analysis, approximate technology, optimization processes, system sensitivity analysis, and artificial interfaces (Sobieszczanski-Sobieski and Haftka 1996), the most important of which is mathematical modeling. The method of dealing with the coupling relationship between systems in mathematical modeling is called multidisciplinary design optimization method. The method to deal with the coupling relationship between systems in mathematical modeling is called multidisciplinary design optimization method, which is the research focus of multidisciplinary design optimization theory. According to whether the optimization is carried out at single-level or at multiple levels, the multi-disciplinary design optimization methods can be divided into single-level optimization algorithm and multi-level optimization algorithm. The single-level multidisciplinary design optimization method does not decompose the original system model, but only carries out optimization at the top level of the system, and achieves subsystem balance through iteration among subsystems or sub-disciplines. Common single-level multidisciplinary design optimization methods include Multidisciplinary Feasible Method (MDF), Individual Disciplinary Feasible Method (IDF), and Successive Approximate Optimization (SAO). The multilevel optimization algorithm is optimized at the system level and in the sub-discipline, and the sub-discipline optimization is conducted around the system optimization. This algorithm not only conforms to people's thinking mode of decomposing a problem into several sub-problems, but also can execute in parallel. Common multi-level optimization algorithms include Collaborative Optimization (CO), Concurrent Subspace Optimization (CSSO) and Bi-Level Integrated Synthesis (BLISS). This chapter introduces the iconic approach and other key technologies for multidisciplinary design optimization.

B. Pan and W. Cui, *Multidisciplinary Design Optimization and Its Application in Deep Manned Submersible Design*, Ocean Engineering & Oceanography 13,
https://doi.org/10.1007/978-981-15-6455-0_2

2.1 Multidisciplinary Design Optimization Modeling

Modeling is the process of abstracting and simplifying real objects. With the development of computer-aided design, most of the modeling in the field of engineering design now adopts digital modeling technology. For example, when designing the aerodynamic performance of an aircraft, a CFD model is established for the shape of the aircraft and the flow field near the body to estimate the speed and maneuverability of the aircraft; when designing the structural performance of the aircraft, a finite element model is established for the main frame members, plates and supporting structures of the aircraft to estimate the force and deformation of the aircraft. The object of multidisciplinary design optimization is the design of complex engineering systems, so the process of modeling is to model the whole process of engineering system design. It not only includes the modeling of various precisions in each subsystem (such as the empirical formula estimation model in the structural subsystem, the overall finite element analysis model), but more importantly, it also includes modeling the coupling relationship between various subsystems. That is, the parameters between the subsystems are passed, and then the subsystems are reorganized through the multidisciplinary design optimization method to establish an efficient design process. The content of multidisciplinary design optimization modeling includes:

2.1.1 Subsystem Modeling

In the development of science, the imbalance of the development of various disciplines leads to the inconsistency of the computational models of each subsystem, and the calculation methods and models with different precisions exist in the same subsystem. Therefore, it is necessary to have a comprehensive understanding of each subsystem involved in the whole complex engineering system and make a technical status assessment, understand the main modeling methods of each subsystem, and obtain the calculation accuracy and calculation time consumption of different modeling methods in each subsystem. At the same time, it is necessary to clarify the relationship between each subsystem and the external and other subsystems, that is, to clarify the input parameters, output parameters and auxiliary parameters of each subsystem.

In addition, when the optimization solver is used for system optimization instead of adjusting design parameters based on manual decision, the model of the subsystem will be transferred multiple times because the optimization algorithm needs to transfer the whole system model for objective function evaluation and constraint function calculation. Therefore, it is necessary to ensure that the models of each subsystem are parameter driven, that is, to achieve parametric modeling and analysis.

Nowadays, the common subsystem modeling methods in the engineering field include:

(1) Simplified estimation methods based on disciplinary theory or empirical formulas, such as the common normative formula method. Its characteristics are based on rigorous theoretical analysis and engineering experience correction. After enough data accumulation and experience modification, this method can have better engineering accuracy and high computational efficiency. However, when the design is beyond the previous experience, this method may cause errors, or even inappropriate.
(2) Fine modeling methods based on the quantification of disciplinary theory, such as finite element models in structural analysis and CFD models in fluid analysis. Based on the current engineering object, this method discretizes the value and then performs a detailed calculation on the actual object of the project, so this method is usually accurate. However, numerical simulation usually consumes a lot of computing resources, and it takes a lot of time on a common design workstation. To achieve parametric modeling and analysis in this method, designers need to not only understand the knowledge of the system, but also have strong programming capabilities. However, when the design changes greatly, the existing parametric model can be invalidated. For example, when the design changes cause the topology of the structure to change, the finite element model of large complex structures such as hull and aircraft frames is difficult to implement automatic modeling.
(3) Database-based comparison and interpolation methods, such as the mother ship method in ship type analysis. This method is directly based on the accumulated design database without the support of strict discipline theory, so it is simple and efficient. However, it has the disadvantage that it requires a large amount of empirical data and is no longer applicable when the design goes beyond the database scope.

2.1.2 Design Process Modeling (System Modeling)

According to the parameter transfer of each subsystem, the diagram of the whole system is established. Then, each subsystem is reorganized by layering, decoupling and approximation, and a new system diagram is established to parallelize and streamline the system design process. These methods of reorganizing the system are called multidisciplinary design optimization. The reorganized system can analyze the key nodes of the whole system (time-consuming subsystem analysis) after several testability design processes, and then reduce the calculation amount by adopting low-precision models and establishing approximate models. When the whole system runs smoothly, it may still encounter problems such as unconvergent optimization and excessive variables, so it is necessary to adopt a new method to restructure the system or to constant the design variables that have little influence on the system indicators through sensitivity analysis.

2.2 Multidisciplinary Design Optimization Method

As mentioned above, the multidisciplinary design optimization method is a method of reorganizing the system and is also a method of processing data transfer among subsystems. The commonly used multidisciplinary design optimization methods are single-level method and hierarchical method. The characteristic of the single-stage method is that the optimization is only carried out in the total system, and each subsystem (or sub-discipline) has only analysis and no optimization, that is, each subsystem can be regarded as a sub-function of the total system analysis, which is the analysis of the total system (to calculate the objective function and the constraint function) to provide state variables. All design variables in the single-stage method are processed and optimized by an optimization solver. Too many design variables can cause high dimensionality in the design space, making the optimization and search difficulty dramatically increase. The multi-level method is optimized at both the system and subsystem levels. Therefore, the design variables of the subsystem need not be transferred to the overall system for unified solution, which overcomes the disadvantage of using a single optimization solver to handle all design variables in the single-stage method. However, it should be noted that the objective function, constraints and design variables of each subsystem optimization in the classification method are different from the objective functions, constraints and design variables of the traditional optimization of the subsystem. In many cases, the difference between the coupling variable calculated by the reduction subsystem and the coupling variable transmitted from the total system is the objective function, that is, the optimization of the subsystem is carried out around the optimization of the total system, rather than only considering the subsystem.

Common single-level methods are:

(1) Multidisciplinary Feasible Method. Hereinafter referred to as the MDF method;
(2) Simultaneous Analysis and Design (SAND), also called the All-At-Once method. Hereinafter referred to as AAO method;
(3) Individual Discipline Feasible. Hereinafter referred to as IDF

Common hierarchical methods are:

(1) Concurrent Subspace Optimization. Hereinafter referred to as CSSO;
(2) Collaborative Optimization. Hereinafter referred to as CO;
(3) Bi-Level Integrated Synthesis. Hereinafter referred to as BLISS;
(4) Analytical Target Cascading. Hereinafter referred to as ATC

The principles of these main methods are described below.

2.2.1 1MDF Method

The Multidisciplinary Feasible Method is to organize subsystems according to the original framework of the system. Then, on the basis of the whole system analysis,

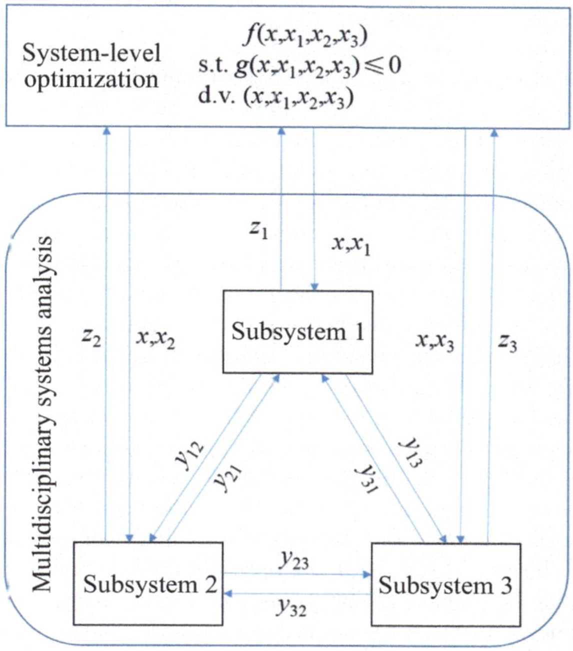

Fig. 2.1 MDF method for three subsystem MDO problems

an optimization solver was added to adjust and optimize the design variables, and the coupling variables were balanced through multiple system analysis iterations. The subsystem organization and parameter transfer of the MDF method with the MDO problem of three subsystems are shown in Fig. 2.1. The system-level optimization solver only deals with design variables, where x represents a common design variable (that is, a variable required by all three subsystems; d is an input parameter), and xi (i = 1, 2, 3) represents a local design of subsystem f Variables (that is, only the input parameters required by subsystem f); f represents the objective function; g represents the constraint function (according to traditional optimization theory, equation constraints can also be easily converted to the inequality constraint form in Fig. 2.1, so this book only discusses inequality constraints). The coupling parameters $y_{ij}(i, j = 1, 2, 3)$ between the subsystems are not processed such as decoupling, but are obtained by setting initial values in the system analysis and then iterating the system analysis multiple times. For example, for any design solution in the optimization process (that is, a set of design known variable x, x_1, x_2, x_3), First assume the initial value of a set of coupling variables $\left(y_{21}^0, y_{31}^0, y_{32}^0\right)$, substituting $\left(y_{21}^0, y_{31}^0\right)$ into subsystem 1 to get $\left(y_{12}^0, y_{13}^0\right)$, then substituting $\left(y_{12}^0, y_{13}^0\right)$ into subsystem 2 to get $\left(y_{21}^1, y_{23}^0\right)$, and then substituting $\left(y_{23}^0, y_{13}^0\right)$ into subsystem 3 to get $\left(y_{31}^1, y_{32}^1\right)$. After such a subsystem iteration, a new set of coupled variable values $y_{21}^1, y_{31}^1, y_{32}^1$ is obtained. If the difference between $y_{21}^0, y_{31}^0, y_{32}^0$ and $y_{21}^1, y_{31}^1, y_{32}^1$ is small (the error is less than the specified value), the subsystems are considered balanced. However, in most cases, the error is relatively large, so $y_{21}^1, y_{31}^1, y_{32}^1$

is used to replace $y_{21}^0, y_{31}^0, y_{32}^0$ for A new round of subsystem iteration, and the difference between $y_{21}^1, y_{31}^1, y_{32}^1$ and $y_{21}^2, y_{31}^2, y_{32}^2$ is compared again. If the difference is still large, $y_{21}^2, y_{31}^2, y_{32}^2$ is substituted into a new round of subsystem iteration to get $y_{21}^3, y_{31}^3, y_{32}^3 \ldots\ldots$ Finally, a set of coupling parameter values (z_1, z_2, z_3) that meet the error requirements is obtained. Then calculate the state parameters (z_1, z_2, z_3) based on this set of coupling variable parameters and design variable values and pass them back to the system layer to provide input parameters for the evaluation of the system-level objective function f and the constraint function g (Note: f And g functions are written as functions of design variables, but the actual input parameters of f and g functions usually include state variables in addition to design variables, just because state variables are also functions of design variables, that is, after the subsystem equilibrium iteration is complete, The value of the state variable corresponding to the design point is unique, so the f and g functions can be considered as a function of the design variable in the final analysis). It can be seen that the success of the subsystem equilibrium iteration depends largely on the selection of the initial values of the coupling variables. If the initial value is close to the final equilibrium value, the equilibrium of the subsystem can be reached after a small number of iterations; if the initial value deviates significantly, it may need to be corrected through multiple iterations. In addition, the reader may have noticed that the premise of subsystem equilibrium iteration is that the design variables are known, that is, for a certain design solution, and the optimization solver usually needs to perform multiple iterative searches on the design space to find the optimal solution. And each design scheme in this optimization process must be balanced and iterated, which means that the MDF method contains two nested iterative processes, that is, the MDF method is a two-cycle method, which will greatly increase the amount of calculation.

系统级优化	System-level optimization
多科学系统分析	Multidisciplinary systems analysis
子系统	Subsystem

The MDF method is the most traditional multidisciplinary design optimization method, and it is also the most primitive processing method for multidisciplinary problems. The MDF method can well deal with the problem of fewer subsystems and tight coupling. Fewer subsystems means fewer subsystem analyses need to be performed in the equilibrium iteration. The tight coupling between subsystems means that the coupling parameters are greatly affected by the subsystems, and the iterative equilibrium converges faster. Moreover, the optimization solver of the MDF method only needs to deal with the design variables of the original MDO problem, and the design space dimension does not expand due to the auxiliary design variables, which makes the optimization solver relatively difficult to solve and easier to pass through fewer optimization iterations Find the optimal solution. The aeroelasticity problem in aviation design is the best example of MDF method. Aeroelastic problems are caused by tightly coupled structural and aerodynamic subsystems. Starting from the initial structure, aerodynamic analysis is performed to obtain the pressure field distribution around the aircraft, especially around the wing; then the pressure field is applied to the structure, and the deformed structure is calculated by the structural subsystem; the deformed structure is then transferred For the aerodynamic subsystem, recalculate the pressure field distribution based on the deformed structure... iterate until it converges. It can be seen that in the aeroelasticity problem, the parameter passed from the structural subsystem to the aerodynamic subsystem is the shape of the structure, and the parameter passed from the aerodynamic subsystem to the structural subsystem is the load.

2.2.2 AAO Method

In order to solve the problem of huge calculation caused by the double-loop structure of the MDF method, auxiliary design variables are introduced to decouple the subsystems, and the subsystems and the overall system are also decoupled. As an example, the three subsystem MDO problems in the MDF method mentioned above are shown in Fig. 2.2.

First introduce two auxiliary design variables: coupling auxiliary design variables $y^u = y_{12}^u, y_{13}^u, y_{21}^u, y_{23}^u, y_{31}^u, y_{32}^u$ and state auxiliary design variables $z^u = z_1^u, z_2^u, z_3^u$, the dimension of the system-level optimized design variables is expanded to $(x, x_1, x_2, x_3, y^u, z^u)$; in In order to ensure that the newly introduced auxiliary design variables are consistent with the calculated values of coupling variables and state variables of each subsystem at the optimal solution, consistency constraints $\| y^u - y \| \le \varepsilon$和$\| z^u - z \| \le \varepsilon$ are added to the system optimization. Among them, ε is the allowable error value (for example, the most common value = 0.001, and sometimes

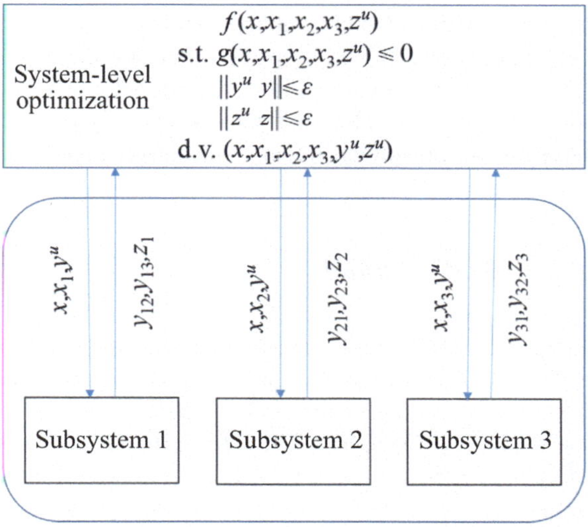

Fig. 2.2 AAO method for MDO problem of three subsystems

the allowable error value ε of the coupling variable y and the state variable z may be different).

Operator $||*||$ is the sum of squares operator, $y = y_{12}, y_{13}, y_{21}, y_{23}, y_{31}, y_{32}$ is the coupling variable value calculated by each subsystem according to the new design variable, and V is the state variable value calculated by each subsystem according to the new design variable. The physical meaning of the consistency constraints is that the system layer needs to introduce auxiliary design variables instead of coupling variables and state variables that are close to the coupling variables and state variables actually calculated by each subsystem (the error is controlled by ε). It can be seen that the optimization solution and the subsystem balance are performed by the same optimization solver at the same time, which is also the origin of the method All-At-Once name.

The advantage of the AAO method is that the subsystems in the system are completely decoupled, that is, the calculation of each subsystem does not have a sequential dependency relationship, and the analysis of each subsystem can be performed in parallel in a distributed computing network. This method is applicable to the case of loose coupling between subsystems and large amount of subsystem analysis and calculation. The disadvantage of the AAO method is that the introduction of new auxiliary design variables instead of coupling variables and state variables will lead to an increase in the dimension of the design space and increase the difficulty of optimization. The increase of constraints brought by the introduction of consistency constraints will also make optimization difficult increase. Therefore, the AAO method is not suitable for the problem of tight coupling between subsystems (more precisely, the AAO method is not suitable for the problem of a large number of coupling parameters and state parameters). In addition, the upper and lower limits of the auxiliary design variables f and 2 cannot be defined, and unbounded optimization algorithms are usually only suitable for simple mathematical problems, and most of them are not suitable for practical engineering problems. Therefore, a common engineering practice is to take an approximate range based on the trial calculation value or the previous calculation result. After optimization, if the optimization point falls on the boundary, the range is increased, but when the optimization converges to the local optimization solution, it may be wrongly judged that the value range is already large enough, resulting in loss of the optimal solution.

2.2.3 IDF Method

In the AAO method, the introduction of coupling-assisted design variables and state-assisted design variables enables decoupling between subsystems and decoupling between the overall system and subsystems. The IDF method can be regarded as an incomplete AAO method. The IDF method framework including the MDO problem of three subsystems is shown in Fig. 2.3. The IDF method also introduces a coupling aid design variable f to decouple each subsystem, but does not introduce a state aid design variable A, which preserves the direct connection between the subsystem and

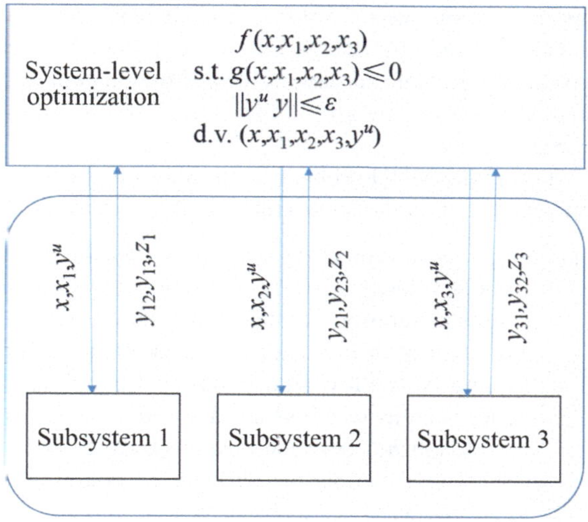

Fig. 2.3 IDF method for MDO problem of three subsystems

the overall system. That is to say, the state variable Z of each subsystem is directly passed to the overall system to evaluate the objective function and constraints, thereby reducing the number of auxiliary design variables and corresponding consistency constraints.

The advantages of the IDF method and the AAO method are similar, that is, the subsystems can calculate in parallel without relying on each other. The IDF method keeps the dimension of the design space smaller than the AAO method by retaining the direct transfer of state variables. However, the number of state variables in many problems is not large, but there are a large number of coupling variables, that is, the number of z^u may be much less than y^u. At this time, compared with the AAO method, the reduction of the IDF method design space dimension may not be obvious.

2.2.4 CSSO Method

The CSSO method was proposed by Sobieszczanski-Sobieski, the founder of MDO, and improved by Renaud and Batill et al. (Yao et al. 2010). The original CSSO method used the global sensitivity equation (which can be seen as a first-order Taylor series expansion approximation method) to construct an approximate relationship between the coupling variables and the design variables. In the process of subsystem optimization, the global sensitivity equation is used to provide approximate values of coupling parameters and state parameters. Since the CSSO method based on the global sensitivity equation needs to perform gradient calculations (numerical difference calculations) with a large amount of calculation, technologies such as response surface and artificial neural network are introduced to replace the global sensitivity equation to form the current CSSO method based on approximate technology. In the approximation-based CSSO method, there is no need to perform optimization within the system, only the design solution generated by each system-level optimization is given, and the state variable value and output coupling variable value corresponding to this subsystem are given. Therefore, it is also called Concurrent Subspace Design (CSD).

Taking the MDO problem with three subsystems as an example, the calculation process based on the approximate CSSO method (CSD method) is:

(1) Given some typical design points (that is, specifying multiple sets of design variable values), perform a system analysis at these design points (including complete iterations of the subsystem balance) to obtain the corresponding coupling variables at each design point value. Based on the corresponding relationship between design points and coupling variables, approximation techniques such as response surface and artificial neural network are used to "fit" the relationship between design variables and coupling variables, and establish an approximate model between coupling variables and design variables $\tilde{y}_{ij} = A_{ij}(x, x_1, x_2, x_3)$: where $i, j = 1, 2, 3$ and $i \neq j$; A_{ij} represents an approximate function.

(2) Perform system layer optimization, as shown in Fig. 2.4. In the process of system optimization, keep design variables unchanged and increase consistency constraints $\|\tilde{y} - y\| \leq \varepsilon$. Where $\tilde{y}_{ij} = A_{ij}(x, x_1, x_2, x_3)$ $(i, j = 1, 2, 3, \text{and } i \neq j)$ are the approximate coupling parameter values calculated by the approximate model, and the opposite y_{ij} are the coupling parameter values obtained after the analysis of each subsystem.
It can also be seen from Fig. 2.4 that the coupling parameters in the input parameters of the subsystem analysis are also substituted by the calculated values of the approximate model. For example, in the analysis of subsystem 1,

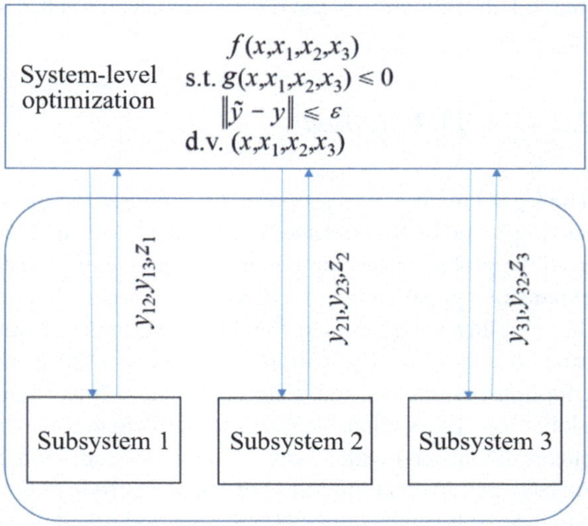

Fig. 2.4 CSSO method (CSD method) for three subsystem MDO problems

the approximate coupling parameters $\tilde{y}_{21}$ and $\tilde{y}_{31}$ calculated by the approximate models A21 and A31 are used instead of the coupling parameters y21 and y31 that were originally provided by the other two subsystems.

(3) Perform a system analysis at the optimal design point and check the results of the subsystem iteration: If the subsystems have reached equilibrium, then the optimization is completed; if the subsystems are unbalanced, then based on the previously calculated data or increase The new system analysis updates the approximate model, improves the accuracy of the approximate model, and then performs a new round of system optimization based on the new approximate model.

Since the CSSO method of this process has only one system optimization, it also belongs to a single-stage method.

In the system analysis of (1), if in addition to the approximate model of the coupling variables $\tilde{y}_{ij} = A_{ij}(x, x_1, x_2, x_3)$, the approximate model of the state variables $\tilde{z}_i = B_i(x, x_1, x_2, x_3)$ can be established, the design variables are divided according to the subsystems, and the design variables optimized at the system level are only the common design variable x. The local design variables of the subsystem are determined by the optimization of each subsystem, and then the hierarchical CSSO method is obtained. In this hierarchical CSSO method, each subsystem performs an objective function that minimizes the error between the approximate value of the state variable and the calculated value (multiple state coupling parameters are reduced to a single objective function by a modulo operator similar to the consistency constraint). For example, the objective function of subsystem 1 can be taken as: $Min \quad (\tilde{z}_1 - z_1)^2$; the constraint condition is the consistency constraint of the output coupling parameters of the subsystem: $(\tilde{y}_{12} - y_{12})^2 + (\tilde{y}_{13} - y_{13})^2 \le \varepsilon$; the design variables are x_1, the parameters x, $\tilde{y}_{21}$ and $\tilde{y}_{31}$ are obtained from the system level. The x_i^{opt} and z_i^{opt} at the optimal design point obtained by the subsystem optimization are passed to the system layer, and these parameters are regarded as the known constant substitute system-level objective function and constraint function, then the system layer can proceed to the next round of design with X System-level optimization of variables… It can be seen that this hierarchical CSSO method also nests optimizations of multiple subsystems within the system-level optimization, so the calculation amount may be larger than the single-level CSSO method. However, the hierarchical CSSO method reduces the design space dimension of the system level and the subsystem level, so the difficulty of its optimization solution will be reduced accordingly.

CSSO method, through the approximate model, skillfully decouples each subsystem so that the analysis or optimization of each subsystem is independent of each other and can be calculated in parallel. However, when the design space dimension is large, a large number of system analysis is needed to obtain enough data points so as to establish an approximate model with sufficient precision. For many large-scale engineering MDO problems, there may be more than dozens or even hundreds of design space dimensions. At this time, it is difficult to conduct a large number of systematic analysis to obtain the data needed for approximate model, which limits the application of CSSO method in large design space dimension problems. With the development of approximation technology, CSSO method is progressing continuously. In the following sections, the response surface, first order approximation, Kriging, artificial neural network and other approximation techniques are introduced.

2.2.5 BLISS Method

Sobieski proposed BLISS method in 1998, which was also based on the global sensitivity equation at the beginning. In this paper, it is referred to as BLISS−98 for short. The basic idea of BLISS−98 is derived from the first order Taylor series expansion of the system- level objective function on the local design variables of the subsystem. Similarly, the MDO problem including the three subsystems mentioned above is taken as an example, the system objective function can be expanded as:

$$f \approx f_0 + D(f, x)\Delta x + D(f, x_1)\Delta x_1 + D(f, x_2)\Delta x_2 + D(f, x_3)\Delta x_3 \quad (2.1)$$

Among them, $f_0 = f(x^0, x_1^0, x_2^0, x_3^0)$ is the value of the objective function at the design point $(x^0, x_1^0, x_2^0, x_3^0)$; $D(f, x)$ is the derivative of the objective function to the common design variable; $D(f, x_i)$ is the derivative of the objective function to the local design variable x_i of subsystem i called sensitivity; it can also be considered to minimize the system objective function f is the first derivative of each term in the minimized expansion (2.1).

The system-level optimized design variables are only the common design variable x, with $D(f, x)\Delta x$ as the objective function. However, the subsystem optimization of subsystem f is performed only for the local design variable xi of the subsystem, and minimizes $D(f, x_i)\Delta x_i$ as the objective function, thereby promoting the convergence of system-level optimization.

Due to the need for a large number of gradient calculations, the calculation amount of the BLISS-98 method is too large to be applicable to many practical engineering problems. Therefore, in the process of system-level optimization, approximate technologies such as response surface are introduced to reduce the calculation amount and improve the smoothness of the system's objective function and constraint function. In 2000, Sobieski et al. reconstructed the optimization problem of the BLISS method, and expressed the objective function of the subsystem optimization as the weighted sum of the state variables output by the subsystem (in the case that the subsystem outputs multiple state variables), called BLISS-2000 method. In the BLISS-2000 method, the objective function of the subsystem optimization is the weighted sum of the state variables, the constraint condition is the consistency constraint of the coupling variables, and the design variable is still a local design variable. Based on the design point data generated in the subsystem optimization process, an approximate model of the state variables and coupling variables in the subsystem is established. Then the system-level optimization can use the approximate models of each subsystem to reduce the amount of calculation. It should be noted that in addition to the shared design variables, the system-level optimized design variables also have weighting coefficients for the state variables of the various subsystems.

Similarly, the BLISS-2000 method is decoupled through an approximate model. Therefore, this method will face the same problem as the CSSO method when dealing with the problem of large dimension of design space. And because the design variables in the system-level optimization increase the weighting coefficient, when the number of state variables of each subsystem is large, it will also cause the design space of BLISS system optimization to be too large. Therefore, many scholars are studying methods to further improve BLISS, such as Lin (2009) and Min (2009).

2.2.6 CO Method

The CO algorithm is the first hierarchical multidisciplinary design optimization method, which was proposed by Koo in 1994. Because the framework of the CO method conforms to the division of labor and collaboration concepts of modern large-scale engineering and the distributed design framework, it has become one of the most widely used multidisciplinary design optimization methods. The CO method divides the MDO problem into system-level optimization and subsystem-level optimization: ① The objective function of the system-level optimization is the system objective function of the original MDO problem; the constraint is the system-level constraint function of the original problem, and the coupling variables and states of each subsystem Constraints on the consistency of the variables; the design variables are the system-level design variables of the original MDO problem (the system-level optimized design variables of the CO method in some literature also include the subsystem design variables) and the newly introduced coupling-assisted design variables (some literature The newly introduced auxiliary design variables include state-assisted design variables. ① The objective function of the subsystem-level optimization is to minimize the residuals of the coupling variables; the constraints are the constraints of the original MDO problem in the subsystem; the design variables are only the local design variables in the subsystem.

The MDO problem including the three subsystems mentioned above is taken as an example, and its CO method framework is shown in Fig. 2.5. The objective function of system-level optimization is the same as f; In addition to the original constraint function g, three consistency constraints J1, J2 and J3 are added to the constraint conditions. In addition to the original common design variable x, $y^u = y^u_{12}, y^u_{13}, y^u_{21}, y^u_{23}, y^u_{31}, y^u_{32}$ are also introduced as auxiliary design variables. In subsystem optimization, x and y^u are obtained from the system level (strictly speaking, each subsystem does not need a complete y^u, but only the components related to the original sub-system. For example, subsystem 1 only needs to be obtained from the system level $\left(x, y^u_{21}, y^u_{31}, y^u_{12}, y^u_{13}\right)$ to optimize the sub-system), and is regarded as a constant in the subsystem optimization process. The objective function is to minimize the difference between the output coupling variable and the

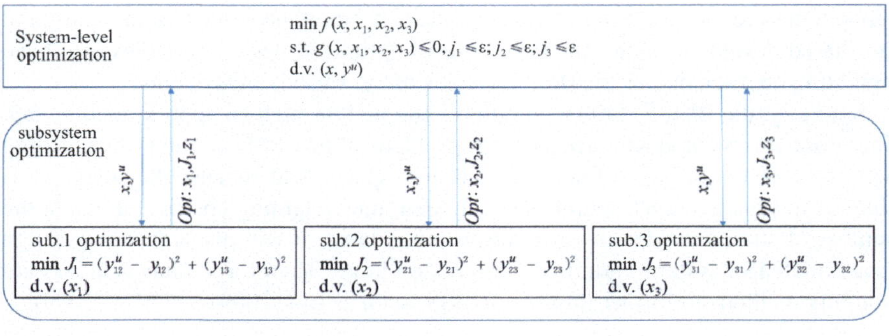

Fig. 2.5 CO method for three subsystem MDO problems

corresponding component of the sub-system. For example, the objective function of subsystem 1 is $J_1 = (y^u_{12} - y_{12})^2 + (y^u_{13} - y_{13})^2$; each subsystem in the original problem has no constraint conditions, the subsystem optimization of the problem is an unconstrained optimization problem. The design variable is the local design variable of each subsystem. For example, the design variable of subsystem 1 is x1.

The optimization process of CO method is as follows:

(1) A round of system analysis was conducted at the initial point of the original MDO problem to obtain the coupling variable value after the balance of each subsystem, which was taken as the initial value y^{u0} of the auxiliary design variable.
(2) Start the system optimization iteration with the initial value of the system-level design variable $\left(x^0, y^{u0}\right)$, and obtain the objective function and constraint value. Among them, the consistency constraint value, the subsystem design variable optimization value, and the subsystem state variable value are obtained by calling each subsystem optimization.
(3) Check whether the constraint function of system-level optimization meets the requirements, and check whether the objective function meets the convergence requirements compared with the previous iteration (the most commonly used convergence criterion is: $\left|f^k - f^{k-1}\right| < \varepsilon$, where ε is the convergence residual, and the default value of commonly used optimization algorithm $\varepsilon = 0.001\frac{n!}{r!(n-r)!}$). If the constraint condition is satisfied and the objective function converges, it is considered to find the best advantage, otherwise the value of the system design variable (x, y^u) is changed, and the next round of system optimization (subsystem optimization is also called during the process) is carried out.

Similar to the AAO method and IDF method, the CO method realizes the decoupling of subsystems by introducing auxiliary design variables and consistency constraints at the system level, enabling each subsystem to be independent and able to be designed in parallel. However, the CO method is different from these single-stage methods. The process of the CO method can be described as: after each system-level design scheme is proposed in the system-level optimization process, the system design variables and coupling variable values are issued to each subsystem in the form of indicators. The subsystem carries out the optimization design of the sub-system based on these system-level indicators, and its goal is to make the output parameters of the sub-system as close as possible to the indicators issued by the system. This process is in line with the idea of division of labor and collaborative design in large-scale engineering design. The CO method maintains the autonomy and modularity of each subsystem to the greatest extent, so it is easily accepted. Moreover, because the design variables for system optimization no longer include design variables for each subsystem, the CO method can effectively reduce the dimension of the design space for system optimization. CO method also has disadvantages, because auxiliary design variables are introduced in system-level optimization for decoupling, when there are many coupling parameters between sub-systems, the spatial dimension of system-level optimization design will increase significantly. In addition, it has been found in practical applications that the CO method cannot obtain convergent optimized solutions for many MDO problems. After research, it is found that system-level consistency constraints cause system-level optimization problems to not meet the Kmush-Kuim-Tucker (KKT) condition, which makes The Grange operator does not exist, which means that the system-level optimization of the CO algorithm cannot use a gradient-based optimization algorithm. The subsystem optimization of the CO method is nested within the system optimization, that is, it belongs to the two-cycle algorithm we mentioned above, so the amount of calculation is large. In addition, when the CO method is applied to a near-linear MDO problem, the nonlinearity of the problem is increased, which also increases the difficulty of optimization.

In order to improve the convergence of the CO method, many scholars have studied the use of penalty functions or augmented multiplier functions to improve the consistency of the system-level consistency method to the objective function, such as Yang (2013). There are also scholars who have introduced approximate methods such as response surface into the CO method, and obtained sufficient data through methods such as experimental design calculations for each subsystem. Based on these data, the output coupling variables and input parameters (design variables and input coupling) The relationship between parameters) is expressed by approximate models. These approximate models are directly called instead of subsystem optimizations when the system is optimized, which can effectively reduce the calculation amount of system optimization. In order to further ensure the accuracy, the approximate model and the exact model can also be called alternately, for example, the variable complexity CO method of alternating the precise model and the approximate model is introduced in the literature (Lin 2009).

2.2.7 ATC Method

The ATC method was proposed by Michelena and KIM et al. of the University of Michigan (Michelena et al. 1999; Kim et al. 2003, 2004). The ATC method was originally aimed at the design of layerable products. The design indicators of the products were decomposed from system indicators into subsystem indicators, and the subsystem indicators were further divided into component indicators… At the same time, each level continuously feedbacks the optimization design results (response feedback) to the upper level, and the upper and lower levels are optimized alternately until the convergence conditions are met.

Assuming that the system-level objective function and constraint function in the MDO problem containing three subsystems above are only functions sharing the design variable x, the framework of ATC method is shown in Fig. 2.6.

In the system-level optimization, the target value of the objective function of the original problem must first be determined, and the difference between the calculated value of the objective function of the original problem and the target value and the sum of the residuals of various consistency constraints as the objective function. In addition to the original system-level constraints, there are consistency constraints for the shared design variable x $\|x_{sub} - x\| = (x_{sub1} - x)^2 + (x_{sub2} - x)^2 + (x_{sub3} - x)^2 \leq \varepsilon_x$, state variable consistency constraints $\left\|z_i^u - z_i\right\| \leq \varepsilon_{z_i}$, and coupling variable consistency constraints $\left\|y_{ij}^u - y_{ij}^{subi}\right\| + \left\|y_{ij}^u - y_{ij}^{subj}\right\| \leq \varepsilon_{y_{ij}}$ where subsystem i feeds back the value of the system-level coupling variables. (Among them, y_{ij}^{subi} is the value of the coupling variable that subsystem i feeds back to the system level. For

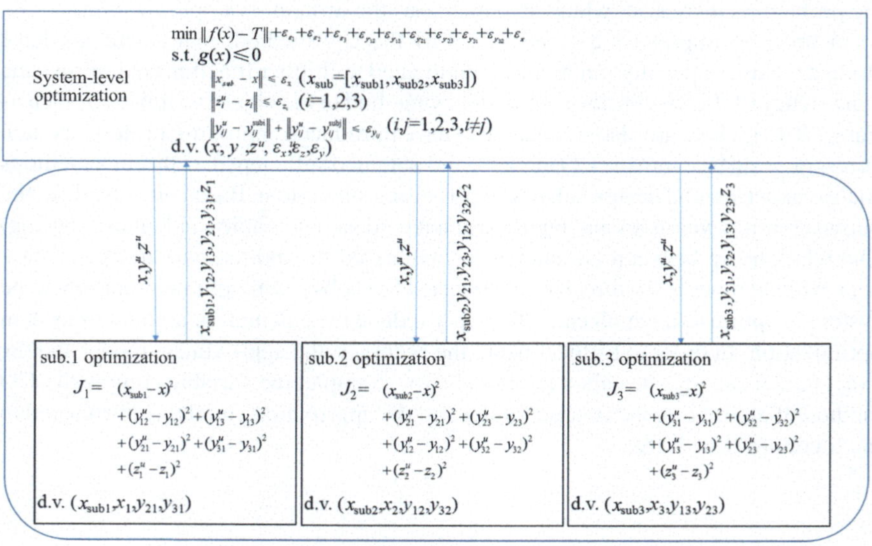

Fig. 2.6 ATC method for three subsystem MDO problems

example: y_{12} returns values from subsystem 1 and subsystem 2 to the system level, and these values are unified by consistency constraints at the system level). In addition to the system-level design variable x, the design variables also include coupling-assisted design variables y^u, state-assisted design variables z^u, and residuals for each consistency constraint.

The objective function of subsystem level optimization is to minimize the differences of common design variables, coupling variables and state variables. In addition to the local design variable xi of the original problem book subsystem, new common auxiliary design variables x_{subi} and input coupling auxiliary design variables y_{ji} ($j \neq i$ that is, the coupling variables passed from other subsystems to subsystem f) are also added to the design variables.

It should be noted that in addition to the decoupling between subsystems, the decoupling between the ATC method also emphasized between system level and subsystem level optimization is also independent of each other, that is in the process of the system level optimization to calculate the objective function and constraint conditions, all subsystem parameters obtained by the last feedback subsystem value substitution, the same subsystem optimization process will be subject to targets for the system level. In this way, the sequence of system-level optimization and subsystem-level optimization is implemented alternately, and the nested structure of the CO method subsystem optimization and system optimization is avoided, which can reduce the calculation amount. However, the ATC method will lead to a large increase in the number of design variables and constraints, especially in MDO problems with many coupled variables. Therefore, in recent years, there have been ATC methods that incorporate consistency constraints into the objective function by using a penalty function method, which reduces the number of constraints and design space dimensions, such as Tosserams et al. (2010) and Dor Mohammad (2013) etc. ATC method and CO method is different, in addition to the embedded system and subsystem optimization iteration from change for alternates, ATC method belongs to the multistage method, and can also process more than two levels (including two stage) of MDO problem, such as the automobile design is divided into the vehicle, systems, subsystems and components, and other design level, and CO in law is a two-stage method, only the system and subsystem; In addition, the ATC method incorporated the residuals of the consistency constraints into the objective function in accordance with the idea of the Lagrange multiplier method of the traditional equity-constrained optimization problem, avoiding the defect that the CO method did not satisfy KKT condition (the basic mathematical basis of the traditional optimization such as Lagrange multiplier method and KKT condition will be introduced below).

2.3 Key Technologies for Multidisciplinary Design Optimization

The principles of seven commonly used multidisciplinary design optimization methods are introduced above. The framework and data transfer flow of these methods are described in detail, and the advantages and disadvantages of each method are analyzed. Reviewing these seven methods, we can find that the cores of these multidisciplinary design optimization methods are decoupling and approximation. The successful solution of the MDO problem depends not only on the modeling using a suitable multidisciplinary design optimization method, but also on the search ability of the optimization solver for the design space (i.e., the ability to optimize). Therefore, this book defines decoupling, approximation, and design space search as key technologies for multidisciplinary design optimization. The following is an in-depth introduction of them, with the purpose of enabling engineering designers to establish and solve appropriate multidisciplinary design optimization models based on the characteristics of specific problems after mastering the three key technologies, instead of being limited to the seven methods mentioned above.

2.3.1 Decoupling

The decoupling here refers to a method of decoupling between subsystems by introducing auxiliary design variables and consistency constraints, and its applications can be seen in the AAO method, IDF method, CO method and ATC method.

Take the simplest example of a problem involving two coupled subsystems (Fig. 2.7). I In calculating coupling variables y_{12}, subsystem 1 needs to obtain coupling variables y_{21} from subsystem 2 in addition to design variables x_1 from the system layer, while subsystem 2 needs to calculate coupling variables y_{21}, and design variables x_2 in turn need to obtain coupling variables y_{12} from subsystem 1. As a result, it has fallen into a logical endless loop.

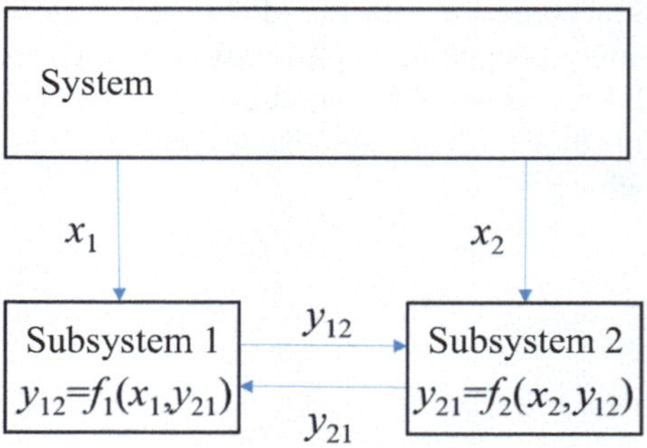

Fig. 2.7 Dual coupling subsystem problem

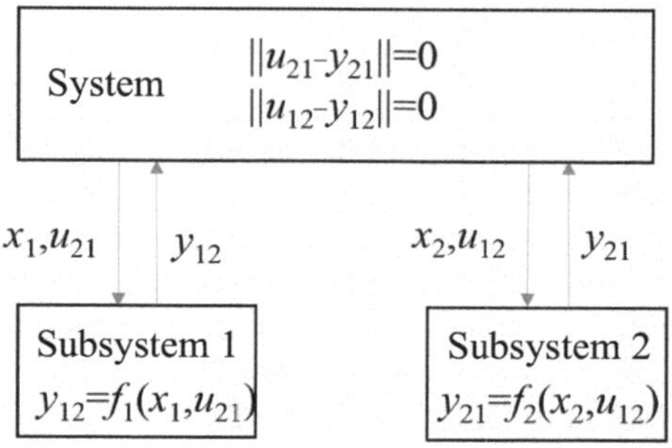

Fig. 2.8 twin system problem after decoupling

In addition to the method used in the MDF method, which assumes the initial value of the coupled variable and then iterates numerically, another method to solve this dead loop is the decoupling method. As shown in Fig. 2.8, adding two auxiliary variables in the system layer u_{21} and u_{12} to replace the coupling variables y_{21} and y_{12}, respectively as input variables of subsystem1 and subsystem 2, two subsystems will be calculated after the coupling variables y_{12} and y_{21} value is returned in the form of system layer, and then the system layer for the coordination of auxiliary variables and the coupling value consistency, namely consistency constraints $||u_{12} - y_{12}|| = 0$ and $||u_{21} - y_{21}|| = 0$. When the consistency constraint is satisfied, it means that the auxiliary variable is equal to the coupling variable, the calculated value of each subsystem substituted by the auxiliary variable is equal to the calculated value directly substituted by the coupling variable. This process transfers the dependency between subsystem 1 and subsystem 2, namely decoupling.

Among them, the consistency constraint is also called compatibility constraint, which is used to ensure the consistency between the newly introduced auxiliary variable and the corresponding variable calculation value. Consistency constraints increase the number of equation constraints in the optimization process of the system layer. For practical engineering problems, the consistency constraints of the equation type are often too strict, so many times the consistency will be introduced by introducing small residuals. Constraints are transformed into inequality constraints. For example, equality-type consistency constraints $\|y^u - y\| = 0$ can be transformed into inequality forms:

$$\left\|y^u - y\right\| \le \varepsilon \tag{2.2}$$

In addition to the most common expression $\|y^u - y\| \le \varepsilon$, there are many equivalent expressions, such as:

(1) $\left|y^u - y\right| \le \varepsilon$

(2) $\left|\frac{y^u}{y} - 1\right| \le \varepsilon$

(3) $\ln\left(\left|\frac{y^u}{y}\right|\right) \leq \varepsilon$

The advantage of decoupling method is that it can remove the dependency among subsystems and make each subsystem independent of each other, so that parallel computing can be carried out and the computing efficiency can be improved. Defect is increased the number of design variables and constraints, increased the difficulty in solving complex optimization algorithm, and decoupling the introduction of the consistency of the constraint conditions may also makes optimization problem does not meet the KKT conditions, make efficient optimization algorithm based on gradient failure, you need to use some math skills to the objective function and constraint conditions of processing.

2.3.2 The Approximate Technology

The approximation technique is to establish an approximate relationship between data based on existing data. Figure 2.9 shows the black box. Just give the box an input x to get an output y, but the functional relationship f in the black box is unknown. In order to explore the functional relationship between the input and output in the black box, given a large number of input $[x_1, x_2, \ldots, x_n]$, the corresponding output $[y_1, y_2, \ldots, y_n]$ is obtained after the black box calculation. It is assumed that the function f is a function with a certain coefficient to be determined, such as the most commonly used response surface approximation technique. It is assumed that f is a quadratic function with undetermined coefficients, and then based on the method of least squares and other methods, known undetermined x and y data points can be used to substitute the undetermined coefficients to obtain an approximate function of f.

In engineering design, the function f in the black box is the model of the design, so approximation technology is often referred to as proxy model or approximation model technology. For the MDO problem, due to the existence of state variables and design variables, coupling variables and design variables, coupling variables and other relationships, so the approximation technology has a wide range of applications: 1. When used to approximate the relationship between the coupling variables between subsystems and design variables, it is called the intermediate approximation; (3) when directly used to approximate the relationship between system-level indicators and design variables, it is called the global approximation.

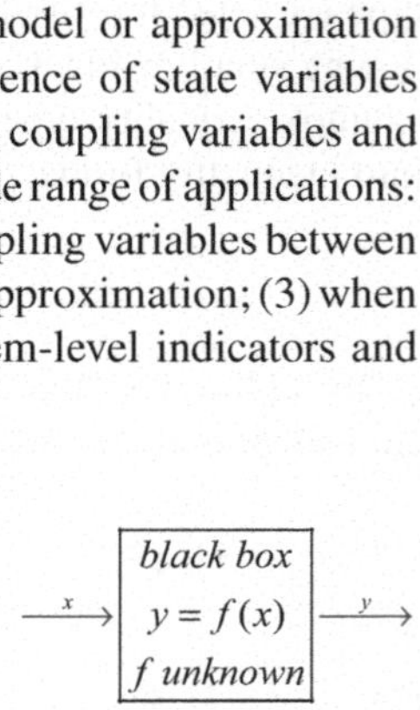

Fig. 2.9 Approximate objects

Approximation techniques are used for other purposes besides establishing unknown functional relationships. For example: in some MDO problem, subsystem model has yet to be clear, but the fine model calculation need to consume a large amount of computing resources, optimization of MDO problem time-consuming or even impossible, then the approximate model was used instead of fine model, quickly get optimal design point, and then through the model validation. In addition, although the development of some discipline analysis tools has been relatively mature and widely used in engineering design, these specialized discipline tools do not consider the connection with other software when they are designed, resulting in programming difficulties, which makes these specialized discipline tools cannot be integrated into the optimization framework. You can assume that these professional discipline tools are black boxes. Based on the data obtained from the black-box analysis, an approximate model can be established, and then the approximate model can be used to replace these professional discipline tools in the optimization framework, so as to realize the indirect integration of professional discipline tools.

It can be seen that the approximation technique involves three steps:

① Obtain sufficient output and input data points; ① Select a suitable approximation model; ③ Determine the coefficients in the approximation model and verify the approximation accuracy.

Step ① is the basis of the approximation technique. The process of obtaining the input-output data points of the black box is the process of testing the black box, and the technique of how to arrange the test to reflect the trend of the output variable with the input variable with the minimum number of trials is called experimental design. After obtaining a sufficient number of reasonably distributed data points, you can use different approximation models to perform regression analysis on these data points, determine the parameters in the approximation model, and then calculate the output of the approximation model at step ① at the known data points and Comparison of known points to verify the accuracy of the approximate model, and sometimes it is necessary to check the difference between the approximate model and the original black box at a new design point. The following will introduce experimental design techniques and common approximation models, including response surface approximation, Taylor expansion approximation, Kriging approximation, and artificial neural networks.

2.3.2.1 Design of Experiment

The essence of Design of Experiment (in this book, DOE for short) is that according to some planning method, the design space is divided into a certain number of grids, and then the responses of these grid nodes are obtained, so that the response can be roughly estimated. Value. In MDO, the most widely used design of experiments is to analyze the sensitivity of system output indicators to input parameters, and to obtain sufficient output-input data points for the establishment of approximate models. Since the development of experimental design technology, it is not limited to providing initial data points for approximate models, and its application scenarios include: ① evaluating the response of design variables (evaluating the impact of design variables on the specified response); ② identifying significant differences between design variables Interaction; ③ Perform a rough analysis of the design space and obtain a rough estimate of the optimal solution, providing initial points and guidance for the optimization solution; ① After the obtained design variables have a large impact on the objective function, they can have a small impact on the objective Elimination of design variables (screening design variables while reducing the dimension of the design space).

First, terms commonly used in design of experiment:

(1) Factor: The design variables that need to be divided into design spaces in the DOE. We can select all design variables or several design variables as factors according to the needs. Generally, the factor is to take all design variables.
(2) Level: The value of the factor. For example, the optimization problem has three design variables x_1, x_2, x_3. We need to study the effect of x_2 on the output. Then take x_2 as a factor and put it in its value range $[x_2^l, x_2^u]$ (where, x_2^l is the lower limit of the value of the factor, and x_2^u is the upper limit of the value of the factor). The three values of x_2 are evenly divided into three $\left[x_2^l, \frac{x_2^l+x_2^u}{2}, x_2^u\right]$: then these three values are the level of x_2.
(3) Response: when the factor takes different levels, the corresponding value of the system output.

Common DOE methods include:

(1) Single parameter design. This method studies the independent effect of factors on the response, that is, when one factor is changed, the other factors remain unchanged. So this method does not include cross-effects between factors. From this process, it can be seen that the number of test points (i.e., the number of analyses) of this method is $1 + N_1 + N_2 + \ldots + N_i + \ldots + N_n$ (where N_i represents the number of levels of the i-th factor), and 1 represents an initialization analysis performed. This method can obtain good results with fewer experimental points when the cross-effects of the factors are relatively small or the factors are independent of each other.
(2) Full factorial design. This method calculates the combination of all levels of all factors, and calculates the response for all nodes after the design space is evenly divided, so its number of test points is $N_1 \times N_2 \times \ldots \times N_i \times \ldots \times N_n$ (where N_i is the number of levels of the ith factor). For example, for a problem with two factors .. and x_2, each factor is divided into 2 levels, you need to calculate 2 $\times$ 2 = 4 test points, that is, the response at (x_1^l, x_2^l), (x_1^l, x_2^u), (x_1^u, x_2^l), (x_1^u, x_2^u).

The information given by this method includes cross-effects, which can get the most design space information if the number of levels is large enough, but the number of test points required for this method will be very large if the design space dimension is high. For example, the model of a ship conceptual design has 25 design variables, each of which only takes 4 levels, then the number of test points (DOE response calculation times) required by this method reaches 425 = 1125899906842624, so the practicability of this method for high-dimensional problems is almost zero.

Table 2.1 $L_8(4^1 \times 2^4)$ orthogonal table

Experiment no.	Factor				
	x_1	x_2	x_3	x_4	x_5
1	x_1^l	x_2^l	x_3^l	x_4^l	x_5^l
2	x_1^l	x_2^u	x_3^u	x_4^u	x_5^u
3	$\frac{2x_1^l+x_2^u}{3}$	x_2^l	x_3^l	x_4^u	x_5^u
4	$\frac{2x_1^l+x_2^u}{3}$	x_2^u	x_3^u	x_4^l	x_5^l
5	$\frac{x_1^l+2x_2^u}{3}$	x_2^l	x_3^u	x_4^l	x_5^u
6	$\frac{x_1^l+2x_2^u}{3}$	x_2^u	x_3^l	x_4^u	x_5^l
7	x_2^u	..	x_3^u	x_4^u	x_5^l
8	x_2^u	x_2^u	x_3^l	x_4^l	x_5^u

(3) Orthogonal design. On the basis of the full factor method, the orthogonal matrix method is used to satisfy the orthogonal test conditions. The method of selecting the partial horizontal points of the partial factors to calculate the response is called the orthogonal design method. Since the development of the orthogonal design method, there have been many standard orthogonal tables available. For example, for a 5-factor problem $(x_1, x_2, x_3, x_4, x_5)$, the factor x_1 has 4 levels $\left(x_1^l, \frac{2x_1^l+x_2^u}{3}, \frac{x_1^l+2x_2^u}{3}, x_2^u\right)$:, and the other factors have 2 levels:$(x_i^l, x_i^u)(i = 2, 3, 4, 5)$, then the problem (where $L_8(4^1 \times 2^4)$ 8 represents the number of trials, 4^1 represents 1 factor with 4 levels, and 2^4 represents 4 factors with 2 levels (Horizontal). The orthogonal table is shown in Table 2.1.
It can be seen that the orthogonal design method is based on the full factorial design method and only calculates the partial horizontal combination of partial factors, which can effectively reduce the number of test points. From the perspective of the division of design space, there are two main purposes for experimental design:
Uniform dispersion: the test points are evenly distributed in the design space, so that each test point is fully representative.
Orderly comparison: it is convenient for the analysis of experimental design results, easy to estimate the main effect and coupling effect of each factor.
Orthogonal design can be thought of as a more "tidily comparable" approach.

(4) Uniform design. In order to improve the uniform dispersion of the test points, Kaitai (1980) proposed a uniform design method. Because of the long length of mathematical theories such as number theory, this article will not introduce them. At present, there are various standard uniform design tables for MDO problems, and related software and modules have been mature. After the development of the uniform orthogonal method, it has the advantages of both the

orthogonal method and the uniform method, and a standard uniform orthogonal table is also available for reference.

(5) Latin square design. In the Latin square design, the same process as the total factorial design method is used first to divide all factors into a certain number of levels, and then randomly combine the levels of these factors to obtain n test points (n is set by man). Because the test points are artificially controlled, the Latin square design is more flexible than the orthogonal design and uniform design. Since the test points are also randomly generated, the distribution of these points is irregular. And in theory, it is not reproducible (in engineering, the reproduction can be guaranteed by fixing the seed of the pseudo-random number).
Based on the test point matrix generated by the Latin square design, the two levels of the factor are swapped to generate a new matrix, and then the distance between the test points is evaluated. If the test points are too close, only one representative test point is retained, and other nearby points are moved or discarded. This process is a process of optimizing the distribution of test points to make it as uniform as possible, so it is also called the optimized Latin method.

(6) Center combination design. The center combination design starts from a 2-level orthogonal table. Based on 2^n test points composed of 2 levels of all n factors, the median of two levels x_i^{l1} and x_i^{l2} of each factor x_i is used to form an intermediate benchmark Point $(x_1^b, x_2^b, \cdots x_i^b, \cdots x_n^b)$ (where $x_i^b = \frac{x_i^{l1}+x_i^{l2}}{2}$). Then based on the confidence region theory (in many cases, the confidence region radius is taken $r_i = \frac{x_i^{l2}-x_i^{l1}}{2}$ directly). At the intermediate reference points, the values of the factors are changed to the lower limit and upper limit of the confidence field one by one (when the value of a factor is changed, the values of other factors are kept as the intermediate reference points), so that the additional 2n test points can be obtained. The new test points can be described by a center point coordinate and the radius of the confidence field, which is convenient for programming. The structure of the center combination design is based on the center reference point, which is suitable for the local design with a certain point as the center, so it is very suitable for establishing the approximate model of the response surface. However, 2 horizontal orthogonal table test points, 1 central reference point, and 2n new points, the total number of test points of this method is $2^n + 1 + 2n$. Therefore, this method is used for high-dimensional problems with too many test points. For example, for the problem of 25 design variables mentioned above, the test points still reach 225 + 1+2 × 25 = 33554483.

2.3.2.2 Empirical Formula Modification

Most disciplines have a set of theoretical formulas with a long history, rigorous derivation and relatively complete. However, because these theoretical formulas are mostly based on simplification and assumptions, the theoretical formulas are not consistent with the functions of complex practical engineering objects, which results in a large calculation error. Therefore, each discipline usually evolves a variety of empirical formulas on the basis of the basic theoretical formulas to reduce the error between the formulas and the actual engineering objects by modifying the coefficients. Because these empirical formulas are consistent with the principles of engineering objects and have been revised many times based on the data of actual engineering products, many of them have become the standard formulas of disciplines or industries. Most of these formulas cannot contain the details of all engineering products, and the calculation accuracy is not as high as the fine model. And we know that in the MDO theory, the phase of system design optimization gains the most is the phase of scheme design. Therefore, these formulas are very suitable for MDO after appropriate modification, which cannot only ensure sufficient accuracy, but also not require a large amount of calculation, and also keep the clear principle of MDO model for subsequent model improvement and upgrade.

Therefore, the adoption of the approximate model should be carried out by the professionals of each discipline. First, evaluate whether the empirical formula can be modified to achieve sufficient accuracy. If the accuracy can meet the requirements of the MDO problem, the empirical formula approximation model is preferred. When the empirical formula does not exist or the accuracy is insufficient, the following other approximation models are used.

2.3.2.3 Response Surface Approximation

In response surface approximation models, polynomials with undetermined coefficients are used to fit real functions. One-to-four-degree response surface models are commonly used. At the same time, in order to reduce the number of coefficients to be determined, the third and fourth mixed terms are usually removed. Therefore, the general expression of the response surface approximation is:

$$y = \tilde{f}(x) = a + \sum_{i=1}^{n} b_i x_i - \sum_{i=1}^{n} c_i x_i^2 + \sum_{1 \le i < j \le n}^{n} d_{ij} x_i x_j + \sum_{i=1}^{n} e_i x_i^3 + \sum_{i=1}^{n} g_i x_i^4 \tag{2.3}$$

where:

y is the response;
$\tilde{f}(x)$ is an approximate function;
$x = (x_1, x_2, \ldots, x_i, \ldots, x_n)$ is the input variable;
a is a constant;
$b = (b_1, b_2, \ldots, b_i, \ldots, b_n)$ is the coefficient of the first term;
$c = (c_1, c_2, \ldots, c_i, \ldots, c_n)$ is a pure quadratic coefficient;
$d = (d_{ij},\ where\ 1 \le i < j \le n)$ is the mixed quadratic term coefficient;
$e = (e_1, e_2, \ldots, e_i, \ldots, e_n)$ is the coefficient of cubic term;

Table 2.2 Correspondence between response surface times and test points

Times of response surface	Number of test points
1次	$n + 1$
2次	$(n + 1)(n + 2)/2$
3次	$(n + 1)(n + 2)/2 + n$
4次	$(n + 1)(n + 2)/2 + 2n$

Note n is the dimension of the input variable

$g = (g_1, g_2, \ldots, g_i, \ldots, g_n)$ is the fourth order coefficient.

When the low order response surface is used, the high order coefficient is zero.

After selecting the number of response surfaces, in order to obtain the coefficients, the number of equations to be calculated must be equal to the number of coefficients to be determined, that is, the number of test points required to construct the response surface approximate model is equal to the number of coefficients to be determined. In general formula (2.3), the test points required when response surfaces of different times are used are shown in Table 2.2. The process of solving these undetermined coefficients is the process of solving the simultaneous equations. After data points are substituted in, it can be seen that the equations of undetermined coefficients are all linear equations. The least-squares estimation of each coefficient can be easily obtained by using multiple linear regression method. Because the difficulty of solving these coefficients is not high, the least square method in traditional optimization can also be directly used to solve them.

The most commonly used in engineering is the secondary response surface. The advantage of this model is that it is continuously differentiable, which can remove the effects of digital noise, smooth the function, and help optimize the call of the solver. The disadvantage is that the response surface model is dealing with highly nonlinear high-dimensional problems (the number of input variables is large) Sometimes the approximation is not ideal.

2.3.2.4 Taylor Series Approximation

Taylor series expansion is usually used to approximate the properties of the function near the expansion point, so it is also a commonly used approximation technique. In addition to obtaining the input and output values of the test points, the Taylor series approximation also requires the derivative information (gradient information) of the function at the design point. The most commonly used Taylor series approximation models are:

(1) Linear Taylor approximation

First, the function value and gradient vector at the expansion point are obtained by the original function calculation (black box calculation). The original function can be approximated as:

$$y = \tilde{f}(x) = y^0 + \sum_{i=1}^{n} g_i(x^0)(x_i - x_i^0) \tag{2.4}$$

where:

y is the response;
$x = (x_1, x_2, \ldots, x_i, \ldots, x_n)$ is the input variable;
$\tilde{f}(x)$ represents an approximate function, and f (x) represents the original function;
x^0, y^0 is the input and output variables of the Taylor expansion point;
$g(x^0) = (g_1(x^0), g_2(x^0), \ldots, g_i(x^0), \ldots, g_n(x^0))$ is the gradient vector of the original function at the test point, and each component of it can be obtained by the forward difference method:
$g_i(x^0) = \left.\frac{\partial f}{\partial x_i}\right|_{x^0} \approx \frac{f(x_1^0,\ldots,x_i^0+\Delta x_i,\ldots,x_n^0)-y^0}{\Delta x_i}$ (Where Δx_i is the differential step size)
A total of at least n + 1 black box calculations are required to obtain the output value and gradient vector at the unfolding point.

(2) Inverse linear Taylor approximation.

$$y = \tilde{f}(x) = y^0 + \sum_{i=1}^{n} g_i(x^0)(x_i - x_i^0)\frac{x_i^0}{x_i} \tag{2.5}$$

The definition in the formula is the same as above.

(3) Mixed linear Taylor approximation.

$$y = \tilde{f}(x) = y^0 + \sum_{i=1}^{n} g_i(x^0)(x_i - x_i^0)\phi_i(x_i, x_i^0) \tag{2.6}$$

where:

$$\phi_i(x_i, x_i^0) = \begin{cases} 1 \; if \; x_i^0 \times g_i(x^o) \geq 0 \\ x_i^0/x_i \; else \end{cases}$$

Other definitions are the same as above.

(4) Two-point Taylor approximation

Through the calculation of the original function, the original function value and the gradient vector at the two expansion points are obtained respectively, and then the approximation of the function between the two points is obtained:

$$y = \tilde{f}(x) = y^{00} + \sum_{i=1}^{n} g_i(x^{00})\frac{(x_i)^{1-p_i}}{p_i}((x_i)^{p_i} - (x_i^{00})^{p_i}) + \frac{1}{2}\varepsilon \sum_{i=1}^{n}((x_i)^{p_i} - (x_i^{00})^{p_i}) \tag{2.7}$$

where:

x^0, y^0 are input and output variables at expansion point 1, respectively. $g(x^0)$ is the gradient vector at expansion point 1;
x^{00}, y^{00} are input and output variables at expansion point 2, respectively. .. is the gradient vector at expansion point 2;
$p = (p_1, p_2, \ldots, p_i, \ldots, p_n)$ is a non-linear exponential vector, ε is the correction coefficient. These two coefficients are replaced by Eq. (2.7) into n + 1 simultaneous equations.

$$\begin{cases} \left.\frac{\partial \tilde{f}}{\partial x_i}\right|_{x^0} - g_i(x^0) = 0 \ (i = 1, 2, \ldots n) \\ \tilde{f}(x^0) - y^0 = 0 \end{cases}$$

Solve this equation

Among them, because the two-point Taylor expansion contains higher-order information, it has wider applicability than the other three approximate models based on linear Taylor expansion (Golovidov 1998). The advantage of Taylor expansion is that it requires fewer calculations of the original function (the first three linear models only need n + 1 calculations, and the two-point model also only needs 2 (n + 1) calculations); the disadvantage is the Taylor approximation Accuracy can only be guaranteed near the expansion point. However, when the design point is far from the deployment point, the error will increase or even no solution.

2.3.2.5 Kriging Approximation

The Kiging approximation (Kiging) was first proposed by the South African geologist Kige in 1951, and was introduced into MDO in 1990. The basic idea of Kriging approximation can be described as: the output value at a certain point can be expressed as a linear combination of the output values of known test points in the neighborhood of the point.

The concepts of expected values, variances, and covariances of random variables are described in the textbook of random mathematics, and their definitions are also given in Chap. 3 of this book. The Kriging method is an interpolation method based on random mathematical theory. In addition to these basic concepts, the Kriging method also uses second-order stationary and eigenhypothesis, which are defined as follows:

(1) Second-order stationary assumption

① The mathematical expectation of y exists in the entire research area and is equal to the constant, that is $\forall x, \forall \Delta x$, there are:

$$E[f(x)] = E[f(x + \Delta x)] = U(\text{常数}) \tag{2.8}$$

The physical meaning is that the spatial variation of the random variable has no obvious trend, but fluctuates around the mean U.

② The covariance of y exists and is stable in the entire study area, that is,

$$\begin{aligned} & Cov\{f(x), f(x+h)\} \\ & = E[f(x)f(x+h)] - E[f(x)]E[f(x+h)] \\ & = E[f(x)f(x+h)] - U^2 \\ & = c(h) \end{aligned} \tag{2.9}$$

$c(h) = E[f(x)f(x+h)] - U^2$ is called the covariance function. The physical meaning of this condition is that the covariance of a random variable does not depend on the absolute position in space, but is determined only by the relative position in space, that is, y has the invariance of c in space.

(2) Eigen Hypothesis
① The mean value of the increment of the random variable is zero:

$$E[f(x) - f(x+h)] = 0 \tag{2.10}$$

② The incremental variance of the random variable exists and is stable:

$$\begin{aligned} & Var[f(x) - f(x+h)] \\ & = E[f(x) - f(x+h)]^2 - \{E[f(x) - f(x+h)]\}^2 \\ & = E[f(x) - f(x+h)]^2 \\ & = 2\gamma(x, h) = \gamma(h) \end{aligned} \tag{2.11}$$

$\gamma(h) = \frac{1}{2}E[f(x) - f(x+h)]^2$ is called the variogram.

Assuming that the output quantity y = f(x) satisfies the second-order stationary hypothesis or eigenhypothesis, and its mathematical expectation is U, the covariance function is c(h), and the variance function is x, then the value of the regional variable y at x can be estimated by the linear combination of the values of the known points in the neighborhood of the point, that is:

$$y = \tilde{f}(x) = \sum_{j=1}^{m} \lambda_j y_j^0 \tag{2.12}$$

where:

M is the number of known data points in the adjacent region;
y_j^0 is the y value at the known data points in the adjacent region;
$\lambda = (\lambda_1, \lambda_2, \ldots, \lambda_m)$ is the weight coefficient.

In order to determine the weighting coefficient γ, an unbiased condition and an optimal estimation condition (also called an estimation variance minimization condition) are used.

(1) Unbiased conditions

$$\begin{aligned} &E[\tilde{f}(x) - f(x)] \\ &= E\left[\sum_{j=1}^{m} \lambda_j y_j^0 - f(x)\right] \\ &= \left(\sum_{j=1}^{m} \lambda_j\right) U - U = 0 \end{aligned} \tag{2.13}$$

The result is:

$$\sum_{j=1}^{m} \lambda_j = 1 \tag{2.14}$$

(2) Optimal estimation conditions

$$\begin{aligned} &Min\ \sigma_E^2 \\ &= E[\{\tilde{f}(x) - f(x) - E[\tilde{f}(x) - f(x)]\}^2] \\ &= E[\tilde{f}(x) - f(x)]^2 \end{aligned} \tag{2.15}$$

Therefore, the process of solving the weight coefficients λ becomes the process of solving the equation-constrained optimization problem:

$$\begin{aligned} &Min\ \sigma_E^2 = E[\{\tilde{f}(x) - f(x)\}^2] \\ &s.t.\ \sum_{j=1}^{m} \lambda_j = 1 \end{aligned} \tag{2.16}$$

When the random variable meets the second order stationary condition, the objective function of the optimization problem can be rewritten by using the covariance function:

$$\begin{aligned} \sigma_E^2 &= E[\{\tilde{f}(x) - f(x)\}^2 \\ &= E\left[\left\{\sum_{j=1}^{m} \lambda_j y_j^0 - f(x)\}^2\right\}\right] \end{aligned}$$

$$= c(x, x) + \sum_{j=1}^{m}\sum_{k=1}^{m}\lambda_j\lambda_k c(x_j^0, x_k^0) - 2\sum_{j=1}^{m}\lambda_j c(x_j^0, x) \tag{2.17}$$

where:

x_j^0 is the input value at the known data point $j(j = 1, 2, \ldots, m)$ corresponding to the output variable y_j^0, and each x_j^0 is an n-dimensional vector.

Using the Lagrange multiplier method, the original optimization problem can be transformed into an unconstrained optimization problem:

$$Min\ F = \sigma_E^2 - 2\mu\left(\sum_{j=1}^{m}\lambda_j - 1\right) \tag{2.18}$$

The partial derivatives of F with respect to $\lambda_j(j = 1, 2, \ldots, m)$ and μ are respectively obtained and set as zero to obtain the Kriging equations:

$$\begin{cases} \frac{\partial F}{\partial \lambda_j} = 2\sum_{k=1}^{m}\lambda_k c(x_j^0, x_k^0) - 2c(x_j^0, x) - 2\mu = 0 \\ \frac{\partial F}{\partial \mu} = -2\left(\sum_{j=1}^{m}\lambda_j - 1\right) = 0 \end{cases} \tag{2.19}$$

After sorting, it can be obtained:

$$\begin{cases} \sum_{k=1}^{m}\lambda_k c(x_j^0, x_k^0) - \mu = c(x_j^0, x) \\ \sum_{j=1}^{m}\lambda_j = 1 \end{cases} \tag{2.20}$$

Solve this m + 1 equation to get the weight coefficient λ and the Lagrange coefficient μ

$$\gamma(h) = c(0) - c(h) \tag{2.21}$$

The Kriging equation is expressed as a function of variation:

$$\begin{cases} \sum_{k=1}^{m}\lambda_k \gamma(x_j^0, x_k^0) + \mu = \gamma(x_j^0, x) \\ \sum_{j=1}^{m}\lambda_j = 1 \end{cases} \tag{2.22}$$

This method is called the simple Kriging method when the expectation of the random variable is known, and the ordinary Kriging method when the expectation is unknown.

It can be seen that the Kriging method is a spatial local interpolation method. This method is based on the variogram theory and uses the known data points of the regionalized variables and the structural characteristics of the variogram to perform a linear unbiased optimal estimation (interpolation) of the unknown points. In MDO, the Kriging method usually breaks an unknown function into a global approximation function and a local deviation function:

$$y = \tilde{f}(x) + \tilde{z}(x) \tag{2.23}$$

where, $\tilde{f}(x)$ is the global approximate model, which usually adopts the response surface approximation model or directly takes as a constant. $\tilde{z}(x)$ is a random process solution with local deviation approximation model, usually take mean value (expectation) is 0, variance is σ^2, and covariance non-zero. Specific function forms $\tilde{z}(x)$ can be taken in various ways, which will not be introduced here. Kriging method only needs at least two data points to establish the model, but engineering experience shows that an ideal Kriging approximation model can be established by taking 10n data points with relatively uniform distribution.

2.3.2.6 Artificial Neural Networks

The basic principle of artificial neural network learning from biological neural network: the processing capacity of each neuron is very low, only received information from the simple reaction, but of a large number of neurons connected with simple processing power network, after all the neurons connected to each other, and by means of the forward or feedback interaction, can become a powerful adaptive system. One important function of neural networks is the ability to learn how things work, not only by learning accurate samples, but also by correcting new data that may be incomplete or noisy. In recent years, the research of neural network in the field of function approximation has been gradually mature, and many new network types have been produced. In addition, neural networks have been widely used in pattern recognition, Figure processing and computer vision, signal processing, time series, medical control, expert systems, power systems, military systems, financial systems, artificial intelligence and optimization.

A neural network consists of the most basic processing unit, the neuron. Each neuron in the organism can only produce simple responses to external stimuli, and the neurons of the artificial neural network also only perform the simplest processing of input parameters. Take a function $y = g(x_1, x_2, x_3)$ containing 3 input variables and 1 output variable as an example. When a single neuron is used to approximate the function g, this neuron is shown in Fig. 2.10.

This neuron can be expressed as:

$$y = \tilde{g}(x) = f(w^{\mathrm{T}}x + b) = f\left(\sum_{i=1}^{n} w_i x_i + b\right) \tag{2.24}$$

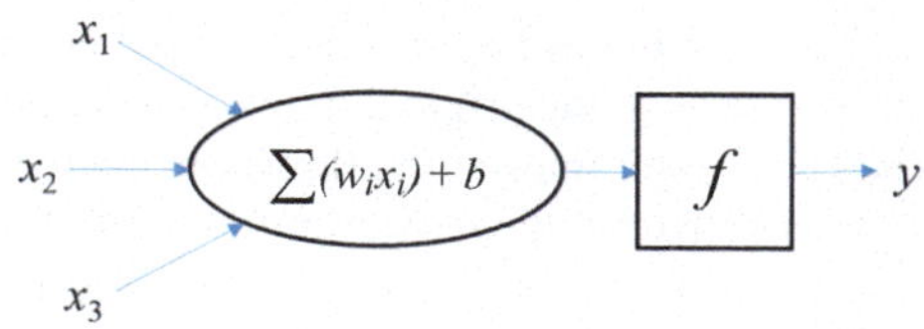

Fig. 2.10 Three input neurons

where:

$x = [x_1, x_2, x_3]^T$ is the input parameter column vector;
$w = [w_1, w_2, w_3]^T$ is the undetermined weight column vector;
$n = 3$ is the dimension of the input variable;
b is the undetermined threshold;
f is called the activation function.

The most commonly used activation Function form is Sigmoid Function, because its Function shape is similar to the signal shape:

$$f(z) = \frac{1}{1 + e^{-z}} \tag{2.25}$$

There are four undetermined coefficients w_1, w_2, w_3, b in Eq. (2.24). The process of training this neuron is to substitute the input and output values of the known data points into the process of obtaining these undetermined coefficients.

Neural network is an adaptive system connected by multiple neurons, among which the most common ones are Back Propagation Neural Network and Radical Basis Function Neural Network:

(1) Back Propagation Neural Network (BPNN)

Back-propagation neural network is a kind of neural network based on error Back Propagation (BP) algorithm to solve the undetermined coefficients of each neuron in the network. Its structure is shown in Fig. 2.11. The whole neural network is mainly composed of three layers: input layer, hidden layer and output layer.

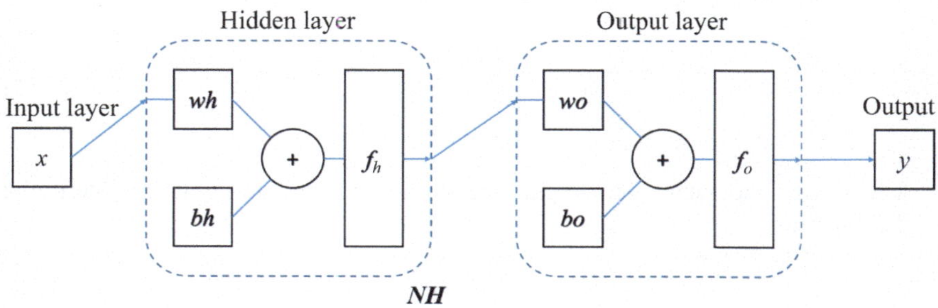

Fig. 2.11 BPNN structure

① The number of neurons in the input layer is usually equal to the number of input variables of the original function n (some algorithms will also add a bias neuron). The weight matrix and threshold vector of the neuron are denoted as wi and bi, respectively; the activation function of the hidden layer neuron is usually an S-shaped function, see Eq. (2.25). The input layer can be seen as a preliminary processing of input parameters.

② Artificially specify the number of hidden neurons NH. Generally, the more neurons in the hidden layer, the higher the fitting accuracy, but the more coefficients to be determined, the longer the training data points and training time required by the neural network. It can be adjusted according to the fitting effect, the amount of existing data and the training time; the weight matrix and threshold vector of hidden neurons are recorded as wh and bh respectively; the activation function of hidden neurons usually also uses S-shaped functions. Each neuron in the hidden layer takes the output values of all neurons in the input layer as input values.

③ The number of neurons in the output layer is usually equal to the number of output variables of the original function (the most common is the case of a single output variable, so the output layer only needs one neuron); the weight and threshold of the neuron are recorded as wo and bo; activation function fo usually uses the simplest linear function, see Eq. (2.26). Each neuron in the output layer takes the output values of all neurons in the hidden layer as input values.

$$f_o(z) = z \tag{2.26}$$

It can also be found from the structure diagram that in the back-propagation neural network, there is no correlation between neurons in the same layer and forward connections between neurons in different layers.

Take the approximate function $y = g(x)$ [where $x = (x_1, x_2, x_3)^T$] containing 3 input variables and 1 output variable as an example. If the hidden layer contains 2 neurons, the number of neurons in the input layer can be determined to be 3 by the input parameter is 3, and the single output parameter also determines that there is only one neuron in the output layer. The approximate model is:

① The input layer. Assume that weight wi and threshold bi are:

$$wi = \begin{bmatrix} wi_{11} & wi_{12} & wi_{13} \\ wi_{21} & wi_{22} & wi_{23} \\ wi_{31} & wi_{32} & wi_{33} \end{bmatrix} \tag{2.27}$$

where: wi_{ij} represents the weight corresponding to the ith input variable in the JTH neuron.

$$bi = [bi_1, bi_2, bi_3] \tag{2.28}$$

Then the output of the JTH (j = 1,2,3) neuron in the input layer is:

$$yi_j = f_i(zi_j) = \frac{1}{1 + e^{-zi_j}} \tag{2.29}$$

$$zi_j = wi^T x + bi_j = \sum_{i=1}^{3} wi_{ij} x_i + bi_j \tag{2.30}$$

② The neurons in the hidden layer take the output $yi = [yi_1, yi_2, yi_3]^T$ of the neurons in the input layer as input parameters, and the weights and thresholds of the two neurons in the hidden layer are assumed to be;

$$wh = \begin{bmatrix} wh_{11} & wh_{12} \\ wh_{21} & wh_{22} \\ wh_{31} & wh_{32} \end{bmatrix} \tag{2.31}$$

$$bh = [bh_1, bh_2] \tag{2.32}$$

The output of the jth (j = 1, 2) th neuron in the hidden layer is:

$$yh_j = f_h(zh_j) = \frac{1}{1 + e^{-zh_j}} \tag{2.33}$$

$$zh_j = wh^T yi + bh_j = \sum_{i=1}^{3} wh_{ij} yi_i + bh_j \tag{2.34}$$

③ The neurons in the output layer take the output $yh = [yh_1, yh_2]^T$ of the hidden layer as the input parameter, and the weight and threshold of one neuron in the output layer are assumed to be:

$$wo = \begin{bmatrix} wo_1 \\ wo_2 \end{bmatrix} \tag{2.35}$$

$$bo \text{ (标量)} \tag{2.36}$$

The output of this single neuron in the output layer is:

$$y = f_o(zo) = zo \tag{2.37}$$

$$zo = wo^T yh + bo = \sum_{i=1}^{2} wo_i yh_i + bo \tag{2.38}$$

It can be seen that the input parameters are continuously passed forward from the input layer, hidden layer to the output layer, and finally the output value is obtained. When there are multiple sets of data points, in order to obtain the weights and thresholds of the neurons in each layer of the network (namely the "training" of the neural network), the errors of each layer are pushed backwards from the calculated value of the network output to the target value, which is called the error back propagation method. The training process of the back-propagation neural network can be divided into two stages: the first stage (the mode forward propagation process), that is, the transfer process of the input information flow through the input layer, the hidden layer and the output layer; In the second stage (error backpropagation process), if the desired output value is not obtained at the output layer, the error signal will be backtransmitted layer by layer along the original path, and the weight and threshold values will be adjusted accordingly. Due to the space limitation, this article will not introduce the detailed training process. In the reverse error propagation, the global optimization algorithm is sometimes used to solve the weights and thresholds. For example, genetic algorithm is used to solve the weights and thresholds in the literature (Xinsheng, 2004), which improves the approximate accuracy of the neural network.

(2) Radical Basis Function Neural Network (RBFNN)

The structure of the Radical Basis Function Neural Network is basically the same as that of the back propagation neural network, which is also composed of the input layer, the hidden layer and the output layer. The input layer and the output layer are the same as the input layer and the output layer of the back propagation neural network, and the main difference lies in the hidden layer, whose structure is shown in Fig. 2.12.

In the neurons in the hidden layer of Radical Basis Function Neural Network:

① The input of the activation function is:

$$z = \|dist\| \times bh \tag{2.39}$$

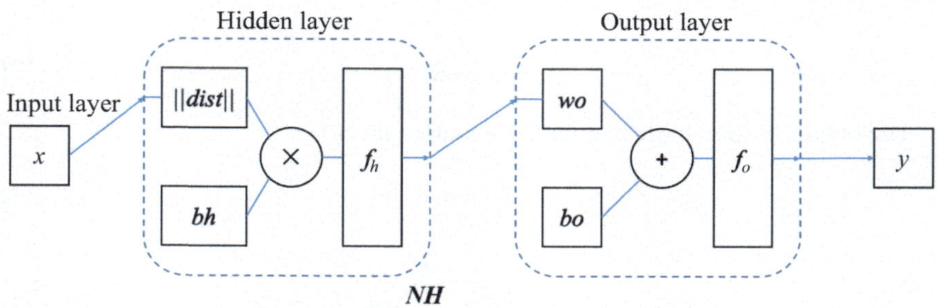

Fig. 2.12 Structure diagram of radical basis function neural network

$$||dist|| = \sqrt{\sum_{i=1}^{m} (wh_i - p_i)^2} \quad (2.40)$$

where:

bh is the threshold of hidden neurons;
wh is the weight of hidden neurons;
p_i is the input of the hidden layer, that is, the output of the output layer. As mentioned above, the number of neurons in the input layer usually takes the same number of dimensions as the input variable of the original function, so in most cases m = n;
$||dist||$ represents the distance (Euclidean distance) between the input vector p_i and the weight vector wh, that is, the radial basis.

② The activation function is usually Gaussian:

$$f_h(z) = e^{-z^2} \quad (2.41)$$

It can be seen that as the radial basis $||dist||$ decreases, the output value of the activation function increases, and when $||dist||$ is 0, the input value is a maximum of 1, so the neurons in the hidden layer can be used to detect whether their input vector is the same as the weight vector Detector (the output value is 1 when the same, the output value is close to 0 when there is a large difference). bh can be used as a coefficient to adjust the sensitivity of the function. When the difference between the input vector and the weight vector is larger, the larger bh, the greater the attenuation of the function, that is, closer to 0.

Radial basis neural network is a neural network based on function approximation theory, and the approximate model of MDO problem is mainly used for function approximation, so radial basis neural network is suitable for MDO problem.

In addition to the commonly used activation functions, the sigmoid, linear, and Gaussian functions are:

$$f(z) = \begin{cases} 1 \ (z \geq z_0) \\ 0 \ (z < z_0) \end{cases} \quad (2.42)$$

$$f(z) = \frac{1}{1 + e^{-az}} \quad (2.43)$$

$$f(z) = \frac{1 - e^{-az}}{1 + e^{-az}} \quad (2.44)$$

(Where α is the specified constant)

Approximate techniques have been used in MDO problems for nearly three decades. From the initial use to approximate the relationship between subsystem state variables and input variables, to the relationship between various variables across the system, such as the relationship between coupling variables and input variables, system indicators and subsystem design The relationship between variables, the relationship between coupling variables of different subsystems, etc. In addition, the approximation technology is combined with an optimization algorithm to perform a large-scale search to find the most likely area for solution, and then use a fine model to conduct a small-scale in-depth search. The approximation technique has also become an important method for reliability analysis and even uncertainty optimization due to its very high calculation efficiency. Approximation technology is being used more and more widely in MDO problems. Therefore, this paper uses approximation technology as one of the key technologies for multidisciplinary design optimization.

2.3.3 Design Space Search

After establishing the multidisciplinary design optimization model, it is necessary to explore the design space to find the optimal design scheme. This process is called design space search or optimization solution. With the increase of problem complexity and design space, the optimal design scheme of most multidisciplinary design optimization problems cannot be obtained by manual guessing, but need to be solved by design space search method. In most cases, the ability to design spatial search methods (i.e. the ability to find the global best) and efficiency (i.e. the speed of finding the best) are the key to the success or failure of multidisciplinary design optimization, so the design spatial search method is also one of the key technologies of multidisciplinary design optimization.

After using the multidisciplinary design optimization method to organize the MDO problem reasonably, the optimization problem at the system level and each subsystem is essentially the same as the traditional optimization problem, except that the design space has more dimensions and the optimization complexity is higher. Therefore, most of the design space search methods follow the traditional optimization methods, and with the application of MDO method in practical engineering problems, it in turn stimulates the birth of new methods. The following is a brief introduction to the basic knowledge of optimization algorithm. The mathematical model of traditional optimization can be uniformly expressed as:

$$\begin{aligned} & Min\ f(x) \\ & s.t.\ g_j(x) \le 0\ (j = 1, 2, \ldots m) \\ & \quad h_k(x) = 0\ (k = 1, 2, \ldots l) \\ & x = (x_1, x_2, \ldots, x_i, \ldots, x_n) \\ & x \in [x^l, x^u] \end{aligned} \tag{2.45}$$

There are three basic concepts of optimization:

(1) Design variable x. xi is a component of the design variable. Design variables refer to the choices made during the design process, and ultimately, individual parameters must be determined. Once the design variables are determined, the design objects are completely determined. The value of each set of design variables is called a design solution. The process of optimizing the solution is to find a design solution that optimizes the performance of the design object. The space described by the coordinate axes of the components of each design variable is called design space, and the number n of design variable components is called the dimension or dimension of the design space, so each design solution corresponds to a point in the design space.
(2) Objective function f. F (x) represents the objective function, the function of the components of the design variable x. Objective function refers to the expected goal to be achieved in the design process. Each objective function represents an important feature of the design, such as performance, quality, volume, cost, benefit, etc. Optimal design is the process of optimizing design variables to achieve the optimal objective function. The most common case is the case where there is only one objective function, which is called the single-objective function problem. The optimization of this problem is called the single-objective optimization. The optimization with more than one objective function is called multi-objective optimization, which usually needs to be combined into a single-objective optimization problem by means of weighted sum.
(3) Constraints. Constraints are restrictions on the value of design variables during the design process. The optimization process of optimization problems with constraints is the process of finding a set of optimal values within the allowable range of design variables to achieve the optimal objective function. Direct restrictions on design variables are called explicit constraints, such as the lower limits x^l and upper limits x^u of design variables; indirect restrictions on design variables are called implicit constraints. For example, m constraints of $g_j(x) \leq 0$ $(j = 1, 2, \ldots, m)$ are imposed by the function gj of the design variable (any inequality constraint Can be transformed into this standard form). In addition, according to the restrictions of the constraints, it is also divided into inequality constraints gj (x) ≤ 0 and equality constraints hk (x) $= 0$.

Generally, system-level optimization and subsystem-level optimization of multidisciplinary design optimization problems can also be expressed by the optimization model of Eq. (2.45). Because the optimization model of formula (2.45) is universal, the various algorithms for solving the model are the optimization algorithms or optimization algorithms we often call. According to the definition of design variables, the process of seeking optimization points is essentially the process of searching the design space. Therefore, the process of solving the optimization design solution is collectively referred to as the design space search process.

The most basic design space search method is analytic method, that is, the research object is described by mathematical equation, and then the optimal solution is directly solved by mathematical analytic method. When the objective function and the constraint function are simple display functions of design variables, the analytic method is the most accurate and efficient method. However, the functions of practical engineering problems are usually more complex and cannot be expressed as the display expressions of design variables.

2.3.3.1 Design Space Search Math Basis

(1) Numerical differentiation. In many design space search methods, the derivative information of a function is used. In many cases, functions do not have explicit expressions. Analytic methods cannot be used to obtain the derivatives. Therefore, numerical methods are needed to solve the approximate value of the function derivatives. In current computer programs, the finite difference method is the most commonly used method.

① Forward difference method:

$$\left.\frac{df}{dx}\right|_{x^0} = \frac{f(x^0+\Delta x)-f(x^0)}{\Delta x} + O(\Delta x) \approx \frac{f(x^0+\Delta x)-f(x^0)}{\Delta x} \tag{2.46}$$

Where: x0 is the derivation point, and Δx is a small change in the variable x (common values such as = 0.001 × 0), which is called the differential step or differential step.

② Backward difference method:

$$\left.\frac{df}{dx}\right|_{x^0} = \frac{f(x^0)-f(x^0-\Delta x)}{\Delta x} + O(\Delta x) \approx \frac{f(x^0)-f(x^0-\Delta x)}{\Delta x} \tag{2.47}$$

③ Center difference method:

$$\left.\frac{df}{dx}\right|_{x^0} = \frac{f(x^0+\Delta x)-f(x^0-\Delta x)}{2\Delta x} + O((\Delta x)^2) \approx \frac{f(x^0+\Delta x)-f(x^0-\Delta x)}{2\Delta x} \tag{2.48}$$

When x is an n-dimensional vector, the partial derivative of the function for the component xi is calculated by the forward difference method:

$$\left.\frac{\partial f}{\partial x_i}\right|_{x^0} = \frac{f(x_1^0, x_2^0 \ldots, x_i^0+\Delta x, \ldots, x_n^0)-f(x^0)}{\Delta x} + O(\Delta x)$$

$$\approx \frac{f(x_1^0, x_2^0 \ldots, x_i^0 + \Delta x, \ldots, x_n^0) - f(x^0)}{\Delta x} \tag{2.49}$$

The backward difference method and the center difference method can be deduced by analogy to obtain:

$$\nabla f(x^0) = \left[\frac{\partial f}{\partial x_1}, \frac{\partial f}{\partial x_2}, \ldots, \frac{\partial f}{\partial x_i}, \ldots \frac{\partial f}{\partial x_n}\right]_{x^0}^T \tag{2.50}$$

Then, when each component of the function's gradient vector [see Eq. (2.50)] is solved by the difference algorithm, the forward difference and backward difference need to call the function to calculate n + 1 times at least, the calculation accuracy and the differential step size, etc. The magnitude of the central difference needs to be called 2n times to calculate the function, and the precision of the calculation can reach the square of the differential step.

④ Complex variable method, It can be seen that when the dimension of the independent variable is large, the finite difference needs to call a large number of functions to calculate the gradient information. In order to reduce the amount of calculation and improve the accuracy, the complex variable method is sometimes used:

$$\begin{aligned} \left.\frac{df}{dx}\right|_{x^0} &= \frac{\operatorname{Im}(f(x^0 + ih))}{h} + O(h^2) \\ &\approx \frac{\operatorname{Im}(f(x^0 + ih))}{h} \end{aligned} \tag{2.51}$$

where:

h is the differentiation step, and the value refers to Δx in the finite difference method;

$x^0 + ih$ represents a complex variable with real part ×0 and imaginary part h;

Im(*) operator means taking the imaginary part.

It can be seen that when solving the function for the gradient vector of n-dimensional vector x, the complex variable method only needs to call the function to calculate n times (the number of times is the same as the forward difference method and the backward difference method), and then the calculation of the square of the differential step size can be obtained Accuracy (equivalent to the central difference method). Of course, the use of the complex variable method requires that the calculation program of the function f (x) supports the input variable to be a complex number, and when the input variable of the function is a complex variable, the memory required by the calculation program is usually more than that of a pure real input variable increase.

(2) One-dimensional search is the simplest and most basic method in the optimization algorithm, and it is also the basis of multi-dimensional search. The quality of the one-dimensional algorithm directly affects the solution speed of the optimization algorithm. Therefore, we need to introduce one-dimensional search before introducing the multi-dimensional search method.
The analytic algorithm for one-dimensional search is based on fermat's theorem: if the function f(x) takes an extreme value at an inner point x^{opt} on the defined interval, and the previous function is differentiable at point x^{opt} then $\left.\frac{df}{dx}\right|_{x^{opt}} = 0$ must occur. When the derivative of a function can be expressed analytically, the best advantage can be obtained by solving the equation with derivative 0. When the derivative of the function cannot be expressed by analytic expression, and the function itself is an implicit function, the analytic method is no longer applicable, and the approximate value needs to be obtained through repeated iteration by numerical method.
The numerical one-dimensional search method assumes that the function is a single valley function. The basic idea is: firstly, the search interval is determined, and then the interval is narrowed down several times according to the interval elimination principle, so as to obtain the approximate optimization point.

① Determine the search interval:
As shown in Fig. 2.13, determining the search interval is the process of determining a range of the independent variable X [^] to include the best features of the function. The principle to determine the search interval is to use the characteristics of the single valley function. The most commonly used method is advance and retreat method, and the specific steps are as follows:

a. Start at the initial point x_0, take the initial step λ;
b. Explore forward, that is, calculate the value of $f(x_0 + \lambda)$;
c. If $f(x_0 + \lambda) > f(x_0)$, then take $\lambda = -\lambda$ (negative step, reverse), and return to step b;
d. If $f(x_0 + \lambda) < f(x_0)$, then take $\lambda = 2\lambda$ (Double the step size, the direction is the same), and return to step b;

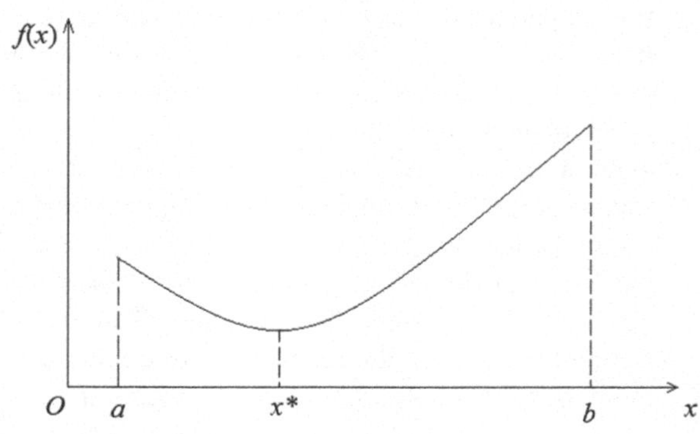

Fig. 2.13 Determine search interval

② Interval elimination: After the interval is determined, the interval elimination is performed. The principle is: assuming that any two points are taken in the interval, and the value of the calculation function at these two points will have the following three cases:

a. $f(a_1) < f(b_1)$, Since the function is a single valley, the minimum point should be within the interval $[a, b_1]$, so the interval $[b_1, b]$ can be eliminated
b. $f(a_1) > f(b_1)$, Similarly, the minimum point should be within the interval $[a_1, b]$, and the interval $[a, a_1]$ can be eliminated;
c. $f(a_1) = f(b_1)$, The minimum point is within $[a_1, b_1]$, and the intervals $[a, a_1]$ and $[b_1, b]$ can be eliminated.

Perform a new round of interval elimination on the remaining interval until the interval is small enough and the function value difference is small enough. According to different A-point-taking methods, the interval elimination method can be divided into golden section method, Fibonacci method, and so on. Another method of interval elimination is the bisection method, that is, in the interval $[a, b]$ containing the extreme point of the single valley function, take the midpoint of the interval $c = \frac{(a+b)}{2}$, calculate the derivative value $f'(c)$ of the function at point c, and judge the rounded interval according to the derivative value:

a. $f'(c) < 0$, the extreme point is to the right of point c, so the interval [a, c] can be eliminated;
b. $f'(c) > 0$, the extreme point is to the left of point c, so the interval [c,b] can be eliminated;
c. $f'(c) = 0$, the extreme point is the point c.

③ The grid method (full search method) divides the interval [a, b] containing extreme points into m equal parts to obtain a vector $\left[x_1, x_2, \ldots, x_i, \ldots, x_{m+1}\right] = \left[a, a + \frac{(b-a)}{m}, \ldots, a + i\frac{(b-a)}{m}, \ldots, a + \frac{(m-1)(b-a)}{m}, b\right]$, calculates the function value of the function at each component x_i of the vector, and then takes the minimum The value point is used as an approximation of the extreme value point.

④ Newton's method (tangent method): Assume that there is a good approximation point x0 near the extreme point of the known function f (x). Since the continuously differentiable function is closer to the quadratic function near the extreme point, it can Near the point x0, the function is approximated by the second-order Taylor expansion of the function, that is,

$$f(x) \approx \phi(x) = f(x_0) + f'(x_0)(x - x_0) + \frac{1}{2} f''(x_0)(x - x_0)^2 \tag{2.52}$$

Assuming that the extreme point of $\phi(x)$ is x_1, Bayes is $\phi'(x_1) = 0$ according to Fermat's theorem, that is,

$$\left.\frac{d\phi}{dx}\right|_{x_1} = f'(x_0) + f''(x_0)(x_1 - x_0) = 0 \tag{2.53}$$

Solve this equation to get the extreme point xi:

$$x_1 = x_0 - \frac{f'(x_0)}{f''(x_0)} \tag{2.54}$$

Thus the iterative formula of Newton's method is obtained:

$$x_{k+1} = x_k - \frac{f'(x_k)}{f''(x_k)} \quad (k = 0, 1, 2, \ldots) \tag{2.55}$$

When Newton's method is used to solve the minimum extreme point, the function requires a first-order continuous derivative and the second-order derivative is greater than zero. Its iterative steps are:

a. Given x_0, ε, δ, let $x_1 = x_0$;
b. Let $x_0 = x_1$
c. Calculate the second derivative of the function at x_0 (the first derivative value is also obtained in the process), calculate $x_1 = x_0 - \frac{f'(x_0)}{f''(x_0)}$
d. Determine whether it meets: $\left|f'(x_0)\right| \le \varepsilon$ or $|x_1 - x_0| \le \delta$;
e. If it is satisfied, the optimization ends, and the extreme point is x_1, otherwise it returns to step b.

The advantage of Newton's method is that the convergence speed is very fast. The disadvantage is that iterative calculation of the second order is required for each iteration.

There are other methods, such as quadratic interpolation and cubic interpolation. For the extreme point of a multivariate function, the most commonly used iteration formula is:

$$x^{k+1} = x^k + \lambda_k d^k \tag{2.56}$$

where:

k represents the k-th iteration;
d^k represents the search direction, usually taking the negative gradient direction;
λ_k is the search step.

After the search direction is determined, and the optimal step size needs to be determined, the problem turns into a one-dimensional search problem:

$$f(x^{k+1}) = f(x^k + \lambda_k d^k) = \varphi(\lambda_k) \tag{2.57}$$

Therefore, the quality of the one-dimensional search algorithm will have an important impact on the multi-dimensional optimization algorithm.

(3) Unconstrained optimization. First consider the simpler case in the optimization model (Eq. 2.45): there are no constraints, that is, mathematically unconditional extreme values.
As can be seen from the above, the process of solving the extremum of the multivariate function is mainly to determine the optimization direction and the search step size. Firstly, it is necessary to analyze the extremum search direction of the multivariate function, which is defined as follows:

$$\left.\frac{\partial f}{\partial d}\right|_{x^0} = \lim_{\Delta d \to 0} \frac{f(x_1^0 + \Delta x_1^0, x_2^0 + \Delta x_2^0, \ldots, x_n^0 + \Delta x_n^0) - f(x_1^0, x_2^0, \ldots, x_n^0)}{\Delta d}$$
$$= \left.\frac{\partial f}{\partial x_1}\right|_{x^0} \frac{\Delta x_1}{\Delta d} + \left.\frac{\partial f}{\partial x_2}\right|_{x^0} \frac{\Delta x_2}{\Delta d} + \ldots + \left.\frac{\partial f}{\partial x_n}\right|_{x^0} \frac{\Delta x_n}{\Delta d}$$
$$= \sum_{i=1}^{n} \left.\frac{\partial f}{\partial x_i}\right|_{x^0} \cos\theta_i \tag{2.58}$$

This is the directional derivative of the multivariate function f in the direction d, where $\cos\theta_i = \frac{\Delta x_i}{\Delta d}$.
Let $d = [\cos\theta_1, \cos\theta_2, \ldots, \cos\theta_n]^T$, then, the directional derivative can be expressed as the product of the gradient vector and d:

$$\left.\frac{\partial f}{\partial d}\right|_{x^0} = \nabla f(x^0)^T d = \|\nabla f(x^0)\| \cos(\nabla f(x^0), d) \tag{2.59}$$

Where, the gradient vector is as described above: $\nabla f(x^0) = \left[\frac{\partial f}{\partial x_1}, \frac{\partial f}{\partial x_2}, \ldots, \frac{\partial f}{\partial x_n}\right]_{x^0}^T$

Modulus of the gradient vector: $\|\nabla f(x^0)\| = \sqrt{\sum_{i=1}^{n} \left(\left.\frac{\partial f}{\partial x_i}\right|_{x^0}\right)^2}$

It can be seen that when $\cos(\nabla f(x^0), d) = 1$, that is, when the gradient direction and the d direction coincide, the value of the direction derivative is the largest and equal to the gradient value. That is, the gradient direction is the direction where the function increases the fastest, and the negative gradient direction is the direction where the function decreases the fastest.
According to the second-order Taylor expansion of the multivariate function and Fermat's theorem, the necessary and sufficient conditions for deriving the minimum value of the multivariate function are: the gradient at the extreme point is zero and the Hessian matrix is positive definite.
Where, Hession Matrix is:

$$G(x^*) = \begin{bmatrix} \frac{\partial^2 f}{\partial x_1^2} & \frac{\partial^2 f}{\partial x_1 x_2} & \cdots & \frac{\partial^2 f}{\partial x_1 x_n} \\ \frac{\partial^2 f}{\partial x_2 x_1} & \frac{\partial^2 f}{\partial x_2^2} & \cdots & \frac{\partial^2 f}{\partial x_2 x_n} \\ \cdots & \cdots & \cdots & \cdots \\ \frac{\partial^2 f}{\partial x_n x_1} & \frac{\partial^2 f}{\partial x_n x_2} & \cdots & \frac{\partial^2 f}{\partial x_n^2} \end{bmatrix}_{x^*} \tag{2.60}$$

Then, to solve the unconstrained optimization problem, the gradient in the extreme sufficient and necessary condition is zero, and then the equations containing n variables and n equations are listed directly:

$$\nabla f(x) = 0 \Rightarrow \begin{cases} \frac{\partial f}{\partial x_1} = 0 \\ \frac{\partial f}{\partial x_2} = 0 \\ \cdots \\ \frac{\partial f}{\partial x_n} = 0 \end{cases} \tag{2.61}$$

Optimal points can be obtained by solving the equations numerically.
However, for practical optimization problems, since the equations in Eq. (2.61) are often non-linear, instead of using numerical methods to solve nonlinear equations, it is better to directly solve the original optimization problems by numerical methods.
The most commonly used numerical method for solving optimization problems is the hill-climbing method shown in Eq. (2.56) above. The mountain climbing method mainly has two steps: ① select the climbing direction, that is, the search direction; ② select the appropriate step size for the search in the determined direction.
The search direction is determined differently from the step generation method, so different unconstrained optimization methods are derived. The procedures of the random direction method and the steepest descent method are described below.

① Random direction method:

a. Specify the initial point x0 and the initial step size λ_0, let x = x^0, $\lambda = \lambda_0$;
b. Generate m n-dimensional random unit vectors, combine x, and λ to calculate m random points xj (j = 1,2,…, m);
c. Calculate the function value at m random points, find the random point x ^ with the smallest function value to generate a feasible search direction: $d = x_L - x$, if xL is not found, shorten the step size to 1f3 and return to step b;
d. Starting from the initial point x, a one-dimensional search is performed along the feasible search direction d with a step size of 1.33λ to find a new point x 'where the objective function no longer drops. Converge if the distance between x 'and x points is less than the specified error or the difference between the function values at the two points is less than the specified error, otherwise return to step b.

The random direction method needs to explore enough directions to find a feasible direction, that is, m must be large enough, especially for high-dimensional problems, which makes the calculation of the random direction method larger.

② The steepest descent method:
Use the negative gradient direction of the multivariate function as the fastest decreasing direction of the function, and use the negative gradient direction as the search direction, thereby obtaining the steepest descent method or gradient method:

$$x^{k+1} = x^k - \lambda_k \nabla f(x^k) \tag{2.62}$$

After determining the direction, only one-dimensional search is needed to determine the optimal step size:

$$\begin{aligned} f(x^{k+1}) &= f\left(x^k - \lambda_k \nabla f(x^k)\right) \\ &= \min f\left(x^k - \lambda \nabla f(x^k)\right) \\ &= \min \varphi(\lambda) \end{aligned} \tag{2.63}$$

According to the Fermat's theorem and the derivation rule of composite functions mentioned above, take $u = x^k - \lambda \nabla f(x^k)$, and then:

$$\begin{aligned} &\varphi'(\lambda) = 0 \\ &\quad \Rightarrow f'(u) \cdot u'(\lambda) = 0 \end{aligned} \tag{2.64}$$

At the optimization point $\lambda = \lambda_k$, and then:

$$\begin{aligned} &\left[\nabla f\left(x^k - \lambda_k \nabla f(x^k)\right)\right]^T \cdot \left(-\nabla f(x^k)\right) = 0 \\ &\Rightarrow \left[\nabla f(x^{k+1})\right]^T \cdot \nabla f(x^k) = 0 \\ &\Rightarrow \left(d^{k+1}\right)^T \cdot d^k = 0 \end{aligned} \tag{2.65}$$

In the steepest descent method, the functions of two adjacent iteration points are perpendicular to each other in the gradient direction. In the process of approaching the minimum iteration, the steepest descent method takes a zigzag route. In places far away from the valley value (minimum value) of the function, each iteration can make the value of the function drop rapidly, while in places near the minimum point, the distance traveled by each iteration is continuously shortened, thus making the convergence speed slow. This is the disadvantage of the algorithm: from the local point of view, the function drops rapidly at each iteration point, but from the global point of view, the function drops in a very tortuous path.

In addition to the mountain climbing method, there are direct search methods such as the simplex method and Cauchy Newton method, which will not be introduced in this book.

(4) Constrained optimization
Before introducing constraint optimization, two commonly used concepts are introduced:

① Convex set: for region (point set)S, if all line segments connecting any two points are contained in S, the region is called convex set. The mathematical expression of convex set is:
For $\forall x_1 \in S, \forall x_2 \in S, \forall \alpha \in 0 \leq \alpha \leq 1$

$$y = \alpha x_1 + (1 - \alpha)x_2 \in S \quad \text{So S is a convex set} \tag{2.66}$$

② A convex function is a function f (x) on a convex set. Connect any two points x0 and x00 in the domain. If the function value of the point on the line segment between these two points is always less than the linear interpolation of f (x0) and f (x00), then the function f (x) is called a convex function.
For $\forall \alpha \in 0 \leq \alpha \leq 1$

$$f\left(\alpha x^0 + (1 - \alpha)x^{00}\right) \leq \alpha f(x^0) + (1 - \alpha) f(x^{00}) \text{ which is always true} \tag{2.67}$$

Then f (x) is a convex function
If f (x) has a continuous first derivative in the domain S, then the sufficient and necessary condition for f (x) to be a convex function is:
For $\forall x^0 \in S, \forall x^{00} \in S$

$$f\left(x^{00}\right) \geq f(x^0) + (x^{00} - x^0)^T \nabla f(x^0) \text{ which is always true} \tag{2.68}$$

Then f (x) is a convex function.
If f (x) has continuous second-order derivatives in the domain S, then the sufficient and necessary condition for f (x) to be a convex function is the semi-positive definite at the Hessian matrix G (x) of formula (2.60).
When both the constraint function and the objective function of a constrained optimization problem are convex functions, the problem is convex programming. The feasible regions of convex programming are all convex sets, and any local optimal solution is a global optimal solution.
Constrained optimization is the most common optimization problem encountered in practical engineering. Constraints are divided into equality constraints and inequality constraints, which are discussed on a case-by-case basis.

① Equality constrained optimization. When the equality-constrained optimization model is:

$$\begin{aligned} &Min\ f(x) \\ &s.t.\ h_k(x) = 0\ (k = 1, 2, \ldots, l) \end{aligned} \tag{2.69}$$

There are usually two ways to deal with this problem: elimination (dimensionality reduction) and Lagrange multiplier (dimensionality elevation).

a. Elimination method. Take the simplest case as an example, the design space is 2 dimensions, that is, x = (xi,x2); There is only one equality constraint, that is, 1 = 1, to get:

$$\begin{aligned} &Min\ f(x_1, x_2) \\ &s.t.\ h(x_1, x_2) = 0 \end{aligned} \tag{2.70}$$

According to the equation of the constraint condition of the equation, $x_1 = \varphi(x_2)$ can be solved, and the variable x1 can be eliminated by substituting into the objective function. Then the original problem turns into a one-dimensional unconstrained optimization problem:

$$Min\ f(x_2, \varphi(x_2)) \tag{2.71}$$

For the n-dimensional problem of Eq. (2.69), the first l design variables of n design variables can be represented by the remaining n-l by l equality constraints:

$$\begin{aligned} &x_1 = \varphi_1(x_{l+1}, x_{l+2}, \ldots x_n) \\ &x_2 = \varphi_2(x_{l+1}, x_{l+2}, \ldots x_n) \\ &\ldots \\ &x_l = \varphi_l(x_{l+1}, x_{l+2}, \ldots x_n) \end{aligned} \tag{2.72}$$

By substituting it into the objective function, the first l design variables can be eliminated and the original problem can be transformed into an n-l dimensional unconstrained optimization problem.

b. Lagrange multiplier method. Unlike the elimination method, which reduces the dimension of the design space, the Lagrangian multiplier method converts the constraint optimization problem into an unconstrained optimization problem by adding design variables.
Similarly, take the simplest problem of Eq. (2.70) as an example. Let the optimal advantage be x*. For any small displacement near x*, d = [dx1, dx2], if the objective function is not wanted to increase, according to the property of the directional derivative in the above, it is necessary to:

$$\nabla f(x^*)^T d = \frac{\partial f}{\partial x_1} dx_1 + \frac{\partial f}{\partial x_2} dx_2 = 0 \tag{2.73}$$

In addition, the constraint condition of the equation at x * must also be satisfied, that is, the direction of d should also be perpendicular to the gradient of h (x *):

$$\nabla h(x^*)^T d = \frac{\partial h}{\partial x_1}dx_1 + \frac{\partial h}{\partial x_2}dx_2 = 0 \tag{2.74}$$

According to Eqs. (2.73) and (2.74):

$$-\frac{dx_2}{dx_1} = \frac{\partial f}{\partial x_1}\Big/\frac{\partial f}{\partial x_2} = \frac{\partial h}{\partial x_1}\Big/\frac{\partial h}{\partial x_2} \tag{2.75}$$

Let $\lambda = -\frac{\partial f}{\partial x_1}\Big/\frac{\partial h}{\partial x_1} = -\frac{\partial f}{\partial x_2}\Big/\frac{\partial h}{\partial x_2}$ Eqs. (2.73) and (2.74) can be transformed into:

$$\frac{\partial f}{\partial x_1} + \lambda\frac{\partial h}{\partial x_1} = 0 \tag{2.76}$$

$$\frac{\partial f}{\partial x_2} + \lambda\frac{\partial h}{\partial x_2} = 0 \tag{2.77}$$

Add the constraint condition h (xi, x2) = 0 to form a system of equations with three unknowns and three equations. Solving this system of equations yields optimization points x * and λ.
For optimization problems involving n-dimensional design space and l equality constraints, the Lagrangian multiplier method can also be derived: The non-zero Lagrangian multiplier λk (k = 1, 2,…, l) is introduced, and the equality constraints of the original problem are incorporated into the objective function to form a new objective function F(x, λ):

$$F(x, \lambda) = f(x) + \sum_{k=1}^{l} \lambda_k h_k(x) \tag{2.78}$$

Then, the unconstrained optimization problem of the new objective function is equivalent to the original equality constrained optimization problem. According to the extreme conditions of unconstrained optimization:

$$\begin{cases} \frac{\partial F}{\partial x_i} = 0\ (i = 1, 2, \ldots, n) \\ \frac{\partial F}{\partial \lambda_k} = 0\ (k = 1, 2, \ldots, l) \end{cases} \tag{2.79}$$

Solve the system of n + 1 equations to get the optimal design point and Lagrange multiplier.

② Inequality constrained optimization. For the inequality-constrained optimization problem, taking the simplest one-dimensional function optimization problem on the interval as an example, the problem can be expressed in a standard form:

$$\begin{aligned} & Min\ f(x) \\ & s.t.\ g_1(x) = a - x \le 0 \\ & \quad g_2(x) = x - b \le 0 \end{aligned} \tag{2.80}$$

By introducing relaxation factor a_1, b_1, inequality constraint can be transformed into equality constraint:

$$h_1(x, a_1) = g_1(x) + a_1^2 = a - x + a_1^2 = 0 \tag{2.81}$$

$$h_2(x, b_1) = g_1(x) + b_1^2 = x - b + b_1^2 = 0 \tag{2.82}$$

Introducing Lagrange multipliers $\lambda_1 \ge 0$ and $\lambda_2 \ge 0$, and incorporating the equality constraints into the objective function, and get:

$$\begin{aligned} & F(x, a_1, b_1, \lambda_1, \lambda_2) \\ & = f(x) + \lambda_1 h_1(x, a_1) + \lambda_2 h_2(x, b_1) \\ & = f(x) + \lambda_1\left(a - x + a_1^2\right) + \lambda_2\left(x - b + b_1^2\right) \end{aligned} \tag{2.83}$$

From the extreme conditions of the unconstrained function, the system of equations is obtained:

$$\begin{cases} \frac{\partial F}{\partial x} = \frac{df}{dx} + \lambda_1 \frac{dg_1}{dx} + \lambda_2 \frac{dg_2}{dx} = \frac{df}{dx} + \lambda_1 + \lambda_2 = 0 \\ \frac{\partial F}{\partial a_1} = 2\lambda_1 a_1 = 0 \\ \frac{\partial F}{\partial b_1} = 2\lambda_2 b_1 = 0 \\ \frac{\partial F}{\partial \lambda_1} = h_1(x, a_1) = a - x + a_1^2 = 0 \\ \frac{\partial F}{\partial \lambda_2} = h_2(x, b_1) = x - b + b_1^2 = 0 \end{cases} \tag{2.84}$$

Since the Lagrangian multiplier and relaxation factor may take zero values, the process of solving this system of equations needs to analyze various possible situations, which will not be expanded in detail here. The final system of equations can be transformed into:

$$\begin{cases} \frac{df}{dx} + \lambda_1 \frac{dg_1}{dx} + \lambda_2 \frac{dg_2}{dx} = 0 \\ \lambda_1 g_1(x) = 0,\ \lambda_2 g_2(x) = 0 \\ \lambda_1 \ge 0,\ \lambda_2 \ge 0 \end{cases} \tag{2.85}$$

The above process can be generalized to an n-dimensional problem, and an optimization problem containing m inequality constraints can be expressed as:

$$
\begin{aligned}
& Min\ f(x) \\
& s.t.\ g_j(x) \le 0\ (j = 1, 2, \ldots, m) \\
& x = (x_1, x_2, \ldots, x_n)^T
\end{aligned}
\tag{2.86}
$$

Introducing m relaxation factors $\bar{x} = [x_{n+1}, \ldots, x_{n+j}, \ldots, x_{n+m}]^T$ and Lagrange multipliers $\lambda = [\lambda_1, \lambda_2, \ldots, \lambda_m]$, the Lagrange function can be obtained:

$$F(x, \bar{x}, \lambda) = f(x) + \sum_{j=1}^{m} \lambda_j \left(g_j(x) + x_{n+j}^2\right) \tag{2.87}$$

Then according to the extreme conditions of the Lagrange function, the system of equations can be obtained:

$$
\begin{cases}
\frac{\partial F}{\partial x_i} = \frac{\partial f}{\partial x_i} + \sum_{j=1}^{m} \lambda_j \frac{\partial g_j}{\partial x_i} = 0\ (i = 1, 2, \ldots, n) \\
\frac{\partial F}{\partial x_{n+j}} = 2\lambda_j x_{n+j} = 0\ (j = 1, 2, \ldots, m) \\
\frac{\partial F}{\partial \lambda_j} = \left(g_j(x) + x_{n+j}^2\right) = 0\ (j = 1, 2, \ldots, m)
\end{cases}
\tag{2.88}
$$

Using a similar derivation to the unary function, the well-known KKT condition can be obtained:

$$
\begin{cases}
\left.\frac{\partial f}{\partial x_i}\right|_{x^*} + \sum_{j \in J} \lambda_j \left.\frac{\partial g_j}{\partial x_i}\right|_{x^*} = 0\ (i = 1, 2, \ldots, n) \\
g_j(x^*) = 0\ (j \in J) \\
\lambda_j \ge 0\ (j \in J)
\end{cases}
\tag{2.89}
$$

That is:

$$\nabla f(x^*) + \sum_{j \in J} \lambda_j \nabla g_j(x^*) = 0 \tag{2.90}$$

Where, j represents the subscript set of constraint conditions in action.

The geometric meaning of KKT condition is that the negative gradient of the objective function at the extremal point of the constrained optimization problem can be expressed as a non-negative linear combination of the gradient of all constraint functions at that point.

We introduced the mathematical basis of one-dimensional search, unconstrained optimization to constrained optimization above. However, in practical engineering applications, many practical difficulties will be faced, such as non-convex feasible regions, discrete variables, non-differentiable functions, multi-objective functions, global optimization, etc. Therefore, the actual optimization algorithm requires a lot of processing skills and engineering experience. These mathematical foundations allow us to understand the kernel and basic concepts of the optimization algorithm, so that we can better select the optimization algorithm and adjust the parameters of the algorithm in the MDO problem.

Most of these common numerical methods are based on extreme conditions, and gradients of functions are used in calculations. Therefore, common numerical optimization algorithms are also called gradient-based optimization algorithms. In addition to these algorithms based on functional properties and calculus theory, algorithms based on other disciplines have appeared in recent years. Among them, genetic algorithms and particle swarm algorithms have been widely used in the field of engineering and have been gradually used to solve MDO problems.

(5) Genetic algorithm. Genetic algorithm is an algorithm based on the theory of evolution and genetics that simulates the evolutionary process of living things. The genetic algorithm starts with a group of randomly generated individuals. This group of individuals is called a population. Each individual in the population has its own genetic code. Individuals in the population generate new individuals through operations such as genetic code crossover and mutation. That is, the offspring, the offspring form a new population; then the fitness of each offspring is evaluated by the adaptability function, that is, the survival of the fittest; the surviving individuals generate the next generation through operations such as crossover and mutation, and then evaluate its fitness... The result is adaptive offspring.

The main operation steps in the genetic algorithm are:

① Coding. The design space data is expressed as genetic space gene coding data, so that each gene coding string data corresponds to a design scheme, and vice versa. Common encodings are binary and decimal.

② Initial population generation. Most of the initial population generation adopts a random method: a specified number of encoding strings are randomly generated, and these encoding strings can be inverted to obtain the same number of corresponding design solutions in the design space, and each encoding string corresponds to an individual.

③ Fitness calculation. The fitness reflects the advantages and disadvantages of each individual. The definition of the fitness function must reflect both the objective function and the constraint conditions. The penalty function method is used to incorporate the constraint conditions into the objective function and then construct the fitness function. Most fitness functions are defined on a positive range.

④ Choose. The selection function is defined with each individual's fitness as the input variable and the individual's probability of participating in reproduction as the output variable. The more adaptive an individual is, the more likely it is to pass on its own code to future generations, which is the principle of survival of the fittest.

⑤ Cross. The process of combining the codes between different individuals to generate new codes is called crossover. The new codes also have the characteristics of parents.

⑥ Variation. The coding of some individuals in the population has a small probability of local changes (usually the probability is 0.001 to 0.01), so that these individuals have new characteristics that were not previously available.
The genetic algorithm has no requirements for the continuity, differentiability, and convexity of the function, and does not require the design space to be a convex set. Therefore, it has been widely used in large-scale engineering optimization problems, and developed a multi-objective genetic algorithm, multi-Island genetic algorithm and many other branches. But genetic algorithms also have disadvantages, that is, the efficiency of genetic algorithms is usually lower than traditional numerical methods, and sometimes "premature" (premature convergence) occurs.

(6) Particle swarm algorithm. Particle swarm optimization is also an optimization method that was inspired by biology. It was proposed by Kennedy and Eberhart in 1995. The original algorithm research was to simulate the predation of bird swarms, a swarm intelligence model called a complex adaptive system. Later, it was found that this model is a good optimization algorithm, so it has gradually developed into today's particle swarm algorithm.
Similar to the genetic algorithm, the particle swarm algorithm also starts from a set of random design schemes, and determines the optimal adjustment through the fitness function. However, particle swarm optimization does not have the operations of encoding, crossover, and mutation of genetic algorithms. It directly evaluates the pros and cons of the design solution through fitness, so it is easier to implement.
The operation steps of the particle swarm algorithm are:

① Randomly generate particle swarms and start iteration;
② Calculate and store the fitness of each particle in the group. For each particle, if the fitness of the particle's current iteration step is better than the fitness of the previous iteration step (historical best fitness), then update the historical best fitness to this iterative fitness;
③ Compare the best fitness of all particles in history, and take the maximum value as the best fitness of the group;
④ Update the speed and position of each particle. The speed and position of each particle are jointly determined by the best fitness of the individual's history and the best fitness of the group. There are many specific speed and position update formulas, which will not be introduced here;
⑤ Iteration steps ② ~ ④ until convergence or reaching the set number of iterations.

From the above steps, it can be seen that particle swarm optimization algorithm has an important feature: by recording the location and fitness of each iteration step, particles are "remembered". This memory ensures that every iteration of pso always increases the fitness of pso (that is, ensures that every iteration optimizes the design scheme and there is no regression), thus improving the efficiency of design space search.

By comparison, particle swarm optimization is simpler and more efficient than genetic algorithms, so it has developed very quickly in just 20 years after its discovery and is used in more and more fields.

In addition to the representative genetic algorithms and particle swarm optimization algorithms introduced above, the design space search methods not based on function derivatives include monte carlo method (stochastic simulation method), pattern search, simulated annealing method, etc., which will not be introduced here.

2.3.3.2 Design Space Search Method Selection

The basic mathematical basis of the design space search method is introduced above, which enables the reader to understand the essence of optimization problems and common concepts, as well as the principle of common optimization algorithms. However, for an engineering designer, especially those who need to design complex and multidisciplinary systems, most of the time there is no need to write the program of optimization algorithm. Therefore, to understand the principle of these algorithms is to better select the appropriate optimization algorithm according to different problems.

So far, no design space search method has been developed that is suitable for all problems. Therefore, only by selecting appropriate methods for different problems can the nature of the problem function be maximized, thereby improving the optimization accuracy and search efficiency. How to choose the right optimization algorithm? It should be comprehensively considered from the following aspects:

(1) Objective function. The first is the number of objective functions. Most optimization algorithms are developed for single objective functions. Currently, the commonly used multi-objective optimization methods are mainly based on weighted sum methods or Pareto solution sets. Among them, Pareto-based solutions are widely used in engineering. Reto's solution set genetic algorithm NSGA and its improved method NSGA-II. Second is the nature of the objective function, such as linearity, convexity, continuity, derivability, and whether it is a least squares problem. After understanding the nature of the objective function, selecting an appropriate optimization algorithm based on its special properties can greatly improve the efficiency of the optimization. On the other hand, if you do not understand the nature of the function and choose an inappropriate optimization algorithm, it will lead to low optimization efficiency and even fail to get the optimization solution.

(2) Constraints. The first is the type of constraint. The most common problem in engineering is the inequality constraint problem, but some problems can be transformed into unconstrained optimization problems. At this time, efficient numerical methods for unconstrained optimization can be used to solve. The second is the nature of the constraint function, such as continuity and differentiability.

(3) The types of design variables. The design variables of most engineering problems are continuous real variables, but sometimes, the design variables are discrete variables. For example, in the design variables commonly used in the optimization of frame structures, the number of support rods can only be an integer. An optimization algorithm capable of supporting discrete variables needs to be selected for solving. Sometimes, a self-programmed preprocessing program (such as the simplest automatic rounding) can also be used to convert discrete variables into equivalent continuous variables, and connect ordinary continuous variable optimization algorithms and discrete variable function solvers.
(4) The scale of the optimization problem. The size of the optimization problem consists of two parts:

① The computation of the objective function and the constraint function. For example, when empirical formulas are used to calculate the objective function (weight, volume, etc.) and constraint function (strain, stress, etc.) in structural optimization design, it usually takes a short time to complete the evaluation of a design solution. However, calculations using fine finite element models or even non-linear finite element models may take hours, days, or even weeks to complete. For small-scale problems, there are usually more optimization methods available. You can focus on obtaining global optimization solutions and high-precision optimization solutions. For large-scale problems, attention should be paid to optimization efficiency. Firstly, a fast convergence algorithm is used to obtain a local optimal solution, even if it is a feasible solution, and then the accuracy of the optimization and whether it is the best optimization is considered.
② Design the dimensions of the space. Most optimization algorithms are sensitive to the dimension of design space, and in many cases the increase of dimension will lead to the surge of optimization calculation. Therefore, in order to improve optimization efficiency, it is usually required to reduce the dimension of design space as far as possible, such as taking parameters insensitive to functions as fixed values. After the final determination of the design space dimension, it is necessary to estimate the call times of the function when selecting the optimization algorithm. Generally speaking, the conventional numerical optimization algorithm based on gradient can converge quickly when applied to the problem of small dimension, while particle swarm optimization and genetic algorithm have better convergence on the problem of large dimension.

Decoupling, approximation, and design space search constitute the key technologies for multidisciplinary design optimization. Therefore, the development of future multidisciplinary design optimization mainly depends on the innovation of these three key technologies. In addition to these three key technologies, the computer technologies required in the practical application of multidisciplinary design optimization include software interface and module integration, user interface, data management and analysis, and so on.

2.4 Optimization and Reliability

In the most common constrained optimization design of engineering, the set of all design schemes that satisfy all constraint conditions is called feasible region, so the optimal solution belongs to feasible region. For practical engineering problems, the system function of constraint conditions is usually the function of mandatory indexes such as safety and rigid requirements. The optimization process not only improves the system's target performance, but also reduces other performance indexes of the system, especially the mandatory indexes represented by the constraint function. Therefore, the optimal design point is located on the boundary of the feasible region defined by these constraints in many cases, that is, the optimal point is located on the constraint conditions in action.

(1) Uncertainty

Actual engineering problems continue to show various natural uncertainty factors: most of the working environment of engineering products is constantly changing, causing the actual load of the engineering system to be inconsistent with the idealized model of the design load; the materials and processing of engineering product components include Various performance differences and quality differences lead to changes in the performance of engineering products; various theories and models used to calculate system performance indicators in the design are based on certain idealized assumptions. There are errors between calculated and actual values. From the perspective of optimization design, there are uncertainties in the optimization model (objective function and constraint function), design variables, and other design parameters of optimal design.

(2) Reliability.

The optimal design points of engineering problems are mostly located on the boundary of feasible region, and the uncertainty change of design parameters or design variables is likely to make the optimal design points fall out of feasible region due to the disturbance. As mentioned above, since most of the constraints that work are important mandatory index requirements such as safety, this means that there is a great possibility that the optimization design point cannot meet the mandatory index requirements of the actual engineering system, that is, the reliability of the optimization design is not up to standard.

(3) Robustness

Like the constraint function, the performance of the objective function will be disturbed due to the uncertainty of the design parameters, etc., and the sensitivity of the objective function to the disturbance of the input parameter at different levels is different. At some design points (the most common is the optimal design point), small changes in input parameters can cause large changes in the value of the objective function (similar to the spike function where the sensitivity is very high at the spike point). As the uncertainty of the design parameters described in (1) exists objectively, it is said that the function at this point is not robust enough. It is necessary to find another design point to optimize the objective function as much as possible while having sufficient robustness.. The objective functions of different optimization problems have different sensitivities to parameters. The objective functions of general engineering products can meet the robustness of input parameters.

The process of design optimization can be considered as the process of finding design limits. The limit design not only optimizes the system performance but also loses the stability of the system. When the external factors or the internal uncertainties of the system are disturbed, the products of the limit design may fail or the performance may be unstable. Therefore, it is necessary to extend the optimal design from the certainty to the uncertainty range. Every design scheme in the uncertain design optimization is no longer a certain point in the design space, but a random vector, that is, a point cloud in the design space. The constraint function is no longer a definite boundary, but a probability of failure, that is, a probability constraint; the value of the objective function corresponding to each design scheme also changes from a point to an interval. At this point, it is necessary to introduce the uncertainty theory and the reliability analysis technology to build the uncertainty optimization model, which will be introduced later in this book.

2.5 Examples of Multidisciplinary Design Optimization

2.5.1 Example 1: A Single Discipline Optimization Problem

Take a simple single-discipline optimization problem-Rosenbrock function as an example. This function is shown in Eq. (2.91), and the function graph in the range of $-1.5 \leq 1.5$ is shown in Fig. 2.14. It can be seen in the figure that the function is continuous, differentiable, non-convex in this range, and the shape is similar to the shallow valley which is slightly deeper but flat in the deep valley. Its special shape makes it a classic function for testing optimization algorithms. The essence of finding the minimum value of f is an unconstrained optimization problem in a two-dimensional design space. According to the above, an unconstrained optimization method can be used to solve it.

$$f = 100(x_2 - x_1^2)^2 + (1 - x_1)^2 \tag{2.91}$$

Because of the special shape of Rosenbrock function, it may be difficult to solve the problem by using the most rapid descent method and other methods, so it is more appropriate to use the method based on direct search for optimization. The optimization results are shown in Table 2.3. It can be seen that the simplex method or cauchy Newton method can be used to find the optimal solution after the function evaluation of 100 orders of magnitude. And Pattern Search method (Pattern Search) after

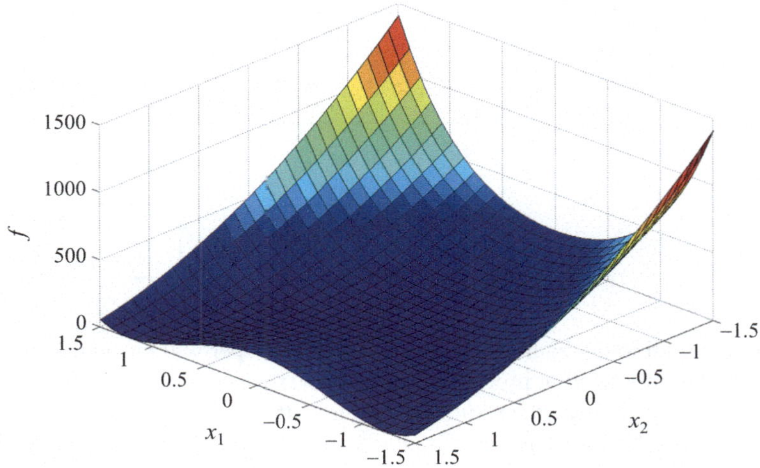

Fig. 2.14 Rosenbrock function graph

Table 2.3 Optimization results of rosenbrock function

Optimization algorithm	Initial point	Optimize point	Function value of optimization goal	Number of function calls
Simplex method	[−1.2, 1]	[1, 1]	0	159
Cauchy Newton method	[−1.2, 1]	[1, 1]	0	138
Pattern search	[−1.2, 1]	[0.9997, 0.9994]	0	11904

10000 time order of magnitude of function evaluation is needed to find the optimal solution, and observe the change of the objective function with the increase of the number of iterations curve and found that the scores of loop optimization algorithm first objective function quickly reduced to less than 0.5, then tens of thousands of times of iteration is used to approach the objective function from 0.5 to 0 (10–7).

2.5.2 Example 2: Two-Disciplinary Optimization Problem

This example from the literature (Sellar et al. 1996), is an optimization problem that encompasses two disciplines. As shown in Fig. 2.15, there are coupling variables between subsystem 1 and subsystem 2. System design variables cover all design variables for both subsystems (that is, no local design variables). Another special feature of this problem is that the state variable of the subsystem is an geometric function of the coupling variable of the subsystem. In the following sections, the various solutions to this problem are introduced. Similar to the detailed derivation process of

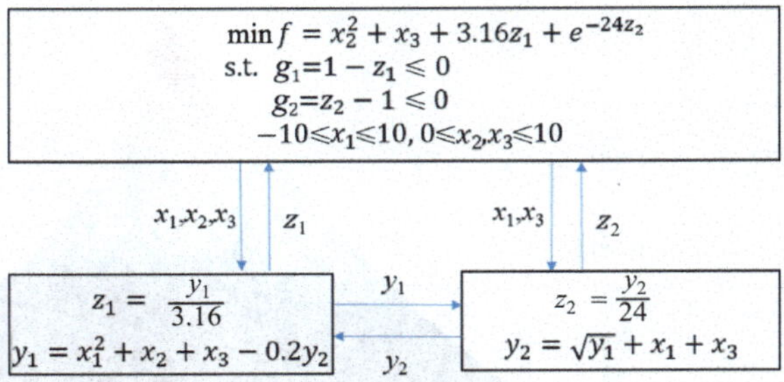

Fig. 2.15 Optimization of two disciplines

examples in traditional mathematical textbooks, this paper will introduce the whole modeling and optimization process in detail through specific Matlab programs, so that readers can reproduce the results according to the program written in this book, so as to deepen the understanding of multidisciplinary design optimization methods.

2.5.2.1 1MDFa Method

First, the MDF method is used, which corresponds to the model of the original problem. The Matlab code of the subfunctions of subsystem 1 and subsystem 2 is shown in Fig. 2.16. You can see that the input parameters of the subsystem 1 function are design variables x = [x1, x2, x3] and coupling variable y2, the output parameters are coupling parameter y1 and state parameter z1; subsystem 2 is also a two-parameter input and a two-parameter output.

```
function [ z1,y1 ] = sub1( x,y2 )
%% State function of subsystem 1
% -- Input --
% x=[x1,x2,x3]
% --Output --
% z = Status parameter
% y = Coupling parameter
%%
y1 = x(1)^2+x(2)+x(3)-0.2*y2;
z1 = y1/3.16;
end
```

```
function [ z2,y2 ] = sub2( x,y1 )
%% State function of subsystem 2
% -- Input --
% x=[x1,x2,x3]
% --Output --
% z = Status parameter
% y = Coupling parameter
%%
y2 = x(1)+x(3)+y1^0.5;
z2 = y2/24;
end
```

Fig. 2.16 MDO example 2: Matlab code for subsystem 1 and subsystem 2 functions

子系统1状态函数	State function of subsystem 1
输入	Input
输出	Output
状态参数	Status parameter
耦合参数	Coupling parameter
子系统2状态函数	State function of subsystem 2

In the MDF method, the system level only transfers design variables to the subsystem level, and the subsystem level returns state parameters to the system level after realizing subsystem balance through multidisciplinary analysis (MDA). Therefore, the multidisciplinary analysis function is an important part of the MDF method. The MDA function of this problem is shown in Fig. 2.17, where the input variable of the function is the design variable $x = [x_1, x_2, x_3]$, the output variable is the state variable $z = [z_1, z_2]$, and the other output variable output is used to record the error of subsystem balance iteration, the number of function calls and other conditions in the debugging, and the value of the quantity does not need to be changed in the optimization. As mentioned above, the multidisciplinary analysis function first take the golden number 0.618 subsystem as a coupling variable to the initial value y1 round iteration (initial iteration), get a more reasonable y1 initial value,

```
function [ z,output ] = MDA( x )
%% Multidisciplinary analysis of MDA
% x=[x1,x2,x3]
%% ----- Initialization iteration ----
y10 = 0.618;
[~,y20]=sub2( x,y10 );
[~,y10]=sub1( x,y20 );
funcCount0 = 1; % Number of initialization calls to each subsystem function
%% -- Start multidisciplinary analysis --
% Subsystem equilibrium iteration is a least squares problem, so lsqnonlin algorithm is most suitable
options = optimoptions('lsqnonlin');% Default optimization algorithm settings
options = optimoptions(options,'Display', 'off');% Modify settings
options = optimoptions(options,'Algorithm', 'levenberg-marquardt');
[y1,~,~,exitflag,output,~,~] = lsqnonlin(@fbalance,y10,[],[],options);
%
[z2,y2]=sub2( x,y1 );
[z1,y1check] = sub1(x,y2);
err_y1 = y1check-y1; % Check the error of y1 for the last time. It is used for debugging and not output.
z = [z1,z2];
%
output.funcCount =...
    funcCount0+output.funcCount+1;% Total number of times each subsystem analysis was invoked
%
%% Subfunction: find the error of coupling parameter y1
    function [err_y1] = fbalance(y1)
        [~,y2cal]=sub2( x,y1 );
        [~,y1cal]=sub1( x,y2cal );
        err_y1 = (y1cal-y1);
    end
end
```

Fig. 2.17 MDO example 2: multidisciplinary analysis functions

then the optimization algorithm is used to replace the traditional simple loop iteration method for balance subsystem, so as to speed up the subsystem iterative convergence speed. The subsystem of MDF method equilibrium iteration process essentially belongs to the least squares problem, so use specifically for this kind of problem of unconstrained least-squares optimization algorithm will obtain better convergence speed, the adopted Matlab optimization toolbox of least-square optimization function lsqnonlin function, readers can also use to write your own optimization algorithm instead.

MDA多学科分析	Multidisciplinary analysis of MDA
初始化迭代	Initialization iteration
初始化调用每个子系统函数的次数	Number of initialization calls to each subsystem function
开始多学科分析	Start multidisciplinary analysis
子系统平衡迭代属于最小二乘问题, 因此采用lsqnonlin算法最合适	Subsystem equilibrium iteration is a least squares problem, so lsqnonlin algorithm is most suitable
默认优化算法设置	Default optimization algorithm settings
修改设置	Modify settings
最后检查一次y1的误差, 调试用, 不输出	Check the error of y1 for the last time. It is used for debugging and not output.
总共调用每个子系统分析的次数	Total number of times each subsystem analysis was invoked
子函数: 求耦合参数y1的误差	Subfunction: find the error of coupling parameter y1

With the multi-disciplinary analysis function, the function can be called directly in the form of a sub-function to obtain the value of the state variable to calculate the system-level objective function and constraint function, thereby calculating the target value and constraint value. The system-level objective function and constraint function are shown in Figs. 2.18 and 2.19, respectively

Fig. 2.18 MDO example 2: MDF method system-level objective function diagram

```
function [ f ] = objsys( x )
%% System level objective function
% -- Input --
% x=[x1,x2,x3]
% --Output--
% f = Objective
%%
% [ z,output ] = MDA( x );
[ z,~ ] = MDA( x ); %Invoking multi-disciplinary system
                     analysis to obtain state variables
f = x(2)^2+x(3)+3.16*z(1)+exp(-24*z(2));
end
```

Fig. 2.19 MDO example 2: MDF method system-level constraint function

```
function [ c,ceq ] = consys( x )
%% System-level constraint function
% -- Input --
% x=[x1,x2,x3]
% --Output--
% c = Inequality constraint
% ceq = Equality constraint, leave blank if not
%%
% [ z,output ] = MDA( x );
[ z,~ ] = MDA( x ); % Invoking multi-disciplinary system
                      analysis to obtain state variables
c(1) = 1-z(1); %g1
c(2) = z(2)-1; %g2
%
ceq = []; %Null assignment without equality constraint
end
```

系统级目标函数 目标 调用多学科系统分析得到状态变量	System level objective function Objective Invoking multi-disciplinary system analysis to obtain state variables

系统级约束函数 不等式约束 等式约束, 没有则留空 没有等式约束赋空值	System-level constraint function Inequality constraint Equality constraint, leave blank if not Null assignment without equality constraint

Therefore, we establish the model of MDF method, which can optimize the main program only by calling the system-level objective function and constraint function like the traditional optimization problem. The main program example is shown in Fig. 2.20. Firstly, the internal point method of Matlab optimization toolbox is adopted for optimization. If the optimization result is not ideal, the sequential quadratic programming method or effective set method can be used.

The comparison between the calculation results of the MDF method and the results of the original literature is shown in Table 2.4. It can be seen that, because the problem is relatively simple and there is only one coupling variable, it is easier to find the subsystem equilibrium solution in multidisciplinary system analysis. The MDF model uses the interior point method, SQP method, or effective set method to converge to the optimal solution of the original problem.

```
clear all
clc
x0=[1,5,2]; % Design Variable Initial Point
xL = [-10,0,0]; % Lower limit of design variables
xU = [10,10,10]; % Upper limit of design variables
options = optimoptions('fmincon');% Default optimization algorithm settings
options = optimoptions(options,'Display', 'off');% Modify settings
% The interior point method is used by default. The optimization results
  did not meet the expectations, you can use the following two methods
% options = optimoptions(options,'Algorithm', 'sqp');
% options = optimoptions(options,'Algorithm', 'active-set');
[xopt,fval,exitflag,output,~,~,~] = ...
fmincon(@objsys,x0,[],[],[],[],xL,xU,@consys,options);
```

Fig. 2.20 MDO example 2 MDF method main program

Table 2.4 MDO example 2MDF method results

Optimization results	Initial point	Optimal point	Function value of optimization goal
Original document	[1, 5, 2]	[1.9776, 0, 0]	3.183390
MDF interior point method	[1, 5, 2]	[1.9776, 0.0003, 0]	3.1834
MDF SQP method	[1, 5, 2]	[1.9776, 0, 0]	3.1834
MDF active set method	[1, 5, 2]	[1.9776, 0, 0]	3.1834

设计变量初始点 设计变量下限 设计变量上限 默认优化算法设置 修改设置 默认采用内点法, 优化结果如果不理想可以使用下面的两种方法	Design Variable Initial Point Lower limit of design variables Upper limit of design variables Default optimization algorithm settings Modify settings The interior point method is used by default. The optimization results did not meet the expectations, you can use the following two methods

2.5.2.2 Decoupling Method

This problem can be decoupled by introducing two coupling aid design variables and two consistency constraints. The functions of the two subsystems are unchanged as shown in Fig. 2.16, and the system-level objective function and constraint function are changed as shown in Figs. 2.21 and 2.22. It can be seen that due to the introduction of auxiliary design variables, the design space has increased from 3 dimensions to 5 dimensional, and the constraint condition adds two consistency constraints written as equality constraints on the basis of the original g1 and g2.

Fig. 2.21 MDO example 2: system-level objective function of decoupling

```
function [ f ] = objsys( x )
%% System level objective function
% -- Input --
% x=[x1,x2,x3,y1,y2]
% --Output--
% f = Objective
%%
% Get state variable
[ z(1),~ ] = sub1( x(1:3),x(5) );
[ z(2),~ ] = sub2( x(1:3),x(4) );
%
f = x(2)^2+x(3)+3.16*z(1)+exp(-24*z(2));
end
```

Fig. 2.22 MDO example 2 system-level constraint function of decoupling method

```
function [ c,ceq ] = consys( x )
%% System-level constraint function
% -- Input --
% x=[x1,x2,x3,y1,y2]
% --Output--
% c = Inequality constraint
% ceq = Equality constraint
%%
% Get state variables and coupling variables
[ z(1),y1 ] = sub1( x(1:3),x(5) );
[ z(2),y2 ] = sub2( x(1:3),x(4) );
% Constraints on the original problem
c(1) = 1-z(1); %g1
c(2) = z(2)-1; %g2
% Consistency constraints
ceq1 = y1-x(4);
ceq2 = y2-x(5);
ceq = [ceq1,ceq2]; % Consistency constraints are
                     written as equality constraints
end
```

得到状态变量	Get state variable
得到状态变量和耦合变量	Get state variables and coupling variables
原问题约束条件	Constraints on the original problem
一致性约束条件	Consistency constraints
一致性约束条件写为等式约束条件	Consistency constraints are written as equality constraints

This optimization model after decoupling of subsystems can be considered as belonging to the IDF model introduced in the previous article, and only one round of optimization is required at the system level. See Fig. 2.23 for the main program of optimization.

```
clear all
clc
y0 = 0.618;% The initial value of the coupling variable
yL = -1e3;% The lower limit of the coupling variable
yU = 1e3;% The upper limit of the coupling variable
%
x0=[1,5,2,y0,y0]; % Design Variable Initial Point
xL = [-10,0,0,yL,yL]; % Lower limit of design variables
xU = [10,10,10,yU,yU]; % Upper limit of design variables
options = optimoptions('fmincon');% Default optimization algorithm settings
options = optimoptions(options,'Display', 'off');% Modify settings
% The interior point method is used by default. The optimization results
  did not meet the expectations, you can use the following two methods
% options = optimoptions(options,'Algorithm', 'sqp');
% options = optimoptions(options,'Algorithm', 'active-set');
[xopt,fval,exitflag,output,~,~,~] = ...
fmincon(@objsys,x0,[],[],[],[],xL,xU,@consys,options);
```

Fig. 2.23 MDO example 2 main program of decoupling method

It should be noted that since the upper and lower limits of the two newly added auxiliary design variables are unknown, the large range [–1000, –1000] is taken here as the range of the two new variables, and the golden ratio 0.618 is taken as the initial point of optimization.

The optimization results of the decoupling method are shown in Table 2.5. It can be seen that the internal point method, SQP method or effective set method can be used in the decoupling method (IDF) model to obtain the optimal solution of the original problem, and the coupling parameter equilibrium values between the two subsystems at the optimal design point y1 = 3.16 and y2 = 3.7553 can also be obtained.

耦合变量初始值	The initial value of the coupling variable
耦合变量下限	The lower limit of the coupling variable
耦合变量上限	The upper limit of the coupling variable

Table 2.5 MDO example 2 optimization results of decoupling method

Optimization results	Initial point	Optimal point	Function value of optimization goal
Original document	[1, 5, 2]	[1.9776, 0, 0]	3.183390
Decoupling interior point method	[1, 5, 2, 0.618, 0.618]	[1,9776, 0.0002, 0, 3.16, 3.7552]	3.1834
Decoupling SQP method	[1, 5, 2, 0.618, 0.618]	[1.9776, 0, 0, 3.16, 3.7553]	3.1834
Decoupling effective set method	[1, 5, 2, 0.618, 0.618]	[1.9776, 0, 0, 3.16, 3.7553]	3.1834

2.5.2.3 Approximation

When using approximation technique for decoupling, it is necessary to obtain enough data through DOE and establish an approximate model between each coupling variable and the design variable based on the data. First, it is necessary to obtain the data of each coupling variable through experimental design and multidisciplinary analysis, in which the system analysis function of coupling parameters are shown in Fig. 2.24.

耦合变量y多学科分析 子函数: 求耦合参数y的误差	Multidisciplinary analysis of the coupling variable y Subfunction: Find the error of the coupling parameter y

It can be seen that, basically, the approximation method is the same as the multidisciplinary analysis function of the MDF method, except that the output variable changes from the state parameter z to the coupling parameter y after the subsystem equilibrium. The experimental design and the establishment and preservation function of the approximate model of the neural network are shown in Fig. 2.25.

Fig. 2.24 MDO example 2 Multi-disciplinary system analysis of coupling parameters of approximate method

```
function [ y,output ] = MDA_y( x )
%% Multidisciplinary analysis of the coupling variable y
% x=[x1,x2,x3]
%% ---- Initialization iteration ----
y10 = 0.618;
[~,y20]=sub2( x,y10 );
[~,y10]=sub1( x,y20 );
[~,y20]=sub2( x,y10 );
y0 = [y10,y20];
%% -- Start multidisciplinary analysis--
% Subsystem equilibrium iteration is a least squares problem,
   so lsqnonlin algorithm is most suitable
options = optimoptions('lsqnonlin');%Default optimization algorithm settings
options = optimoptions(options,'Display', 'off');% Modify settings
options = optimoptions(options,'Algorithm', 'levenberg-marquardt');
[y,~,~,exitflag,output,~,~] = lsqnonlin(@fbalance,y0,[],[],options);
%
%% Subfunction: Find the error of the coupling parameter y
    function [err_y] = fbalance(y)
        % y = [y1,y2]
        [~,y1cal]=sub1( x,y(2) );
        [~,y2cal]=sub2( x,y(1) );
        err_y = [y1cal-y(1),y2cal-y(2)];
    end
end
```

Fig. 2.25 MDO example 2 approximation DOE and establish coupling parameter approximation Model

```
clear all
clc
xL = [-10,0,0]; % Lower limit of design variables
xU = [10,10,10]; % Upper limit of design variables
%% DOE
% Sample generation
Ndoe = 1000;
rng('default');%The pseudo-random number generation algorithm is fixed
               so that the results of each calculation can be repeated
x = zeros(Ndoe,length(xL));
for ind_1 = 1:length(xL)
    u = rand(Ndoe,1);
x(:,ind_1) = xL(ind_1)+u*( xU(ind_1)-xL(ind_1) );
end
clear u
% Calculate function value
Ny = 2; % Number of coupling variables
y = zeros(Ndoe,Ny);
for ind_2 = 1:Ndoe
y(ind_2,:) = MDA_y(x(ind_2,:));
end
%% Approximate
N_hide = 10; % Number of hidden neurons
net = fitnet(N_hide); %BP Artificial neural networks
net = train(net,x',y'); % Training network
save Ynet.mat net % Save the network for optimization
```

样本生成 伪随机数生成算法固定,使得每次的计算结果可重复 函数值计算 耦合变量个数 近似 隐层神经元个数 人工神经网络 训练网络 保存网络供优化使用	Sample generation The pseudo-random number generation algorithm is fixed so that the results of each calculation can be repeated Calculate function value Number of coupling variables Approximate Number of hidden neurons Artificial neural networks Training network Save the network for optimization

Among them, by the random method, experiment design to generate 1000 sample points within the scope of design variable values, and then through the analysis of the coupling parameter of the system function to obtain two coupling parameters corresponding data points, on the basis of using the artificial neural network of neurons in hidden layer has 10 coupling parameter approximation model is set up, and stored in Ynet. The mat files. Then, 200 new sample points were randomly generated and calculated and compared respectively through the coupling parameter system analysis function and the established neural network. It was found that the error of yi and was in the order of 0.05, and the approximate precision was considered to have met the requirements.

After the approximate model is established, the coupling parameters required for the calculation of each subsystem can be directly calculated using the approximate model, so as to achieve the decoupling of the subsystem analysis. The subsystem function after the approximate model becomes as shown in Fig. 2.26 and as shown in Fig. 2.27, the y2 required for the calculation of subsystem 1 is directly obtained through the neural network, and the same is true for subsystem 2.

输入耦合变量通过已经建立的近似模型计算	Input coupling variables are calculated using an established approximation model

The objective function and constraint function of system-level optimization are the same as that of MDF method, and the main program of optimization is also the same as that of MDF method. The optimization results are shown in Table 2.6. It can be seen that the approximate error leads to a small deviation in the optimization solution, but such a small error is acceptable from an engineering perspective.

Fig. 2.26 MDO example 2 subsystem 1 function after introduction

```
function [ z1 ] = sub1( x )
%% State function of subsystem 1
% -- Input --
% x=[x1,x2,x3]
% -- Output --
% z = Status parameter
%% Input coupling variables are calculated
    using an established approximation model
load Ynet.mat
y2 = net(x');y2 = y2(2);
%%
y1 = x(1)^2+x(2)+x(3)-0.2*y2;
z1 = y1/3.16;
end
```

Fig. 2.27 MDO example 2 introduces the function of subsystem 2 after approximation

```
function [ z2 ] = sub2( x )
%% State function of subsystem 2
% -- Input --
% x=[x1,x2,x3]
% -- Output --
% z = Status parameter
%% Input coupling variables are calculated
    using an established approximation model
load Ynet.mat
y1 = net(x');y1 = y1(1);
%%
y2 = x(1)+x(3)+y1^0.5;
z2 = y2/24;
end
```

Table 2.6 MDO example 2 approximation optimization results

Optimization results	Initial point	Optimal point	Function value of optimization goal
Original document	[1, 5, 2]	[1.9776, 0, 0]	3.183390
Approximate interior point method	[1, 5, 2]	[1.9577, 0.003, 0]	3.1841
Approximate SQP method	[1, 5, 2]	[1.9578, 0, 0]	3.1841
Approximate effective set method	[1, 5, 2]	[1.9578, 0, 0]	3.1841

In this case, the state parameters of each subsystem are directly expressed as functions of design variables through the approximate model. Then, the approximate model of state variables can be directly invoked during system optimization without invoking each subsystem.

References

DorMohammadi S, Rais-Rohani M (2013) Exponential penalty function formulation for multilevel optimization using the analytical target cascading framework. Struct Multi Optim 47(4):599–612

Golovidov O (1998) Flexible Implementation of approximation concepts in an MDO framework. In: AIAA 98, p 4959

Kaitai F (1980) Uniform design—application of number theory in DOE. J Appl Math 3(4). (方开泰. 均匀设计——数论方法在试验设计的应用[J]. 应用数学学报,1980,3(4).)

Kim HM, Michelena NF, Papalambros PY, Jiang T (2003) Target cascading in optimal system design. J Mech Des (Transactions of the ASME) 125(3):474–480

Kim H, Ragon S, Soremekun G et al. (2004). Flexible approximation model approach for bi-level integrated system synthesis. Reston, VA, USA. AIAA 2004-4545

Lin X (2009) Research and application of aircraft MDO process and related technologies. PhD thesis of National university of defense science and technology (许林. 飞行器MDO过程及相关技术研究与应用. 国防科学技术大学博士学位论文. 2009.)

Michelena N, Papalambros P, Park H et al (1999) Hierarchical overlapping coordination for large-scale optimization by decomposition. AIAA J 37(7):890–896

Min M (2011) The method of collaborative optimization of two-stage integrated system and its application in the conceptual design of deep-sea space station. PhD thesis of Shanghai Jiaotong University (赵敏. 两级集成系统协同优化方法及其在深海空间站总体概念设计中的应用. 上海交通大学博士学位论文. 2009

Sellar RS, Batill SM, Renaud JE (1996) Response surface based, concurrent subspace optimization for multidisciplinary system design. AIAA Paper, 714

Sobieszczanski-Sobieski J, Haftka RT (1996) Multidisciplinary aerospace design optimization: survey of recent developments. 34th AIAA Aerospace Sciences Meeting and Exhibit, Reno, Nevada, AIAA Paper No. 96-0711, p. 32

Tosserams S, Kokkolaras M, Etman LFP, Rooda JEA (2010) Nonhierarchical formulation of analytical target cascading. J Mech Des (Transactions of the ASME) 132

Xinsheng L (2004) Application of combination optimization algorithm of neural network and genetic algorithm. J Guizhou Univer: Nat Sci 21(2):179–184. (赖鑫生. 神经网络结合遗传算法优化应用[J]. 贵州大学学报: 自然科学版, 2004, 21(2): 179–184.)

Yang Z (2013) Research on optimization design of multidisciplinary collaborative design process. Master's thesis of shenyang university of technology (张阳. 多学科协同设计过程优化设计研究. 沈阳理工大学硕士学位论文. 2013.)

Yao W, Chen XQ, Ouyang Q, Wei YX (2010) A Concurrent subspace optimization procedure based on multidisciplinary active regional crossover optimization. In 51st AIAA/ASME/ASCE/AHS/ASC structures, structural dynamics, and materials conference < BR > 18th. Orlando, Florida. AIAA, pp 2010–3082

Chapter 3
Uncertainty Theory

In 1657, Huygens first studied randomness as a type of uncertainty, and in 1836 referred to the term "uncertainty". The widely accepted classification method of uncertain information is to divide the uncertain information into "Aleatory Uncertainty" (AU) and "Epistemic Uncertainty" (Oberkampf et al. 2004). Figure 3.1 illustrates the relationship between knowledge mastery and uncertain information (Guo and Du 2007). With the development of science and technology, the understanding of a natural phenomenon is getting deeper and closer to the essence of the phenomenon, so the cognitive uncertainty will gradually decrease with the increase of knowledge, but many natural phenomena have inherent Occasionally, sometimes the phenomenon is inconsistent with the conclusions drawn from the existing scientific observations, which reflects the inherent uncertainty of natural phenomena.

未知 现有知识 完全知识 知识 认知不确定性 固有不确定性 不确定性	The unknown Current knowledge Full knowledge knowledge Cognitive uncertainty Inherent uncertainty uncertainty

The definition of uncertainty is: contingent uncertainty refers to objective and irreducible uncertainty; Cognitive uncertainty refers to subjective uncertainty caused by lack of knowledge, which can be reduced (subjective and reducible uncertainty that stems from a lack of knowledge regarding input data) (Helton 1997). Accidental uncertainty describes the characteristics of random changes inherent in real systems or environments (Oberkampf et al. 2001; Helton et al. 2006; Oberkampf and Helton

B. Pan and W. Cui, *Multidisciplinary Design Optimization and Its Application in Deep Manned Submersible Design*, Ocean Engineering & Oceanography 13,
https://doi.org/10.1007/978-981-15-6455-0_3

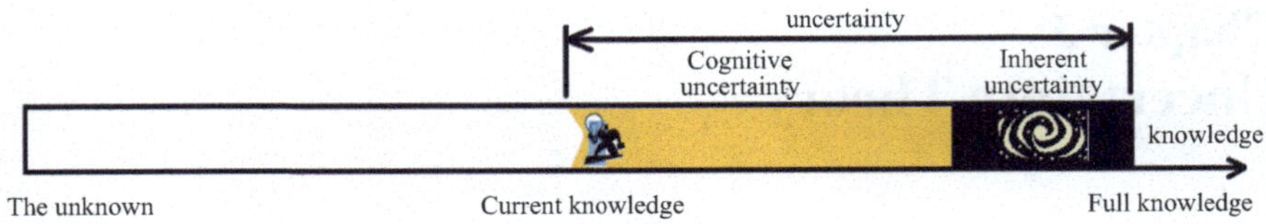

Fig. 3.1 The relationship between cognition and uncertainty

2002), so it is often referred to as inherent uncertainty, such as wave loads in ship design. Cognitive uncertainty, sometimes called subjective uncertainty, is caused by lack of knowledge or incomplete information. Of course, the division itself is controversial. If the nature of the universe is known, accidental uncertainty can be reduced. Assuming that the nature of the universe is unknowable, accidental uncertainty can be assumed to be unreducible beyond a certain point. The famous debate between Einstein and Bohr about quantum mechanics is essentially a debate between these two world views. Existing scientific theories are built on the basis of some assumptions, which may not be completely true in practical applications. In other words, there is always a certain difference between the object of scientific research and the reality, which is reflected in the difference between the analysis model and the actual engineering structure in engineering design. This kind of cognitive uncertainty caused by the difference between the analysis model itself and the actual object is often referred to as model cognitive uncertainty (Agarwal 2004). The method to reduce the cognitive uncertainty of the model depends on the progress of scientific theories and designers to include enough details in the modeling. Uncertainty caused by the lack of data or insufficient information of uncertain parameters that cannot accurately describe their statistical characteristics is usually classified as cognitive uncertainty, which is generally referred to as parametric cognitive uncertainty. At present, the cognitive uncertainty in engineering reliability research mostly refers to the parameter cognitive uncertainty, and the cognitive uncertainty in this book usually refers to this kind of uncertainty.

Nowadays, computer is playing an increasingly important role in engineering design, but in the computer calculation, there will always be rounding error, decimal co., LTD., the convergence criterion Uncertainty leads to the calculation of Uncertainty, such as this kind of Uncertainty is called Numerical Uncertainty (Numerical Uncertainty, NU) (Agarwal 2004), sometimes referred to as the simulation of Uncertainty. Numerical uncertainties cannot be ignored in computational fluid dynamics analysis (CFD), but they are not usually involved in general engineering design. Even if they are involved, such uncertainties are often approximated as random variables. Therefore, the research objects of this book focus on accidental uncertainties and cognitive uncertainties.

The most popular way to describe the inherent uncertainty is Probability Theory. The founder of Classical Probability Theory was Swiss mathematician Bernoulli, who established the first limit theorem in Probability Theory, Bernoulli's law of large Numbers. In 1933, a.n.kolmogarov summarized the basic properties and relations of events and their probabilities with the ideas of set theory and measure theory, established the axiomatic system of probability, and laid the theoretical foundation of probability theory (Durrett 1996; Xiao et al. 2002), making probability theory officially a science (Billingsley 1995). Classical probability theory is the most well developed uncertainty theory in the theoretical system. It has become a required basic course in most universities, and there are numerous references, such as Agarwal (2004), Durrett (1996), Shutiei (2000), Billingsley (1995) and Cui (1990). The first section of this chapter gives the basic concepts of classical probability theory and briefly discusses its limitations.

At present, Evidence Theory (Shafer 1976), Fuzzy Set Theory (Zadeh 1965, 1978), Possibility Theory (Klir 1999, 2004), interval theory (Moore 1962, 1966), etc. are the widely used description methods for cognitive uncertainty. Fuzzy set theory was proposed by Zadeh in 1965. In fuzzy set theory, cognitive uncertainty information is described by fuzzy variables and membership functions that represent the credibility of uncertain information (Klir 2004; Dubois and Prade 1988; Savoia 2002). At present, because the axiom system of fuzzy theory is not as perfect as probability theory, its application has been limited. The possibility theory is based on the fuzzy theory. Dubois points out the Probability-Possibility Consistent Principle, that is, the probability of an event occurring is not less than the probability of the event (Probability) (Dubois and Prade 1988). Possibility theory is only initially established. Compared with probability theory, its theoretical system and development depth are far from enough (De Cooman 1997). Interval theory was first established by Moore in 1962 in order to track the upper and lower limits of an accurate solution under the condition that the number of digits retained after the decimal point is limited in computer numerical calculation (Jacobsen 2002).

With the in-depth development of the uncertainty description theory and the accumulation of experience in the practical application of the reliability design theory, people have summarized the applicable scope of the existing uncertainty description theory. Different uncertainty description theories are applicable to describe different uncertain information, as shown in Table 3.1 (Bifeng 2006). Although conceptually uncertainties can be classified differently, for a specific engineering problem or a specific design parameter, the uncertainty is the result of multiple factors and in many cases cannot be quantified separately. Therefore, in order to overcome the shortcoming of single Uncertainty description Theory, many of the scientists working on a can be used to describe Uncertainty by accident, can also be used to describe the unity of the cognitive Uncertainty information description Theory (Unifying Uncertainty Handling, UUHT). For example, Dempster (1967), Cui and Blockley (1990) combine Probability Theory with Interval Theory to obtain Interval Probability Theory (IPT), and use Interval Probability Theory as a method to measure the credibility of evidence in knowledge system. Weichselberger proposed some problems that need to be studied when using interval probability theory as a unified uncertainty description method (Weichselberger 2000). Zadeh proposed the concept of extending fuzzy set theory into generalized uncertainty theory (Zadeh 2006). Cui introduces some concepts of fuzzy mathematics into classical Probability Theory, overcomes the shortcomings of classical Probability Theory and expands the ability

Table 3.1 Description theory of common uncertain information and its applicable scope (Bifeng 2006)

Uncertainty Analysis (UA) method	Uncertain representation measures	Theoretical method	Application condition
UA method based on probability	Probability distribution	Probability and mathematical statistics	A large sample is uncertain
UA method based on fuzzy theory	Fuzzy membership degree	Fuzzy mathematics	cognitive uncertainty
UA method based on the theory of possibility	Possibility measure	Possibility theory	cognitive uncertainty
UA method based on evidence theory	Reliability	Evidence theory	cognitive uncertainty
A method based on interval analysis	Distribution interval	Interval mathematics	No sample, bounded
A method based on convex set model	Mean deviation ratio	Convex analysis	No sample, bounded

of classical Probability Theory without modifying the axiom system of classical Probability Theory. The expanded Probability Theory is called Generalized Probability Theory (GPT) (Cm 1993a, b). In the general theory of probability, the possibility of a certain event occurs is considered to be subjective believe in the possibility of the size, rather than the real possibility of the incident, so as to make the event probability includes subjective uncertainty, which makes the generalized probability theory can deal with uncertainty by accident, not only can deal with cognitive uncertainty. In generalized probability theory, the probability of an event can be expressed as interval number, fuzzy number or real number, depending on the depth and breadth of information. Thus, the generalized probability theory not only retains the perfect axiomatic system of classical probability theory, but also overcomes the limitations of classical probability theory, which will be introduced in detail in Sect. 3.6 of this chapter. In this book, we treat uncertainty as a natural phenomenon and apply the theory of generalized probability to deal with it, without further discussing what kind of uncertainty it belongs to. The research on unified uncertainty description method is one of the frontiers of reliability design theory.

3.1 Classical Probability Theory

In order to distinguish traditional probability theory from interval probability theory and generalized probability theory, the original probability theory is called classical probability theory. Classical probability theory treats uncertain information as a random variable. When there is enough data, statistical methods can be used to obtain the random characteristics of random variables, such as mean value, variance, correlation, etc. Classical probability theory is based on probability space and has the following basic concepts (Zhentao 2007; Shutie 2000):

(1) Sample space Ω refers to the set of all possible outcomes of a randomized trial.
(2) Random events are events that can be determined by the results of randomized trials. Random events are a subset of the sample space.
(3) The event domain Ξ (or event family) is an event family composed of all observable events (events with reasonable probabilities), and Ξ is the power set of Ω.
(4) If the aggregate function P (A) (0 ≤ P (A) ≤ 1) defined on the event domain Ξ satisfies the following three conditions:

① Non-negative: $\forall A \in \Xi, P(A) \geq 0$ is true;
② Normative: $P(\Omega) = 1$;
③ Countably additive: if $A_i \in \Xi (i = 1, 2, \ldots)$ and incompatible with each other, then $P(\bigcup_{i=1}^{\infty} A_i) = \sum_{i=1}^{\infty} P(A_i)$. P (A) is the probability of event A.

(5) Probability space refers to triples: (Ω, Ξ, P)。
(6) For a continuous random variable X, its probability density function (PDF) $f(x)$ is defined as the derivative of its probability distribution function, and the statistical characteristic parameters of X are defined as:

① Cumulative Distribution Function (CDF):

$$F(x) = P(X \leq x) = \int_{-\infty}^{x} f(t)dt \tag{3.1}$$

② Mean:

$$EX = \int_{-\infty}^{+\infty} x \times f(x)dx \tag{3.2}$$

③ Variance:

$$DX = \int_{-\infty}^{+\infty} (x - EX)^2 \times f(x)dx \tag{3.3}$$

④ Standard deviation: $\sqrt{DX}$
⑤ Variable coefficient: $\frac{\sqrt{DX}}{EX}$

(7) By the random variable X defined in the same probability space; (f = 1, 2,... ‘) is called a random vector, then F(x1, x2..., xn) = P (X1 acuities were X1, X2

X2, or less..., Xn ≤ Xn) is the joint distribution function of random vector X, and its derivative is the joint probability density function of random vector X, which is:

$$F(x_1, x_2, \ldots, x_n) = \int_{-\infty}^{x_n} \int_{-\infty}^{x_{n-1}} \cdots \int_{-\infty}^{x_1} f(y_1, y_2, \ldots, y_n) dy_1 dy_2 \ldots dy_n \tag{3.4}$$

(8) For any $k < n$ and $1 \leq i_1 < i_2 \ldots < i_k \leq n$, the distribution of the subvectors $(X_{i_1}, X_{i_2}, \ldots, X_{i_k})$ of the random vector X is called the edge distribution of X. The most studied is k = 1, that The edge distribution of Xi and the edge distribution of Xi are:

$$F_{X_i}(x_i) = F(+\infty, x_i, \ldots, +\infty) = \int_{-\infty}^{+\infty} \cdots \int_{-\infty}^{x_i} \cdots \int_{-\infty}^{+\infty} f(y_1, y_2, \ldots, y_n) dy_1 dy_2 \ldots dy_n \tag{3.5}$$

The corresponding edge probability density function of Xi is:

$$fx_i(x_i) = \int_{-\infty}^{+\infty} \cdots \int_{-\infty}^{+\infty} \int_{-\infty}^{+\infty} \cdots \int_{-\infty}^{+\infty} f(y_1, \ldots, y_{i-1}, y_{i+1}, \ldots, y_n) dy_1 \ldots dy_{i-1} dy_{i+1} \ldots dy_n \tag{3.6}$$

Then, the mean value of Xi:

$$EX_i = \int_{-\infty}^{+\infty} x_i f_{X_i}(x_i) dx_i \tag{3.7}$$

The variance of X_i is:

$$DX_i = \int_{-\infty}^{+\infty} (x_i - EX_i)^2 f_{X_i}(x_i) dx_i \tag{3.8}$$

The covariance of Xi and Xj is:

$$\begin{aligned} Cov(X_i, X_j) = \rho_{ij}\sqrt{DX_i \times DX_j} &= E[(X_i - EX_i)(X_j - EX_j)] \\ &= E(X_i \times X_j) - EX_i \times EX_j \end{aligned} \tag{3.9}$$

In Eq. (3.9), ρ_{ij} is called the correlation coefficient between X_i and X_j. When X_i and X_j are independent of each other, $\rho_{ij} = 0$.

(9) Let Y = g (X) be a function of the random vector X, then the distribution function of Y is:

$$F_Y(y) = P(g(X) \leq y) = P(g(X) \in (-\infty, y])$$
$$= P(X \in g^{-1}(-\infty, y])) = \int\int_{g^{-1}(-\infty,y]} \mathrm{f}(x_1, x_2, \ldots, x_n)d_{x_1}d_{x_2}\ldots d_{x_n} \tag{3.10}$$

(10) The mean and variance of Y = g (X) are:

$$EY = \int_{-\infty}^{+\infty}\ldots\int_{-\infty}^{+\infty} \mathrm{g}(\mathrm{x}_1, \mathrm{x}_2, \ldots, \mathrm{x}_\mathrm{n}) \times \mathrm{f}(x_1, x_2, \ldots, x_n)d_{x_1}d_{x_2}\ldots d_{x_n} \tag{3.11}$$

$$DY = \int_{-\infty}^{+\infty}\ldots\int_{-\infty}^{+\infty} (\mathrm{g}(\mathrm{x}_1, \mathrm{x}_2, \ldots, \mathrm{x}_\mathrm{n}) - \mathrm{EY})^2$$
$$\times \mathrm{f}(x_1, x_2, \ldots, x_n)d_{x_1}d_{x_2}\ldots d_{x_n} \tag{3.12}$$

When there are enough samples of uncertain data, classical probability theory can describe uncertain information more accurately. For example, when determining the yield strength of a certain metal material in structural design, it is necessary to use sampling to extract enough metal materials of this type to be processed into samples, and then perform batch tensile tests. After statistical analysis of the experimental results, Determine which probability density function is used to approximate these data with the least error (for such uncertainties, the most commonly used is the normal distribution), and at the same time, the corresponding mean parameters and variances can be determined. However, in practice, often the number of samples is not enough. At this time, using random variables to describe the uncertainty information will cause a large error.

3.2 Fuzzy Theory

In fuzzy theory, all objects are called the domain U. According to classical set theory, A common subset A on the domain U refers to A group composed of some elements in U. In classical set theory, it is clear whether each element belongs to A, that is, the element either belongs to A or does not belong to A. Therefore, classical set theory is also called clear set theory. However, in practice, the membership relationship is not as clear as defined by the classical set theory in many cases. For example, the common concepts such as "danger", "heavy" and "high" are obviously no longer applicable to the classical set theory, so the fuzzy theory extends the classical set theory to the fuzzy set. The following are some basic concepts of fuzzy theory including fuzzy sets.

(1) For a given subset U of a given universe U, for any element n of U, a number between [0,1] is specified to correspond to it, then this number is called the membership of u to A. The mapping from U to the set of all memberships is a function $\mu_{\hat{A}}(u)$, then function $\mu_{\hat{A}}(u)$ is a membership function of Â, and set Â is a fuzzy set.

The size of the membership function $\mu_{\hat{A}}(u)$ reflects the degree of u's membership to the fuzzy set Â. The closer the value of $\mu_{\hat{A}}(u)$ is to 1, the higher the degree of u's subordination to Â; the closer the value of $\mu_{\hat{A}}(u)$ is to 0, the lower the degree of $\mu_{\hat{A}}(u)$'s subordinate to Â. For the selected threshold $\lambda(0 \leq \lambda \leq 1)$, when $\mu_{\hat{A}}(u) \geq \lambda$, u is regarded as belonging to Â, otherwise it is not Â. The non-binary nature of membership functions makes a fuzzy set Â a "flexible" set with a "moving boundary".

It can be seen from this definition that the classic set is a special case of fuzzy sets. Whether the element u belongs to a classic set A is represented by a membership function:

$$\mu_A(u) = \begin{cases} 1 & u \in A \\ 0 & u \notin A \end{cases} \tag{3.13}$$

It can be seen that the classic set is a special case where the membership function has only two values, 0 and 1. For fuzzy sets Â, the value of membership function $\mu_A(u)$ is no longer just 0 or 1, but can be any value on [0,1].

This definition has many membership functions, and any function with a range between [0,1] can be defined as a membership function. In order to define meaningful membership functions, Zadeh defines the convexity and boundedness of membership functions, and the consistency of membership functions in practical use.

(2) Boundedness: For any $\alpha \in (0, 1]$, if the set $\{u : \mu_A(u) \geq \alpha\}$ is bounded, the membership function is bounded. Since the range of the membership function is [0,1], according to the definition of boundedness of ordinary functions, membership functions are always bounded. Please refer to the literature (Zadeh 1965) for detailed information on the boundedness of membership functions.
(3) Convexity: for any $\alpha \in (0, 1]$, if the set $\{u : \mu_A(u) \geq \alpha\}$ is convexity, then the membership function is convexity. Notice that the convexity of the membership function is different from that of the general function. The convexity of the membership function is equivalent to:

$$\mu_A(\lambda u_1 + (1 - \lambda)u_2) \geq \min(\mu_A(u_1), \mu_A(u_2))$$

(4) Consistency: There is a unique point um, which makes $\mu_A(u_m) = 1$, and point um is also called the point of maximum membership.
(5) For a non-empty set Θ, its power set is P (Θ), if the set function Pos satisfies:

① $Pos\ (\Theta) = 1$;
② $Pos(\phi) = 0$;

③ For any set $\{A_i\}$ in P(Θ), $Pos(\cup_i A_i) = \sup_i(Pos\{A_i\})$ is true.

Then, Pos is a measure of likelihood (Nahmias 1978), where sup means taking the upper bound (Supremum), and Pos (A) refers to the highest probability among all the values that make event A true. It should be noted that there are still some differences in the definition of the possibility measure. For example, De Cooman (1997) has a different definition of the possibility measure than the aforementioned concept.

(6) The possibility space consists of triples [Θ, P (Θ), Pos].
(7) A fuzzy variable is a function that maps from the probability space [Θ, P (Θ), Pos] to the real number set R. The n-dimensional fuzzy vector is a function from the probability space [Θ, P (Θ), Pos] to the $\mathbf{R}^n$-dimensional real vector space.

For continuous fuzzy variables, the boundedness and convexity of membership functions are equivalent to:

① $\lim_{x\to-\infty} \mu(x) = \lim_{x\to\infty} \mu(x) = 0$;
② When $x \leq x_m$, $\mu(x)$ is a non-decreasing function, and when $x \geq x_m$, $\mu(x)$ is a non-increasing function.

Coupled with the consistency requirements in practical use, it can be seen that the shape of the membership function should be as shown in Fig. 3.2, that is, with x_m as the boundary, the membership function can be divided into left ($\mu_{X,\,L}(x)$) and right (µX, R (x)), take, L(x) non-decreasing, take, (x) non-increasing.

When a fuzzy variable is transformed from a random variable, two principles should be followed:

① Probability-possible consistency principle: the probability of an event is not less than the probability of the event (Dubois and Prade 1988);
② The system designed according to fuzzy theory is more conservative than the system designed according to probability theory, and the membership function should choose the system with the least conservative quantity (Du 2006).

Although the above concepts specify some properties of membership function, they do not give a clear definition of membership function. Therefore, it is often necessary to select membership function according to subjective opinions in practical

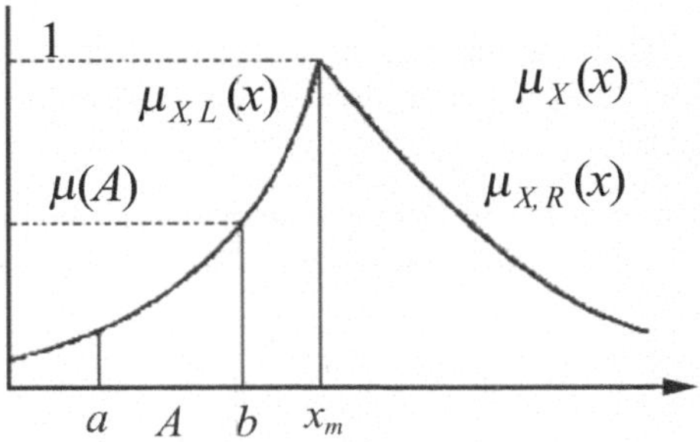

Fig. 3.2 Properties of membership functions

use, which brings a lot of trouble to the application of fuzzy theory in engineering practice.

In Zadeh's fuzzy set theory, in addition to the possibility measure Pw, the impossibility of the opposite of the event is also defined as the necessity measure Nec of the event. If ξ is a fuzzy variable and its membership function is $\mu(x)$, $r \in R$, which is a real number, then the possibility and necessity of fuzzy event $\{\xi \geq r\}$ are as follows:

$$Pos\{\xi \geq r\} = \sup_{u \geq r} \mu(u) \tag{3.14}$$

$$Nec\{\xi \geq r\} = 1 - Pos\{\xi < r\} = 1 - \sup_{u < r} \mu(u) \tag{3.15}$$

From this definition, it can be seen that even if the possibility of a fuzzy event is equal to 1, the event may not be true; similarly, the necessity of a fuzzy event is equal to 0, and the event may also occur. In order to facilitate practical applications, Liu Baoding introduced a credibility measure (Liu 2002):

$$Cr\{\xi \geq r\} = \frac{1}{2}(\mathrm{P}os\{\xi \geq r\} + \mathrm{N}ec\{\xi \geq r\}) \tag{3.16}$$

That is, the credibility measure Cr of a fuzzy event is equal to the average value of the possibility and necessity of the fuzzy event. It can be found that Cr is self-dual, which greatly facilitates decision-making in practical applications. When the Cr of a fuzzy event is equal to 1, the event must occur; and when the Cr of a fuzzy event is equal to 0, the event must not occur.

On the basis of the credibility measure, Liu Baoding also introduced a credibility distribution function and a credibility density function, which correspond to the probability distribution function and probability density function in probability theory, respectively. Based on this, a complete axiomatic system based on the credibility theory of credibility measures has been developed.

(8) For fuzzy variable ξ, if function $\Phi[-\infty, +\infty] \rightarrow [0, 1]$ satisfies any $x \in [-\infty, +\infty]$:

$$\Phi(x) = Cr\{\xi \leq x\} \tag{3.17}$$

Then the function Φ is the credibility distribution of the fuzzy variable ξ, which represents the credibility of the value of the fuzzy variable ξ not greater than x (Liu 2002).

(9) Φ is the credibility distribution of the fuzzy variable ξ. For any $x \in [-\infty, +\infty]$, if the function $\phi[-\infty, +\infty] \rightarrow [0, +\infty]$ satisfies:

$$\Phi(x) = \int_{-\infty}^{x} \phi(y)dy \tag{3.18}$$

Then, the function ϕ is the credibility density function of the fuzzy variable ξ.

(10) Based on the credibility measure, the expected value of the fuzzy variable ξ is defined as Liu (2004) as long as there is a finite value in the integral $\int_0^{+\infty} Cr\{\xi \geq r\}dr$, $\int_{-\infty}^{0} Cr\{\xi \leq r\}dr$:

$$E[\xi] = \int_{0}^{+\infty} Cr\{\xi \geq r\}dr - \int_{-\infty}^{0} Cr\{\xi \leq r\}dr \tag{3.19}$$

(11) If the expected value of the fuzzy variable ξ is $E[\xi]$, then the variance of the fuzzy variable ξ is:

$$V[\xi] = E[(\xi - E[\xi])^2] \tag{3.20}$$

(12) $\xi_1, \xi_2, \ldots, \xi_m$ are all fuzzy variables, and if it is true for any subset $B_1, B_2, \ldots, B_m$ of R in the real field.

$$\mathrm{P}os\{\xi_i \in B_i, i = 1, 2, \ldots, m\} = \min_{1 \leq i \leq m} \mathrm{P}os\{\xi_i \in B_i\} \tag{3.21}$$

Then, $\xi_1, \xi_2, \ldots, \xi_m$ which are independent of each other.

(13) ξ and η are independent fuzzy variables, and the expected value is limited, then the expected value operator has the property of a linear operator (Liu 2004), which is true for any real numbers a and b

$$E[a\xi + b\eta] = aE[\xi] + bE[\eta] \tag{3.22}$$

Taking the common trapezoid membership function (Fig. 3.3) as an example, according to the above definition, the credibility $Cr\{\xi \geq x\}$ can be written as:

Fig. 3.3 Commonly used trapezoid membership function

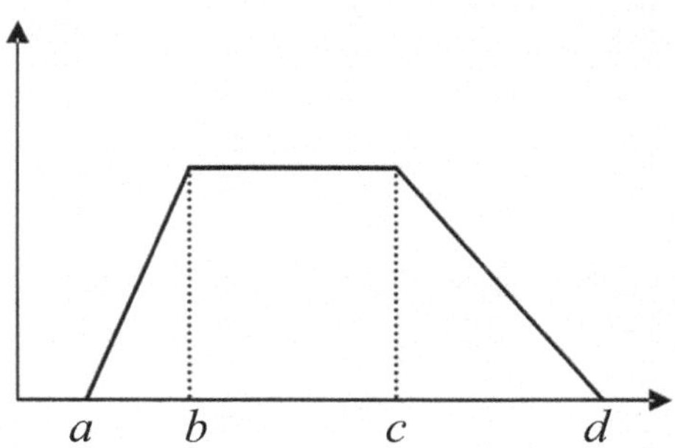

$$Cr\{\xi \geq x\} = \begin{cases} 1 & x < a \\ (2b - a - x)/2(b - a) & a \leq x < b \\ 1/2 & b \leq x \leq c \\ (d - x)/2(d - c) & c < x \leq d \\ 0 & \text{else other} \end{cases} \quad (3.23)$$

Possibility theory is based on fuzzy theory, so many concepts are derived from the concept of fuzzy mathematics, which will not be introduced here. Possibility theory and fuzzy set theory are collectively referred to as uncertain information processing methods based on fuzzy theory. For detailed information about more concepts and their application in engineering, please refer to the literature (Chen et al. 2000; Zimmermann 1991; Sakawa 1993; Rao and Chen 1998; Tonon and Bernardini 1990; Jensen and Sepulveda 2000; Maglaras et al. 1997; Hall and Lawry 2001; Mourelatos and Zhou 2005; Liu and Rao 2005; Huang and Li 2005).

3.3 Convex Set Theory

Convex set theory is a theory used in the case of incomplete information. The convex set models commonly used in convex set theory include interval models and ellipsoid models (Ben-Haimand Elishakoff 1990). Convex set theory requires less information to describe uncertainty information than probability theory and possible theory. Generally, analysis in convex set theory needs to find the worst case of the system (Worst Case) and analyze the worst case. When the convex set theory uses interval models, the methods in interval analysis theory can be used for analysis. Strictly speaking, interval analysis theory is a kind of convex set theory. Because of its wide application, it is often mentioned as an independent theory.

3.4 Interval Theory

Interval theory is a theory that uses interval variables to represent uncertain variables. The concept of interval is widely used in engineering. For example, the R inch of a mechanical part is marked as: $r = 5^{+0.002}_{-0.001}$, then the uncertain R inch variable r can be expressed by the number of intervals as:

$$r = [4.999, 5.002] \quad (3.24)$$

The basic concepts of interval theory and its operations are as follows:

(1) For a given real number $\underline{x}, \overline{x} \in R$, and $\underline{x} \leq \overline{x}$, then:

$$X^I = [\underline{x}, \overline{x}] = \{x \in R | \underline{x} \leq x \leq \overline{x}\} \quad (3.25)$$

X^I is called a bounded closed interval, referred to as the interval, where $\underline{x}$ and $\overline{x}$ are the lower and upper endpoints of the interval, and X^I is a real interval number. If $\underline{x} = -\overline{x}$, then the interval X^I is a symmetric interval. The set of all intervals on the real number field R is denoted as 1 (R).

(2) Median of the interval: $X^C = m(X^I) = (\underline{x} + \overline{x})/2$
(3) Interval radius (also called interval uncertainty): $\Delta X = rad(X^I) = (\overline{x} - \underline{x})/2$
(4) Intervals are represented by median and radius: $X^I = X^C + \Delta X[-1, 1]$
(5) Intervals $X^I = [\underline{x}, \overline{x}]$ and $Y^I = [\underline{y}, \overline{y}]$ contain: $X^I \subseteq Y^I \Leftrightarrow \underline{y} \leq \underline{x} \leq \overline{x} \leq \overline{y}$
(6) If the interval has a common part, the common part is called the intersection: $X^I \cap Y^I = [\max(\underline{x}, \underline{y}), \min(\overline{x}, \overline{y})]$
(7) If the interval has a common part ($X^I \cap Y^I \neq \varphi$), the union of the interval is $X^I \cup Y^I = [\min(\underline{x}, \underline{y}), \max(\overline{x}, \overline{y})]$
(8) The four algorithms of the interval $X^I = [\underline{x}, \overline{x}] \in I(R)$ and $Y^I = [\underline{y}, \overline{y}] \in I(R)$ are:

$$X^I + Y^I = [\underline{x} + \underline{y}, \overline{x} + \overline{y}] \tag{3.26}$$

$$X^I - Y^I = [\underline{x} - \overline{y}, \overline{x} - \underline{y}] \tag{3.27}$$

$$X^I \times Y^I = [\min(\underline{x} \cdot \underline{y}, \underline{x} \cdot \overline{y}, \overline{x} \cdot \underline{y}, \overline{x} \cdot \overline{y}), \max(\underline{x} \cdot \underline{y}, \underline{x} \cdot \overline{y}, \overline{x} \cdot \underline{y}, \overline{x} \cdot \overline{y})] \tag{3.28}$$

$$\begin{aligned} X^I/Y^I &= [\underline{x}, \overline{x}]/[\underline{y}, \overline{y}] = [\underline{x}, \overline{x}] \times [\frac{1}{\overline{y}}, \frac{1}{\underline{y}}] \\ &= [\min(\underline{x}/\underline{y}, \underline{x}/\overline{y}, \overline{x}/\underline{y}, \overline{x}/\overline{y}), \max(\underline{x}/\underline{y}, \underline{x}/\overline{y}, \overline{x}/\underline{y}, \overline{x}/\overline{y})] \end{aligned} \tag{3.29}$$

The four operations that can be deduced from (8) satisfy the combination and exchange laws, but the distribution laws may not hold. In addition, if the interval number is used as the denominator, it must not contain zero values, otherwise special treatment is required.

(9) Let $X_i^I = [\underline{x_i}, \overline{x_i}] \in I(R)(i = 1, \ldots, n)$, then:

$$X^I = [\underline{X}, \overline{X}] = \begin{bmatrix} X_1^I \\ X_2^I \\ \cdots \\ X_n^I \end{bmatrix} = \begin{bmatrix} [\underline{x_1}, \overline{x_1}] \\ [\underline{x_2}, \overline{x_2}] \\ \cdots \\ [\underline{x_n}, \overline{x_n}] \end{bmatrix} \tag{3.30}$$

Composes an interval vector, where $\underline{X} = (\underline{x_1}, \underline{x_2}, \ldots, \underline{x_n})^T, \overline{X} = (\overline{x_1}, \overline{x_2}, \ldots, \overline{x_n})^T$. The interval vector is an n-dimensional cuboid in R^n. The set of all interval vectors on R^n is denoted by $I(R^n)$.

(10) Let $a_{ij}^I = [\underline{a_{ij}}, \overline{a_{ij}}] \in I(R)(i = 1, \ldots, n;\ j = 1, \ldots, n)$ be the matrix of $n \times n$ consisting of these n^2 real interval numbers:

$$A^I = [\underline{A}, \overline{A}] = (a^I_{ij}) = \begin{bmatrix} a^I_{11} & a^I_{12} & \dots & a^I_{1n} \\ a^I_{21} & a^I_{22} & \dots & a^I_{2n} \\ \dots & \dots & \dots & \dots \\ a^I_{n1} & a^I_{n2} & \dots & a^I_{nn} \end{bmatrix} \quad (3.31)$$

A I is called an interval matrix. Among them:

$$\underline{A} = (\underline{a_{ij}}) = \begin{bmatrix} \underline{a_{11}} & \underline{a_{12}} & \dots & \underline{a_{1n}} \\ \underline{a_{21}} & \underline{a_{22}} & \dots & \underline{a_{2n}} \\ \dots & \dots & \dots & \dots \\ \underline{a_{n1}} & \underline{a_{n2}} & \dots & \underline{a_{nn}} \end{bmatrix} \quad (3.32)$$

$$\overline{A} = (\overline{a_{ij}}) = \begin{bmatrix} \overline{a_{11}} & \overline{a_{12}} & \dots & \overline{a_{1n}} \\ \overline{a_{21}} & \overline{a_{22}} & \dots & \overline{a_{2n}} \\ \dots & \dots & \dots & \dots \\ \overline{a_{n1}} & \overline{a_{n2}} & \dots & \overline{a_{nn}} \end{bmatrix} \quad (3.33)$$

The set of all interval vectors on $R^{n\times n}$ is written as $I(R^{n\times n})$.

(11) Similar to the number of intervals, the median and radius (uncertainty) of the interval vector and interval matrix are:

$$X^C = (\underline{X} + \overline{X})/2 \quad (3.34)$$

$$\Delta X = (\overline{X} - \underline{X})/2 \quad (3.35)$$

$$A^C = (\underline{A} + \overline{A})/2 \quad (3.36)$$

$$\Delta A = (\overline{A} - \underline{A})/2 \quad (3.37)$$

(12) Let $X^I = [\underline{X}, \overline{X}] \in I(R^n)$, $A^I = [\underline{A}, \overline{A}] \in I(R^{n\times n})$, then interval vectors and interval matrices can be expressed as:

$$X^I = X^C + \Delta X[-1, 1] \quad (3.38)$$

$$A^I = A^C + \Delta A[-1, 1] \quad (3.39)$$

(13) Let $X^I = X^C + \Delta X[-1, 1] \in I(R^n)$, $A^I = A^C + \Delta A[-1, 1] \in I(R^{n\times n})$, is an arbitrary real number and $U = (U_1, \dots, U_n)^T \in R^n$ is an arbitrary real vector, then the following three formulas hold:

$$\alpha X^I = X^I\alpha = X^C\alpha + \Delta X|\alpha|[-1, 1] = \alpha X^C + |\alpha|\Delta X[-1, 1] \quad (3.40)$$

$$A^I U = A^C U + \Delta A|U|[-1, 1] \tag{3.41}$$

$$U^T A^I = U^T A^C + |U|^T \Delta A[-1, 1] \tag{3.42}$$

(14) Let $f : R^n \to R$, if there exists an interval value mapping $F : I(R^n) \to I(R)$ that satisfies $F([x_1, x_1], \ldots, [x_n, x_n]) = f(x_1, \ldots, x_n)$ for any $x_i \in X_i^I (i = 1, \ldots, n)$, then F is said to be an interval extension of the function f.

It can be seen that $F(X^I)$ is a function of the interval vector $X^I = (X_1^I, \ldots, X_n^I)$. Let $X^I, Y^I \in I(R^n)$, and $X^I \subseteq Y^I$, if $F(X^I)$, then F has monotonicity. From this, it can be proved that all four operations have monotonicity, and since any rational function is a combination of four arithmetic operations with a finite number of intervals, the function F has monotonicity in the rational interval.

(15) The interval function F is the interval extension of the function F. If F contains monotonicity, the upper and lower limits of the range of F can be obtained through the $F(X^I)$ approximation. The error value of the approximation has not been determined yet, but for special cases, the following conclusions can be drawn:

① If each independent variable in the f expression only appears once, then:

$$\{f(x_1, \ldots, x_n)|\forall x_i \in X_i^I, i = 1, \ldots, n\} = F(X_1^I, \ldots, X_n^I) \tag{3.43}$$

② When using the subdistributive property, we should add and subtract first, then multiply and divide.

After the interval theory is established, its algorithm is constantly developing, but it is still difficult to apply interval analysis to practical engineering problems because the operation of interval Numbers only meets the commutative law and associative law, and the correlation of interval Numbers will lead to the expansion of interval operations. One Method to solve the distributive law calculation problem of interval theory is Vertex Method (VM) (Dong and Shah, year unknown), and the computation amount of this Method increases exponentially as the number of interval dimensions increases.

3.5 Evidence Theory

Unlike classical probability theory, which uses a single probability to measure uncertainty, evidence theory uses the concepts of Belief and Plasibility to describe uncertainty. Credibility and likelihood are determined by the varying degrees of evidence supporting the argument. Similar to the domain of fuzzy set theory, in the theory of evidence, the entire object in question is called the Universal Set U, also called the Frame of Discernment. The basic concepts of evidence theory are as follows.

(1) For the frame of discernment U, its power set is2^U. If the function $m : 2^U \rightarrow [0, 1]$ satisfies:

a. $m(A) \geq 0 \ \forall A \in 2^U$;
b. $m(\varphi) = 0$;
c. $\sum m(A) = 1 \ for \ all \ A \in 2^U$

Then, the function m is called Basic Probability Assignment (BPA) of event A, where Ø represents the empty set, and the power set 2^U represents all possible subsets of U. Event A is a subset of U, then m (A) represents the degree of evidence support for event A.

(2) The credibility Bel (A) and the likelihood W (A) of event A can be calculated using the following formulas:

$$Bel(A) = \sum_{B|B \subseteq A} m(B) \tag{3.44}$$

$$Pl(A) = \sum_{B|B \cap A \neq \varphi} m(B) \tag{3.45}$$

It can be seen that Bel(A) is the sum that fully agrees with all views of event A, while $Pl(A)$ is the sum that fully agrees with and partially agrees with all views of event A. Therefore, Bel(A) and $Pl(A)$ are the lower limit and upper limit of the possibility of event A, respectively, and the relationship between them is shown in Fig. 3.4.

That is, the following formula holds:

$$Pl(A) + Bel(\overline{A}) = 1 \tag{3.46}$$

Sometimes, there are different sources of evidence, so Rules of Combination (ROC) are needed. Common Rules of Combination include: The Dempster Rule of Combination (TDRC); Discount + Combine Method (DCM); Yager's Modified Dempster's Rule (YMDR); Inagaki's Unified Combination Rule (IUCR); Zhang's Center Combination Rule (ZCCR); Dubois and Prade's Disjunctive Consensus Rule (D&PDCR); Mixing the or Averaging (MA). The most common is TDRC, which assumes that evidence is independent of each other and ignores conflicts between different evidence.

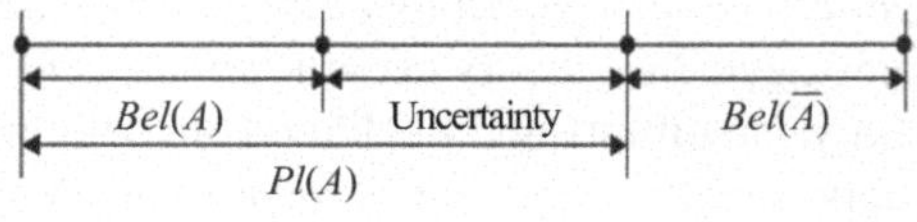

Fig. 3.4 Relationship between likelihood and reliability

$$m(A) = \frac{\sum\limits_{B \cap C = A} m_1(B)m_2(C)}{1 - \sum\limits_{B \cap C = \varphi} m_1(B)m_2(C)}, \quad A \neq \varphi \tag{3.47}$$

Among them, m1 and m2 are two completely independent evidences. From the assumption of TDRC, it can be seen that TDRC is not suitable for situations where there is more conflict between the evidences.

More evidence theory and its application can be referred to in the literature (Bae et al. 2002, 2004, 2006; Agarwal et al. 2004).

3.6 Broad Probability Theory

There is no doubt that classical probability theory is the most widely used theory of uncertainty information description, but its shortcomings have been criticized by many scholars. For example, classical probability theory is based on two-valued logic, that is, the relationship between an element and a set can only belong to or does not belong. It cannot express "the degree to which an element belongs to a set…", and this concept is exactly the fuzzy set theory An important advance. Many concepts in nature cannot be expressed in statistical mathematics, especially in the category of linguistics, such as "high", "short", "fat", "thin", etc. Most of these concepts contain subjective will, so they are vague.

Classical probability theory is also supported by many scholars, such as literature (Henrion 1986; Spiegelhalter 1986; Spiegelhalter 1987; Stallings 1977; Tribus 1979; Watson 1987) of probability theory can deal with fuzzy information theory and probability theory, fuzzy set theory and evidence than the existing defects, etc., this paper explores the thought with other theoretical description of the uncertainty parameters can use probability theory is described, and the widely application of the theory of probability to prove its irreplaceable status in the theory of dealing with uncertain information. Based on the above situation, Cui extends and extends some concepts in classical probability theory without violating the axiom system of classical probability theory, so that the theory can deal with non-statistical uncertain information, which is called generalized probability theory (GPT) (Cui 1993a, b). Here is a brief introduction to some of the concepts that distinguish GPT from classical probability theory.

(1) The definition of probability.

In classical probability theory, probability is often defined as the frequency of events that occur objectively in multiple random experiments. For most events, people observe and judge whether an event occurs, so probability inevitably comes with human subjective judgment. The definition of probability in GPT will use the definition of subjective probability, that is, probability is the degree of confidence in the occurrence of an event (Degree of Belief). The occurrence of an event in this definition is no longer just objective, but with the subjective opinion of the observer, that is, with human cognition and judgment, so the probability in the GPT can describe both the inherent uncertainty information and the Describe cognitive uncertainty information. The most extreme case is that there is no experimental data and only rely on human experience and cognition to describe uncertainty. At this time, the accuracy of the probability is very low, such as "someone grows high". And if there is a large amount of statistical test data for an event, such as flipping a coin, then the statistical analysis can be directly based on the data to obtain uncertain information, thereby reducing human subjective factors to a very low level. For most events, it is necessary to combine experimental data with human subjective judgments.

(2) Single parameter representation of fuzzy event probability.

Zadeh uses a single parameter to represent the probability of fuzzy event A (Zadeh 1978):

$$P_A(X \leq x_0) = \int m_A(x)p(x)dx \tag{3.48}$$

Where $m_A(x)$ is a feature function (membership function) and p (x) is a probability density function.

Then:

$$p_A(x) = \frac{m_A(x)p(x)}{\int m_A(x)p(x)dx} \tag{3.49}$$

It is called A-Qualified PDF (A-Qualified PDF, QPDF).

When PDF in classical probability theory is replaced with QPDF, the formula in classical probability theory can be used to calculate the probability of fuzzy events.

(3) Interval number representation of fuzzy event probability.

$$P(A) = [S_n(A), S_P(A)] \tag{3.50}$$

Where $S_n(A)$ and $S_P(A)$ are the lower and upper bounds of the probability of event A, respectively. $S_P(A) - S_n(A)$ is the uncertainty of the probability of event A. The smaller $S_P(A) - S_n(A)$ is, the more knowledge of A; the larger $S_P(A) - S_n(A)$ is, the less knowledge of A.

The probability of a counter event for event A is:

$$P(\overline{A}) = [1 - S_P(A), 1 - S_n(A)] \tag{3.51}$$

According to the interval theory in Sect. 3.4 of this chapter, the four operations of interval numbers do not satisfy the distribution law. In order to overcome this shortcoming, the vertex method needs to be used for calculation.

Let the interval number X = [a, b], and its function is:

$$Y = f(X) = \{f(x)|x \in [a, b]\} \tag{3.52}$$

Then Y is also an interval number. If the function is continuously monotonic on $[a, b]$, then

$$Y = [\min(f(a), f(b)), \max(f(a), f(b))] \tag{3.53}$$

Similarly, interval vector

$$X^I = [\underline{X}, \overline{X}] = \begin{bmatrix} X_1^I \\ X_2^I \\ \cdots \\ X_n^I \end{bmatrix} = \begin{bmatrix} \underline{x_1}, \overline{x_1} \\ \underline{x_2}, \overline{x_2} \\ \cdots \\ \underline{x_n}, \overline{x_n} \end{bmatrix}$$

If the function $Y = f(X^I) = \{f(x)|x \in [\underline{X}, \overline{X}]\}$ is monotonically continuous in an n-dimensional rectangular region, and the coordinate of the j-th vertex of the n-dimensional rectangle is $c_j (j = 1, \ldots, 2^n)$, then:

$$Y = [\min(f(c_1), \ldots, f(c_{2^n})), \max(f(c_1), \ldots, f(c_{2^n}))] \tag{3.54}$$

If the function has m extreme points in an n-dimensional rectangular area, let the coordinates of the k-th extreme point be:

$$Y = [\min(f_c, f_e), \max(f_c, f_e)] \tag{3.55}$$

Among them, $f_c = (f(c_1), \ldots, f(c_{2^n}))$, $f_e = (f(e_1), \ldots, f(e_m))$ is a vector consisting of the values of the function at all vertices and all extreme points. It can be seen that the calculation volume of the vertex analysis method increases exponentially with the increase of the dimension of the variable.

(4) Fuzzy number representation of fuzzy event probability.

The interval number of the probability of the fuzzy event indicates that there is a sudden change between "possible" and "impossible", and when the fuzzy number is used to represent the probability of the fuzzy event, the transition between "possible" and "impossible" will be smooth.

For the probability of fuzzy events represented by fuzzy numbers, it is still possible to use a method similar to the vertex analysis method, but at this time, multiple thresholds of membership functions need to be calculated. Literature (Möller et al. 2000; Shih et al. 2003) introduced the "a-cut method" (α-cut method).

(5) Dependency (Cui 1993a, b)

$$\gamma = P(A \cap B)/\min(P(A), P(B)) \tag{3.56}$$

$\gamma = 0$ means that A and B are independent of each other; $\gamma = 1$ means that A and B are the most dependent, that is, or.

Then:

$$P(A \cap B) = \gamma \min(P(A), P(B)) \tag{3.57}$$

$$P(A \cup B) = P(A) + P(B) - \gamma \min(P(A), P(B)) \tag{3.58}$$

Because the above concept is derived from the following inequality:

$$\max(0, P(A) + P(B) - 1) \le P(A \cap B) \le \min(P(A), P(B)) \tag{3.59}$$

Therefore, the dependency degree includes various possible situations. The independent operators in classical probability theory and the related operators in fuzzy mathematics are special states of dependence.

① Minimum dependencies:

$$\gamma = \max(0, P(A) + P(B) - 1)/\min(P(A), P(B)) \tag{3.60}$$

$$P(A \cap B) = \max(0, P(A) + P(B) - 1) \tag{3.61}$$

$$P(A \cup B) = \min(1, P(A) + P(B)) \tag{3.62}$$

② Statistically independent:

$$\gamma = \max(P(A), P(B)) \tag{3.63}$$

$$P(A \cap B) = P(A)P(B) \tag{3.64}$$

$$P(A \cup B) = P(A) + P(B) - P(A)P(B) \tag{3.65}$$

③ Maximum dependency (this is the state of the largest and smallest operator in fuzzy theory):

$$\gamma = 1 \tag{3.66}$$

$$P(A \cap B) = \min(P(A), P(B)) \tag{3.67}$$

$$P(A \cup B) = \max(P(A), P(B)) \tag{3.68}$$

From the definition above, it can be known that the correlation coefficient in classical probability theory only indicates the state of linear dependency, and the actual nonlinear dependency cannot be described by the correlation coefficient. In addition, the statistical independence does not mean that there is no dependency between the two, but only a special dependency state. In practical applications, the strength of "if A is established, B can be determined" is used to determine which dependency definition is adopted: if A is established, then the possibility of B being established is small, and the minimum dependency definition should be adopted; if A is established, then It is very likely that B is established, and the maximum dependency definition should be adopted; if the relationship between A and B is not so clear, other definitions can be used, such as statistically independent of each other.

In generalized probability theory, probability is defined as the probability of events, which can be expressed by single parameter, interval number or fuzzy number according to the number of known information. More importantly, generalized probability theory is the extension of the main concepts of classical probability theory, which preserves the complete axiom system of classical probability theory, and some reliability analysis methods and reliability design methods that have been widely used in engineering can be used. Therefore, the generalized probability theory is of great significance to the development of engineering reliability design.

Classical probability theory is the most widely used and accepted theory. The complete axiom system and the characteristic of describing uncertain information accurately make the probability theory widely used in the field of engineering. However, the accuracy of classical probability theory is based on obtaining enough data, but not enough statistical data can be obtained in all cases. When information is not sufficient, often need according to individual subjective experience and knowledge to describe the uncertain information, this is the starting point for the development of fuzzy theory, and the fuzzy theory in order to expand the applicability, make its axiom system incomplete, lost not sure the accuracy of the information description, so it is the rationality and the possibility of application in the engineering remains controversial. If there is no subjective knowledge of certain uncertain information, only the approximate range of the information is known, then the interval theory will become an ideal descriptive theory. In the generalized probability theory, the accuracy and accuracy of uncertain information description theory always contradict each other. A good theory must be based on the obtained information and deal with the balance between accuracy and accuracy. While keeping complete axiom system on the basis of classical probability theory, using the definition of subjective probability, extended to describe the originally only real events probability can be described with interval number and fuzzy number, at the same time, by introducing the concept of dependency between two events, make the theory of probability and can handle any type of uncertainty information.

References

Agarwal H (2004) Reliability based design optimization: formulations and methodologies. PhD thesis, Notre Dame

Agarwal H, Renaud JE, Preston EL et al (2004) Uncertainty quantification using evidence theory in multidisciplinary design optimization. Reliab Eng Syst Saf 85(1–3):281–294

Bae HR, Grandhi RV, Canfield R (2002) Uncertainty quantification of structural response using evidence theory. In: 43rd AIAA/ASME/ASCE/AHS/ASC structures, structural dynamics, and materials conference. Denver, Colorado
Bae HR, Grandhi RV, Canfield RA (2004) Epistemic uncertainty quantification techniques including evidence theory for large-scale structures. Comput Struct 82(13–14):1101–1112
Bae HR, Grandhi RV, Canfield R (2006) Sensitivity analysis of structural response uncertainty propagation using evidence theory. Struct Multi Optim 31:270–291
Ben-Haim Y, Elishakoff I (1990) Covex methods of uncertainty in applied mechanics. Elsevier, Amsterdam
Bifeng S (2006) Aircraft reliability engineering. Northwestern polytechnical university press. (宋笔锋. 飞行器可靠性工程. 西北工业大学出版社, 2006.)
Billingsley P (1995) Probability and measure. Wiley-Interscience Publication
Chen YH, Wang WJ, Chiu CH (2000) New estimation method for the membership values in fuzzy sets. Fuzzy Set Syst 112:521–525
Cui WC (1990) Uncertainty analysis in structural safety assessment. PhD thesis, Department of Civil Engineering, University of Bristol
Cui WC (1993a) Generalized probability theory for handling uncertainty: the foundation. Technical report. China Ship Scientific Research Center
Cui WC (1993b) Generalized probability theory for handling uncertainty: the theory. Technical report. China Ship Scientific Research Center
Cui WC, Blockley DI (1990) Interval probability theory for evidential support. Int J Intell Syst 5(2):183–190
De-Cooman G (1997) Possibility theory I: the measure-and integral-theoretic groundwork. Int J Gen Syst
Dempster AP (1967) Upper and lower probabilities induced by a multivalued mapping. Ann Math Stat 38:325–329
Du L (2006) Reliability-based and possibility-based design optimization using inverse analysis methods. PhD thesis of graduate college of the university of Iowa
Dubois D, Prade H (1988) Possibility theory: an approach to computerized processing of uncertainty. Plenum Press, New York
Durrett R (1996) Probability: theory and examples. Duxbury Press
Guo J, Du X (2007) Sensitivity analysis with mixture of epistemic and aleatory uncertainties. AIAA J 45(9):2337–2349
Hall J, Lawry J (2001) Imprecise probabilities of engineering system failure from random and fuzzy set reliability analysis. In: 2nd international symposium on imprecise probabilities and their applications. Ithaca, New York
Helton J (1997) Uncertainty and sensitivity analysis in the presence of stochastic and subjective uncertainty. J Stat Comput Simul 57:3–76
Helton JC, Johnson JD, Oberkampf WL, Storlie CB (2006) A sampling-based computational strategy for the representation of epistemic uncertainty in model predictions with evidence theory. Comput Method Appl Mech Eng 196(37):3980-–998
Henrion M (1986) Uncertainty in AI: is probability epistemologically and heuristically adequate. In: Expert systems and expert judgement. Proceedings of the NATO advanced research workshop in Porto. Portugal
Huang HZ, Li HB (2005) Perturbation finite element method of structural analysis under fuzzy environment. Eng Appl Artif Intell 18(1):83–91
Jacobsen HS (2002) Representation and calculation of economic uncertainties: intervals, fuzzy numbers, and probabilities. Int J Prod Econ 78:91–98
Jensen HA, Sepulveda AE (2000) Use of approximation concept in fuzzy design problem. Adv Eng Softw 31:263–273
Klir G (2004) Generalized information theory: aims, results, and open problems. Reliab Eng Syst Saf 85(1–3):21–38

Klir G, Wierman M (1999) Uncertainty-based information-elements of generalized information theory. Physica-Verlag, Heidelberg, New York

Liu B (2002) Theory and practice of uncertain programming. Physica-Verlag, Heidelberg

Liu B (2004) Uncertainty theory: an Introduction to its axiomatic foundations. Springer, Berlin

Liu Q, Rao SS (2005) Fuzzy finite element approach for analysis of fiber-reinforced laminated composite beam. AIAA J 43(3):651–661

Maglaras G, Nikolaidis E, Haftka RT et al (1997) Analytical-experimental comparison of probabilitic and fuzzy set based methods for designing under uncertainty. Struct Optim 13:69–80

Möller B, Graf W, Beer M (2000) Fuzzy structural analysis using α-level optimization. Comput Mech 26:547–565

Moore RE (1962) Interval arithmetic and automatic error analysis in digital computing. PhD Dissertation, Stanford University

Moore RE (1966) Interval analysis. Prentice-Hall, Englewood Cliffs, NJ

Mourelatos ZP, Zhou J (2005) Reliability estimation and design with insufficient data based on possibility theory. AIAA J 43(8):1696–1705

Nahmias S (1978) Fuzzy variable. Fuzzy Sets Syst 1:97–101

Oberkampf W, Helton J (2002) Investigation of evidence theory for engineering applications. AIAA 2002–1569. In: 43th AIAA/ASME/ASCE/AHS/ASC structure, structural dynamics and material conference, 4th non-deterministic approaches forum. Denver, Colorado

Oberkampf W, Helton J, Sentz K (2001) Mathematical representation of uncertainty. AIAA 2001–1645. In: 42th AIAA/ASME/ASCE/AHS/ASC structures, structural dynamic and materials conference and exhibit, 2001AIAA non-deterministic approaches forum. Seattle, WA

Oberkampf W, Helton J, Joslyn C et al (2004) Challenge problem: uncertainty in system response given uncertain parameters. Reliab Eng Syst Saf 85:11–19

Rao SS, Chen L (1998) Numerical solution of fuzzy linear equations in engineering analysis. Int J Numer Meth Eng 43:391–408

Sakawa M (1993) Fuzzy sets and interactive multiobjective optimization. Plenum Press

Savoia M (2002) Structural reliability analysis through fuzzy number approach, with application to stability. Comput Struct 80:1087–1102

Shafer G (1976) A mathematical theory of evidence. Princeton University Press, NJ

Shih CJ, Chi CC, Hsiao JH (2003) Alternative α-level-cuts methods for optimum structural design with fuzzy resources. Comput Struct 81:2579–2587

Shutie X, Minpin Q, Jun Y (2000) Random mathematics. Higher education press. ISBN 7-04-008990-4. (萧树铁, 钱敏平和叶俊. 随机数学. 高等教育出版社, 2000, ISBN 7-04-008990-4.)

Spiegelhalter DJ (1986) A statistical view of uncertainty in expert systems. In: Gale W (ed.) Artificial intelligence and statistics. Addison-Wesley

Spiegelhalter DJ (1987) Probabilistic expert systems in medicine: practical issues in handling uncertainty. Stat Sci 2(1):3–44

Stallings W (1977) Fuzzy set theory versus Bayesian statistics. IEEE Trans Syst Man Cybern SMC-7, 216–219

Tonon F, Bernardini A (1990) Multiobjective optimization of uncertain structures through fuzzy set and random set theory. Comput-Aided Civil Infrastruct Eng 14:119–140

Tribus M (1979) Comment on fuzzy sets, fuzzy algebra and fuzzy statistics. Proc IEEE 67:11–68

Weichselberger K (2000) The theory of interval-probability as a unifying concept for uncertainty. Int J Approx Reason 24:149–170

Xiao S, QIAN M, Ye J (2002) Random Mathematics [M]. Beijing: Higher Education Press. (萧树铁, 钱敏平, 叶俊. 随机数学[M]. 北京: 高等教育出版社, 2002.)

Zadeh LA (1965) Fuzzy sets. Inf Control 8:338–353

Zadeh LA (1978) Fuzzy sets as a basis for a theory of possibility. Fuzzy Sets Syst 1:3–28

Zadeh LA (2006) Generalized theory of uncertainty (GTU)—principal concepts and ideas. Comput Stat Data Anal 51:15–46

Zhentao L (2006) Research on analysis method of uncertain structure. PhD thesis of Xidian university. (梁震涛. 不确定性结构的分析方法研究. 博士学位论文, 西安电子科技大学, 2007.)

Zimmermann HJ (1991) Fuzzy set theory and its applications. Kluwer Academic Publishers

Chapter 4
The Analysis of System Reliability

4.1 Reliability

The ability of an engineering system or products completing the prescribed functions under the prescribed condition and within the prescribed time is called reliability. In the design stage, prescribed functions of an engineering system usually can be expressed as the function of input variable x, which usually can be called the state function $g(X)$. The value of the state function can judge whether functions of an engineering system can meet the design requirements.

The most common judging logic is two-valued logic. When the state function g (X) exceeds a certain status (a certain allowed value), an engineering system can't meet a function requirement prescribed as the design. The particular state is called the limit state of the function. If an engineering system can conform to functional requirements specified as the design, then the system can be judged as "reliable", conversely it can be judged as "invalid" or "unreliable". Usually, the reliability of an engineering system is divided as the zero point of a state function.

Reliable:

Invalid:

$$\begin{cases} Reliable : g(X) \geq 0 \\ Invalid : g(X) < 0 \end{cases} \tag{4.1}$$

Then the judging function of two-valued logic (also called the indicative function) can be written as:

$$I(g(X)) = \begin{cases} 1 & 当 g(X) < 0 \\ 0 & 当 g(X) \geq 0 \end{cases} \tag{4.2}$$

B. Pan and W. Cui, *Multidisciplinary Design Optimization and Its Application in Deep Manned Submersible Design*, Ocean Engineering & Oceanography 13,
https://doi.org/10.1007/978-981-15-6455-0_4

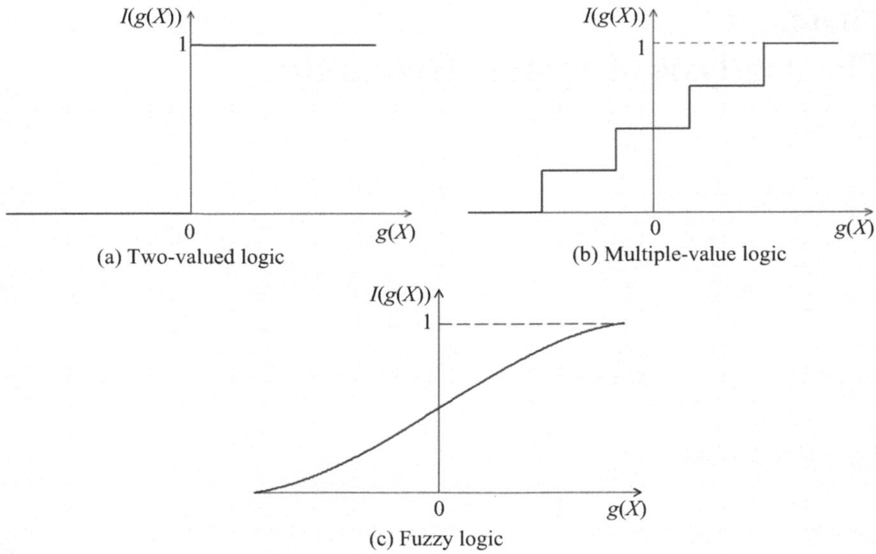

Fig. 4.1 Different logic function (indicative function)

In the two-valued logic, the state function g(X) is divided into two different areas under the limit state, as Fig. 4.1a, while the state function value near the limit state has no essential difference. The sudden change doesn't conform to engineering reliability. Therefore, people proposed to avoid the sudden change, using the logic of smooth transition to judge. Multi-value logic divides the value range of functional indicators of the system into multiple intervals. For example, in the common five-valued logic, the system state is not simply divided as the reliable and invalid, but divided into five states: excellent, good, middle, poor, very poor. The judging function of multi-value logic is still not smooth completely, but it exists sudden changes at the boundary between intervals, as shown in Fig. 4.1b, if division intervals of value range of functional indicators are many enough, then it can be judged that the transition of functions is smooth, for instance, polyline approximation smooth curve. The fuzzy logic achieves judgment through the introduction of smooth membership functions, as shown in Fig. 4.1c. Currently, there are no unified methods by which a suitable membership function can be selected. Up to now, reliability judging logic widely applied in real project design is still two-valued logic. This book will introduce two-valued logic, theories, program systems and fuzzy logic. It doesn't talk about how to ascertain the membership function in the fuzzy logic.

二值逻辑	Two-valued logic
多值逻辑	Multiple-value logic
模糊逻辑	Fuzzy logic

4.2 Reliability Analysis Methods

Reliability analysis is closely related to the uncertainty description theory of uncertainty variables. When the uncertainty variable x is expressed by different uncertainty description theory, reliability analysis methods of the system state function g(x) vary. Uncertainty variables are expressed as the random numbers in the probability theory, reliability analysis is called probability reliability analysis. Uncertainty variables are expressed as fuzzy numbers in the fuzzy theory, reliability analysis is called fuzzy reliability analysis. When uncertainty variables are described as interval numbers, the vertex analysis method is usually adopted to make the interval analysis. The concept of the vertex analysis method is simple. Its concept: When there exists a monotonic relation between the function value and interval variables, poles are on the boundary of interval variables; when there exists a more complex relation between the function value and interval numbers, to adopt DOE, optimization algorithm, etc. to conduct design space search to find the extreme point. After finding the extreme point, to judge whether a system is reliable through judging whether the value of a state function exceeds the permission interval.

Reliability Analysis Methods mentioned by us, include probability reliability analysis methods and fuzzy reliability analysis methods.

(1) Probability reliability analysis methods

In the probability theory widely applied in engineering application, the uncertainty variable $X = (x_1, \ldots, x_n)$ is viewed as the random variable that follows certain probability density distribution, $(i = 1, \ldots, n)$ is called basic random variable, then the state function value of the system Z = g(X) is also the random variable. At this time, the failure probability of the system P_f can be expressed as:

$$P_f = \int \cdots \int I(g(X)) f_X(x) dx_1 \ldots dx_n \tag{4.3}$$

In which,

$$I(g(X)) = \begin{cases} 1 & 当g(X) < 0 \\ 0 & 当g(X) \geq 0 \end{cases}$$ It is demonstrative or it is a judging logic function.

fx(x) is the joint probability density function of the random variable X.

If the failure probability of a system is P_f, then the system reliability R = 1-P_f.

The calculation of failure probability is a process of multidimensional integration. If adopting rapid and highly effective multidimensional numerical direct integration methods, it generally reaches 7 dimensions (Wei and Weicheng 1997). It is very difficult to calculate the integral in the real issue, so various methods of approximately evaluating reliability were created (Moffitt 2010), which are called reliability analysis methods, also called reliability evaluation methods.

Currently, the first-order second moment method in the reliability index method is widely used. The basic concept of first-order second moment method was proposed firstly by Erlannichin of the Soviet Union. Hasofer and Lind put forward a first-order second moment method applied in the normal distribution in 1974–checking point method. Rackwitz and Fiessler promoted the checking point method to any distribution type. It shows stable and reliable characteristics in the most application cases, so it is recognized generally. The checking point method has been adopted by the Joint Committee on Structural Safety. The checking point method is also called JC method (Hasofer and Lind 1974; Palle and Michael 1982; Madsen et al. 1986).

Monte Carlo Method (another probability reliability analysis method. It is also called random simulation method or random sampling method) is also widely used. Monte Carlo Method sources from the Buffon experiment of the 18th century. The concept: to get the statistical feature value of the research object through the statistical experiment, and take the statistical feature value as the approximate solution of the object. The basis of Monte Carlo Method is the law of large numbers. When the times of random experiments are many enough, Monte Carlo Method will be converged to the true solution. Advantages of Monte Carlo Method: intuitive, accurate, easy to realize, highly nonlinear problems can also be solved accurately through this method and it is easy to realize parallel computing. Disadvantages of Monte Carlo Method: low efficiency, large amount of computation, especially under the condition of very high reliability (Hammersley and Handscomb 1964; Rubinsrein 1981). In order to improve the efficiency of random simulation, reduce sampling times, there emerged various improved Monte Carlo methods, for example, Descriptive Sampling (DS), (Saliby 1990; Ziha 1995), Importance Sampling (IS) (Harbitz 1986; Melchers 1989; Dong Cong 1996), etc. Of course, applicability and stability of reduced sampling algorithms is smaller than that of original Monte Carlo methods (Crude Monte Carlo, CMC). It is required to ascertain the use of Monte Carlo methods according to sampling space and systematic structure. For large computation amount of engineering system, for example, finite element analysis of structure systems, large computation amount of random simulation method is unbearable. With the development of the computing capability of computers, Monte Carlo methods are widely used.

To make the algorithm of solving checking point in the reliability indicator method easier to converge, performance evaluation method is created (Youn and Choi 2004; Youn et al. 2005a, b; 2004). Detailed introductions of the method are shown in the following text. In addition, quadratic polynomial response surface, neural network and other technologies are introduced to reduce the amount of computation by Monte Carlo method (Faravelli 1989; Cruse 2001; Liu 2000; Myers and Montgomery 1995; Qu et al. 2001; Zhu 2011).

Solving multidimensional integral is the essence of reliability analysis, so fast integration methods (Wei and Weicheng 1997), eigenvalue dimension reduction methods (Youn et al. 2007), etc. are developed. In recent years, Polynomial Chaos Expansion (PCE) Method is also used in the uncertainty analysis of random systems. Wiener firstly adopted Hermite polynomials as the base function to simulate random variables as the random process of Gaussian distribution (Wiener 1938). Xiu, etc. extended the base function to more orthogonal polynomials, thus making the Polynomial Chaos

Expansion Method applied in more distribution types, for example, uniform distribution, Beta distribution, Gama distribution, etc. (Xiu 2010; Xiu and Karniadakis 2002). The basic concept of Polynomial Chaos Expansion Method: to use the sum of chaos of orthogonal polynomials containing independent random variables to approximate the random process. Arbitrary Polynomial Chaos Expansion (APCE) (Oladyshkin and Nowak 2012) that can simulate the arbitrary distribution function and reduce computation amount will be the development trend. The polynomial chaos expansion method has been applied in the hydrodynamics (Zhao et al. 2009; Ghisu et al. 2008; Lurati et al. 2008; Ghisu et al. 2010), composite materials (Manan and Cooper 2008, 2009a, b), flexible systems (Ting et al. 2011), landing analysis of Mars probes (Gaofeng and Shengying 2012), etc. Currently, polynomial chaos expansion methods still exist an obvious limitation: computing amount increases rapidly with the increase of random variables. When there are many random variables in a system, computing amount of polynomial chaos expansion method is more than Monte Carlo method and its stability is lower than Monte Carlo method, so Polynomial chaos expansion method is mostly applied in simple systems having smaller amount of uncertain variables. Its application in the large complex engineering system still needs to be researched. Readers interested in it can refer to the document (Oladyshkin and Nowak 2012; Gaofeng and Shengying 2012) or other relevant documents.

Current probability reliability evaluation methods include following 5 types:

① Reliability Index Approach, RIA, including:

a. First Order Reliability Method (FORM), JC algorithm, etc.
b. Second Order Reliability Method (SORM).
c. RIA Method directly adopting General Sequential Quadratic Programming (SQP) to search MPP.

② Random Simulation Method (Monte Carlo Method):

a. Crude (Original) Monte Carlo, CMC.
b. Improved Monte Carlo Method: descriptive sampling, stratified sampling, importance sampling, etc.

③ Performance Measure Approach, PMA, including:

a. Mean Value (MV)
b. Advanced Mean Value (AMV)
c. Hybrid Mean Value (HMV)
d. Enriched Hybrid Mean Value (HMV +)

④ Approximate model method:

a. Response Surface Method (RSM)
b. Neural network method, etc.

⑤ Polynomial Chaos Expansion (PCE).

Detailed introductions of representative algorithms:

(2) Fuzzy reliability analysis method

When judging logic function is the fuzzy membership function or when adopting the fuzzy theory to describe uncertainty variables, reliability analysis can be called fuzzy reliability analysis.

⑥ Fuzzy logic reliability analysis:

In the fuzzy reliability analysis, the analyzer still adopts the probability theory to describe uncertainty variables. In terms of judging logic function, users adopt the fuzzy logic function to replace traditional two-valued logic. Because logic function is the fuzzy membership function, Most Probable Point (MPP) under the two-valued logic will no longer exist. RIA and PMA algorithms based on MPP analysis will not be used. Only algorithms based on sampling (for instance, random simulation method) can be used.

⑦ Reliability analysis of fuzzy variables:

In the reliability analysis of fuzzy variables, set the parameter $X = (x_1, \ldots, x_n)$ (X is a fuzzy variable) and a state variable of a system (a index) $Z = g(X)$ (a fuzzy variable), require that the possibility of failure of a state variable $\leq \alpha_t$, that is,

$$\mu(g(X) < 0) \leq \alpha_t \tag{4.4}$$

In which, α_t is the specified failure probability (or allowed failure probability). Generally, α_t is set as 0.05, μ is the membership function.

Common fuzzy variable reliability analysis methods:

a. α cut set method (α-cut or level-cut): conduct reliability analysis of state function in the probability confidence level α
b. Multilevel-cut set method: Conduct the analysis of problems at the probability confidence level α to improve the accuracy of a solution when nonlinearity g(X) is higher;
c. Advanced performance evaluation method: Based on the PMA method in the classic probability theory, Choi, etc., extend PMA, which is also applied in the fuzzy reliability analysis (Du 2006). They find that it can meet the engineering needs in terms of the accuracy and computing efficiency, but theoretically, MPP will no longer exist if fuzzy numbers are adopted. The practicality of the method still needs to be verified.
d. Fuzzy simulation method: It develops on the basis of credibility theory. Liu B's series of documents introduce function expectation value, function optimism value, fuzzy simulation method of constrained conditions (Liu and Iwamura 1998; Liu 1999, 2000, 2002, 2006, etc.). Documents (Xiaoyu 2006; Shaojuan 2007; Yufu 2006) also introduce it.

The development of fuzzy theory is not improved. There are no unified methods and standards that can be used in ascertaining the application of fuzzy logic function or membership function. Fuzzy reliability analysis methods can't be used in engineering. Computing amount usually exceeds the acceptable range. The following text makes preliminary introductions of fuzzy reliability analysis through simple cases of computing.

This book makes detailed introductions of representative algorithms (for instance, reliability index method, performance evaluation method and random simulation method) that are commonly used in engineering, and makes comparisons of computing errors and computing efficiency of reliability analysis algorithms through typical computing cases.

4.2.1 Reliability Index Algorithm

RIA algorithm transforms reliability into reliability indexes to show reliability. Reliability index is the ratio of the mean of state variable Zs to the standard deviation, that is:

$$\beta = \frac{\mu_Z}{\sigma_Z} = \frac{E[g(X)]}{\sqrt{Var[g(X)]}} \tag{4.5}$$

Relations between reliability index and probability of failure P_f/reliability R

$$P_f = \Phi(-\beta) = 1 - \Phi(\beta) \tag{4.6}$$

$$R = \Phi(\beta) \tag{4.7}$$

In which, Φ shows standard positive distribution function.

Hasofer and Lind gave the physical meaning of reliability indexes (Hasofer and Lind 1974). Reliability index β refers to the shortest distance from the coordinate origin to limit state surface g(X) = 0 in standard positive space. The point that the shortest distance corresponds in the limit state is called design checking point (short name: checking point). Because a system is most likely to become invalid at this point, the checking point is also called the most probable point (MPP). Only finding MPP, reliability index β can be calculated. MPP is the index that can measure reliability, which should be searched. RIA and PMA algorithms are two common algorithms used to search MPP.

MPP is in the limit state surface g(X) = 0. In the definition of RIA, the process of solving of MPP is described as the solution of nonlinear constrained optimization.

$$\begin{aligned} &\min\beta = |V| \\ &s.t. \quad g(V) = 0 \end{aligned} \tag{4.8}$$

Principles of RIA algorithm: In the limit state surface [g(V) = 0 hypersurface], to search the point that is nearest to the mean point. The point is MPP. The distance between MPP and the mean point means reliability index β. RIA algorithm adopts linear programming to solve the optimization issue, so the applicable scope of RIA algorithm is limited. Optimization algorithms are improved. Non-linear and global optimization algorithms can be used to solve, for example, secondary planning, genetic algorithm, particle swarm optimization, etc.

The introduction of global optimization algorithms greatly increases computation amount, so various methods used to reduce RIA algorithm computing amount are researched, for example, expansion or approximation method, which simplifies a state function into linear or auasilinear function. In the RIA algorithm, first-order Taylor series of a state function are expanded, which is called first order second moment method (FOSM). Rackwitz and Fiessler promoted FOSM to arbitrary distribution type, giving an iterative algorithm that can steadily find MPP in the most limit state functions, namely checking point method. Advantages of checking point method: stable, reliable. Due to the above merits, the method is widely used. It has been adopted by the Joint Commission of Structure Security (JCSS). The method is also named JC method (Hasofer and Lind 1974; Palle and Michael 1982; Madsen et al. 1986). Steps of JC algorithm:

(1) Assuming the initial checking point $X^{*(0)} = (x_1^{*(0)}, \ldots, x_n^{*(0)})$. set the mean as the initial point, namely $x_i^{*(0)} = \mu_{x_i}$
(2) Set the checking point $p^* = x^*$, conduct equivalent normalization of non-normal random variables and calculate corresponding equivalent normal distribution parameter μ_i^N, σ_i^N.
(3) Calculate the reliability index.

① When random variables are independent:

$$\beta = \frac{g(x_1^*, \ldots, x_n^*) + \sum_{i=1}^{n} \left(-\frac{\partial g}{\partial x_i}|_{P^*}(\mu_{x_i} - x_i^*)\right)}{\sqrt{\sum_{i=1}^{n} \left(\frac{\partial g}{\partial x_i}|_{P^*}\sigma_{x_i}\right)^2}} \tag{4.9}$$

② When random variable relation coefficient matrix is ρ (symmetric matrix):

$$\beta = \frac{g(x_1^*, \ldots, x_n^*) + \sum_{i=1}^{n} \left(-\frac{\partial g}{\partial x_i}|_{P^*}(\mu_{x_i} - x_i^*)\right)}{\sqrt{\sum_{i=1}^{n} \left(\rho_i \frac{\partial g}{\partial x_i}|_{P^*}\sigma_{x_i}\right)^2}} \tag{4.10}$$

(4) Calculate importance coefficient (direction coefficient)

① When random variables are independent of each other,

$$\alpha_i = \cos(\theta_{x_i}) = \frac{-\frac{\partial g}{\partial x_i}|_{P^*}\sigma_{x_i}}{\sqrt{\sum_{i=1}^{n}\left(\frac{\partial g}{\partial x_i}|_{P^*}\sigma_{x_i}\right)^2}}; (i = 1, \ldots, n) \quad (4.11)$$

② When random variable relation coefficient matrix is ρ (symmetric matrix),

$$\alpha_i = \cos(\theta_{x_i}) = \frac{-\rho_i\frac{\partial g}{\partial x_i}|_{P^*}\sigma_{x_i}}{\sqrt{\sum_{i=1}^{n}\left(\rho_i\frac{\partial g}{\partial x_i}|_{P^*}\sigma_{x_i}\right)^2}}; (i = 1, \ldots, n) \quad (4.12)$$

(5) Calculate the new checking point

$$x_i^* = \mu_{x_i} + \beta\sigma_{x_i}\alpha_i; (i = 1, \ldots, n) \quad (4.13)$$

(6) Judgment convergence: $|\beta_k - \beta_{k-1}| \leq \varepsilon$ is viewed as computation convergence (in which, ε is the required smallest reliability index computation error), stop iterating; Otherwise, return (2), continue to iterate.

In the definition formula of reliability indexes, if all random variables obey normal distribution, adopt improved Rosenbluth method to calculate the mean and standard deviation of failure state function (Cui Weicheng, etc., 1998). By this method, to avoid the computation of derivation of functions, in addition, to achieve the accuracy of expansion of more than two-order Tylor levels.

4.2.2 Performance Evaluation Method

PMA (an individual reliability analysis method) can't be used to analyze state functions whose reliability is unknown. Mostly, this method is used in the double circulation RBDO algorithm whose reliability is known.

$$\begin{aligned} &\min g(V) \\ &s.t. \quad \beta = |V| = \beta^t \end{aligned} \quad (4.14)$$

In which, β^t is target reliability.

Principles of PMA algorithm: To find the point that makes the state function value smallest on the hypersphere $\beta = |V| = \beta^t$.

It is difficult for limit state surface to capture complex state functions. Through RIA method, the solution meeting the constrained conditions in the formula (4.8) can't be found sometimes. Constrained conditions in PMA refer to the Hypersphere that are easy to capture state functions. PMA method is more stable than RIA, which has been verified in many cases. The book (written by Youn and Choi 2004) can be referred.

In order to speed up the convergence rate and convergence property, Youn and Choi, etc. researched a series of improved PMA method, for example, HMV + method mentioned in the document (Youn et al. 2004).

4.2.3 *Random Simulation Method*

4.2.3.1 Monte Carlo Method

The theoretical basis of random simulation method is the law of Large Numbers. If the joint probability density function of random vector X is f(X), then adopt random number generation method to generate enough samples (N stands for the quantity of samples), and put these samples into the state function g(X) to get the value. $g(X) \leq 0$ means a failure. According to the law of large numbers, the failure probability of g(X) (that is, the probability of $g(X) \leq 0$) can be approximate as:

$$P_f = \lim_{N \to \infty} \frac{m}{N} \tag{4.15}$$

In which, m stands for the times of $g(X) \leq 0$; m = sum (f(g)). f(g) stands for logic function value (Fig. 4.1). In the two-valued logic, I(g) = o or 1. The method is called as the original Monte Carlo method (CMC). Computing steps of Monte Carlo method:

① Construct the probability statistics models of uncertain variables.
② According to the probability statistics model and random number generation algorithms, to draw enough random numbers. To conduct the simulation test calculation of state function in the generated sampling point to get enough state function value samples;
③ Conduct the statistical analysis of state function values to ascertain the probability distribution, quantile and digital features.

The random number generation algorithm is used in the Monte Carlo. In order to ensure the repeatability of computation results, it is required to fix random number generation algorithms and random number seeds. The mostly common random number generation algorithm in engineering is Mersenne Twister algorithm whose breed is 0. Refer to computer-related documents to know more random number breed algorithms, specific algorithms of generating random numbers will not be introduced here.

In the real computation, N value is limited. Based on the variance analysis, it can be known that the estimated value of N is:

$$N = \frac{Z_{\alpha/2}^2}{P_f^\circ \varepsilon^2} \tag{4.16}$$

In the above formula, P_f° stands for the estimated value of failure probability, which can be derived from Form method.. $\varepsilon = \frac{|P_f - P_f^\circ|}{P_f}$ is the allowable failure probability error. 1-α is confidence degree; $Z_{\alpha/2}$ is the confidence limit. Reliability is highly demanded in the engineering system, so P_f° is small, n is big. For example, $P_f^\circ = 10^{-3}$, $\varepsilon = 5\%$, $Z_{\alpha/2} = 2$ (that is,$1 - \alpha = 0.95$), conduct $N \approx 1.6 \times 10^6$ experiments to get the estimated value of failure probability. Due to the large computation amount, the method is hard to be applied in engineering. Therefore, many improved sampling methods are developed, for example, LHS, TOA, IS, etc. (Saliby 1990; Ziha 1995; Harbitz 1986; Melchers 1989; Dong Cong 1996).

4.2.3.2 Importance Sampling Algorithm

Basic principles of importance sampling algorithm: Through increasing opportunities of g(x) ≤ 0, to improve CMC efficiency, reduce the computation amount. It focuses on invalid areas to sample. If most of the sampling points fall into the invalid area, the convergence speed of failure probability computation will speed up.

Introduce an additional probability density function $h_\gamma(x)$, transform formula (4.3) into:

$$\begin{aligned} P_f &= \int \cdots \int I(g(X)) \frac{f_X(x)}{h_\gamma(x)} h_\gamma(x) dx_1 \ldots dx_n \\ &= E\left[I(g(X)) \frac{f_X(x)}{h_\gamma(x)} \right] \end{aligned} \tag{4.17}$$

N stands for the times of random sampling (from $h_\gamma(x)$). Sample vector $x_i (i = 1, \ldots, N)$ then the approximate value of P_f is:

$$P_f^* = \frac{\sum_{i=1}^{N} \left[I(g(x_i)) \frac{f_X(x_i)}{h_\gamma(x_i)} \right]}{N} \tag{4.18}$$

Then, the variance of P_f^* is:

$$Var(P_f^*) = \frac{E\left(I(g(x_i)) \frac{f_X(x_i)}{h_\gamma(x_i)} \right)^2 - P_f^2}{N} \tag{4.19}$$

It can be seen that the variance will decrease when the shape of $f_X(x)$ is close to that of $h_\gamma(x)$. The general method of constructing $h_\gamma(x)$ is normal distribution;

the mean value of $h_\gamma(x)$ is generally design checking point (MPP) x^*, so the first step of importance sampling algorithm is to use RIA or PMA method to find MPP point; however, there is no uniform conclusion for the variance method of $h_\gamma(x)$. The popular method: If the original basic random variable is subject to normal distribution, the variance of $h_\gamma(x)$ is 1; if the original basic random variable is subject to non normal distribution, the variance of $h_\gamma(x)$ is 1-3 times of the variance of the original basic random variable (Shisong et al. 1998). In this book, after the calculation of some examples, it is found that variance selection has a great influence on the results, but now there is no unified opinion on variance selection in the field of reliability research.

The Importance Sampling Procedure Using Design points (ISPUD) program is the well-known reliability analysis software in importance sampling algorithm. It was first jointly developed by Schueller, Bucher, Bourgund and other Austrian structural experts (Schueller et al. 1987). Later, the program was transplanted from VAX machine to PC machine (Weicheng 1990). In this book, the ISPUD program was rewritten by using Matlab and the following improvements were made:

① In the Nataf algorithm for solving the correlation coefficient of random variables, the Fzero and Fminbnd functions in Matlab are used to directly solve the nonlinear formula [see formula (4.20)], so that Nataf algorithm is not limited by the distribution type supported by the original empirical formula, and ISPUD can be used for almost all common two-parameter probability distribution pattern.

$$\rho_{ij} = \int_{-\infty}^{\infty}\int_{-\infty}^{\infty}\left(\frac{y_i-\mu_i}{\sigma_i}\right)\left(\frac{y_j-\mu_j}{\sigma_j}\right)\varphi_2(z_i, z_j, \rho'_{ij})dz_i dz_j \tag{4.20}$$

In above formula: $y_i = F_{X_i}^{-1}(\Phi(z_i))$, $y_j = F_{X_j}^{-1}(\Phi(z_j))$, F_{X_i} is the distribution function on the right of random variable (Matlab function is cdf, and $F_{X_i}^{-1}$ is its inverse function, and Matlab function is icdf), Φ is the standard normal distribution function (cdf), $\varphi_2(z_i, z_j, \rho'_{ij})$ is the two-dimensional normal density function with mean point (z_i, z_j) and correlation coefficient ρ'_{ij}. For known ρ_{ij}, it is necessary to solve ρ'_{ij} through this formula.

② Parallel computing is used to speed up the computing speed during stochastic analog computation.

③ All the input parameters supporting the state function are random variables or only some of them are random variables.

4.2.4 Calculating Example of Reliability Analysis

In this book, JC method, original Monte Carlo method, and importance sampling algorithm routine are used to calculate three classic examples, so as to compare the characteristics and calculation errors of these algorithms.

4.2.4.1 Example 1

Lose efficacy
Limitation
Reliability

$$G(X) = X(1) \times X(2) + X(3) \begin{cases} < 0 & \text{失效} \\ = 0 & \text{极限} \\ > 0 & \text{可靠} \end{cases} \tag{4.21}$$

In which:

$X(1) \sim \text{logn}(12.5197, 0.1245)$ (Unless otherwise specified in this book, logn denotes lognormal distribution);

$X(2) \sim \text{logn}(-7.1052, 0.0998)$;

$X(3) \sim ev(-103.9438, 19.5259)$ (Unless otherwise specified in this book, ev denotes the minimum distribution)。

The correlation coefficient matrix is as follows:

$$\rho = \begin{bmatrix} 1 & 0.4 & 0 \\ 0.4 & 1 & 0 \\ 0 & 0 & 1 \end{bmatrix}$$

The reliability of the problem is less than 0.999. According to the formula (4.16) above, it needs millions of Monte Carlo random simulation to get the reliability meeting the engineering accuracy (5%). To ensure sufficient accuracy, 10 million (10^7) times of sampling are taken in this random simulation method. See Table 4.1 for comparison of calculation results.

Table 4.1 Example 1 calculation results

Reliability analysis algorithm	Reliability
10^7 stochastic simulation	0.9931
JC	0.9907
1000 times of sampling in IS procedure of this book	0.9907

Table 4.2 Example 2 calculation results

Reliability analysis algorithm	Reliability
10^7 stochastic simulation	0.8887
JC	0.9987
1000 times of sampling in IS procedure of this book	0.9145

The state function of example 1 is a weak nonlinear function, and the calculation results of JC algorithm, CMC method, and importance sampling algorithm have little differences, which is consistent with JC algorithm using the first expansion (linear expansion) of the state function as the approximate solution at the checking point.

4.2.4.2 Example 2

$$G(X) = 3 + X(1) - \frac{1}{6}\sum_{i=2}^{10} X^2(i) \begin{cases} < 0 & 失效 \\ = 0 & 极限 \\ > 0 & 可靠 \end{cases} \tag{4.22}$$

$X(i) \sim N(0, 1)$ (Unless otherwise specified in this book, N denotes the normal distribution) is mutually independent ($i = 1, \ldots, 10$). The comparison of calculation results is shown in Table 4.2.

In example 2, the state function is a nonlinear function (quadratic function). The error of JC algorithm and CMC algorithm reaches $\frac{0.9987-0.8887}{0.8887} = 12.4\%$, while the error of 1000 times of sampling for importance sampling algorithm and 10^7 times of sampling for CMC algorithm is only $\frac{0.9145-0.8887}{0.8887} = 2.9\%$. It can be seen that when JC algorithm is used for nonlinear state function, the approximation method of simplifying limit state function into linear function in the algorithm will lead to large error and the accuracy of importance sampling algorithm can be close to CMC algorithm when the parameter selection is appropriate.

4.2.4.3 Example 3

$$G(X) = min\begin{pmatrix} 4X(4) + X(1) \\ 4X(5) + X(1) + X(2) \\ 4X(6) + X(1) + X(2) + X(3) \end{pmatrix} \begin{cases} < 0 & 失效 \\ = 0 & 极限 \\ > 0 & 可靠 \end{cases} \tag{4.23}$$

In above formula:

X(1)∼ev(—8.1998,3.1188)(Unless otherwise specified in this book, ev denotes the minimum distribution);
X(2)-ev(-5.6499,2.3391);

Table 4.3 Example 3 calculation results

Reliability analysis algorithm	Reliability
10^7 times of stochastic simulation	0.9990
JC	0.9993
1000 times of sampling in IS procedure of this book	0.9983

X(3)-ev(—4.5999,1.5594);
X(4)~logn(2.1817,0.1764)(Unless otherwise specified in this book, logn denotes lognormal distribution);
X(5)-logn(2.5533,0.15295);
X(6)-logn(2.6954,0.1590).

The correlation coefficient matrix:

$$\rho = \begin{bmatrix} 1 & 0.5 & & & & \\ 0.5 & 1 & & & & \\ 0.2 & 0.5 & 1 & & & \\ 0 & 0 & 0 & 1 & & \\ 0 & 0 & 0 & 0.3 & 1 & \\ 0 & 0 & 0 & 0 & 0.3 & 1 \end{bmatrix}$$

The comparison of calculation results is shown in Table 4.3.

For the system failure in example 3, the state function of the system consists of several sub-state functions. The sub-state functions that work at different points are different, so the state function can be regarded as the problem of non-smooth. For each sub state function in this example is still a linear function, the calculation results among JC algorithm, CMC algorithm, and IS algorithm are not very different, which shows that JC algorithm does not require high smoothness of state function, but it is sensitive to the non-linearity of state function.

The program of JC, CMC, and IS algorithms is verified by these three examples. These three reliability analysis algorithm programs will be used as the sub modules of RBDO algorithm hereinafter.

4.3 Introduction of Fuzzy Reliability Analysis

In the three examples of reliability analysis in the previous section, the failure of the state function and the reliability judgment logic both are two-valued logic. When using the reliability analysis algorithm of sampling class, such as CMC algorithm, the calculation method and program can be used for fuzzy reliability analysis based on fuzzy logic with slight changes. If the MPP point of two-valued logic is taken as the approximate MPP point of fuzzy logic, the IS algorithm can also be used for

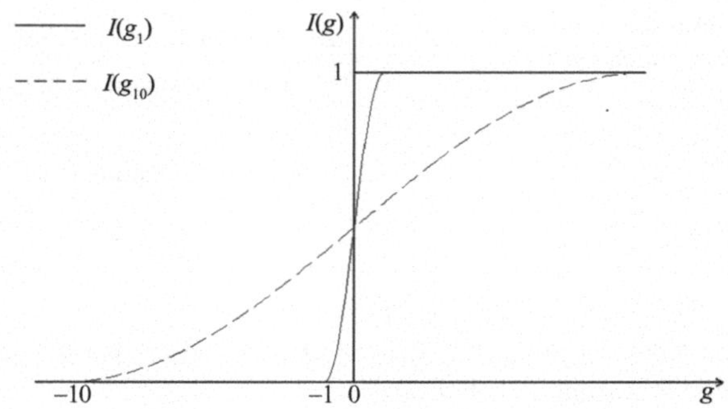

Fig. 4.2 Two membership functions selected

approximate fuzzy reliability analysis, but the rationality of this assumption is worth considering, so the error of IS algorithm may be large.

In this paper, the example 2 in the previous section is selected to introduce the fuzzy reliability analysis. In the original example, the two-valued logic is used, and the zero point of the function value is taken as the watershed to distinguish the failure domain and the reliability domain, that is, the logic judgment function at the zero point of the function value is mutational, while in the fuzzy logic, the logic judgment function (similar to the membership function in the fuzzy theory) is continuous and smooth, as shown in Fig. 4.1. At present, there is no unified method to select logical functions, and the selection of logical functions is closely related to the specific problems, so it is hard to establish unified rules and standards of logical functions. Two membership functions with different transition smoothness are selected for demonstration, as shown in Fig. 4.2 and formula (4.24) and (4.25). After replacing the two-valued logic function in CMC and IS algorithm program with the new logic function, the fuzzy reliability analysis of example 2 can be carried out.

$$Ig_1 = \begin{cases} 1 & \text{when} \quad g > 1 \\ \frac{1}{2} + \frac{1}{2}\sin\left(\frac{2\pi}{4}g\right) & \text{when} \; -1 \le g \le 1 \\ 0 & \text{when} \quad g < -1 \end{cases} \tag{4.24}$$

$$Ig_{10} = \begin{cases} 1 & \text{when} \quad g > 10 \\ \frac{1}{2} + \frac{1}{2}\sin\left(\frac{2\pi}{40}g\right) & \text{when} \; -10 \le g \le 10 \\ 0 & \text{when} \quad g < -10 \end{cases} \tag{4.25}$$

The CMC method using 10^7 times of sampling, IS algorithm using 1000 times of sampling, 2000 times of sampling, and 3000 times of sampling are used to analyze the fuzzy reliability of example 2 with membership function of I (g_1) and I (g_{10}) respectively. The results are compared with the calculation results based on two-valued logic in the previous section, as shown in Table 4.4.

Table 4.4 Example 2 results of fuzzy reliability analysis

Logic function	Algorithm	Reliability
Two-valued logic	10^7 times of CMC	0.8887
	1000 times of IS	0.9145
	10^7 times of CMC	0.8755
$I(g_1)$	1000 times of IS	0.8703
	2000 times of IS	0.4677
	3000 times of IS	1.0444
	10^7 times of CMC	0.6147
$I/(g_{10})$	1000 times of IS	0.5985
	2000 times of IS	0.3775
	3000 times of IS	0.7257

It can be seen from the calculation results and comparison in Table 2–5 that:

(1) By using the CMC method with 10^7 times of sampling in three logic functions of two-valued logic, $I(g_1)$, and $I(g_{10})$, it can be found that the selection of membership function has a great influence on the reliability calculation results: When the "steep" membership function $I(g_1)$ is selected, the fuzzy reliability analysis results are similar to those of two-valued logic (the difference is $\frac{0.8887-0.8755}{0.8887} \approx 1.5\%$); but when the "gentle transition" membership function $I(g_{10})$ is selected, the results of fuzzy reliability analysis are much smaller than those of two-valued logic (the difference is up to $\frac{0.8887-0.6147}{0.8887} \approx 30\%$). In practical engineering, the boundary between system failure and non failure is not as clear as that of two-valued logic. Failure usually occurs gradually rather than suddenly, so appropriate fuzzy logic is judgment function that is more consistent with the actual situation. It can be seen from the result of this example that the result of reliability analysis of traditional two-valued logic is larger than that of fuzzy logic, which means that the reliability of actual system is not as high as that of two-valued logic, so from the engineering safety and on the premise of proper logic function, fuzzy logic reliability analysis is more able to ensure safety.

(2) The result of 1000 times of IS algorithm is close to that of 10^7 times of CMC algorithm, but the error between 2000 times and 3000 times of IS algorithm, and 10^7 times of CMC algorithm is large, even when the logic function is $I(g_1)$, the IS algorithm with 3000 times of sampling gets reliability>1. Examples 1 and 3 are also calculated in this book (the results will not be listed in this book). It is found that the error of IS algorithm is large, and the reliability is more than 1 for many times. Therefore, it is only a coincidence to judge that the result IS algorithm with 1000 times of sampling is close to that of CMC method. For the MPP point of two-valued logic is used as the sampling center in the fuzzy reliability analysis of IS algorithm, it leads to a large error. This in turn shows that the CMC method is the only one with more reasonable theory and more accurate calculation accuracy in the fuzzy reliability analysis, and the RIA, PMA and IS algorithms

that need to use MPP points are all invalid. However, the calculated amount of CMC is usually unacceptable in practical engineering, so it is necessary to use high-precision approximate model to replace the time-consuming real engineering calculation model.

References

Cong D (1996) Important sampling based reliability analysis for non-linear system. Strength Environ 2:60–64. (董聪. 非线性系统可靠性分析的重要抽样法. 强度与环境, 1996, 2, pp. 60–64.)

Cruse TA (2001) Issues and strategies in probabilistic structural modeling. In: Proceedings 6th national turbine engine high cycle fatigue conference

Cui WC (1990) Uncertainty analysis in structural safety assessment. PhD thesis, Department of Civil Engineering, University of Bristol

Du L (2006) Reliability-based and possibility-based design optimization using inverse analysis methods. PhD thesis of graduate college of the university of Iowa

Faravelli L (1989) Response surface approach for reliability analysis. J Eng Mech 115(12)

Gaofeng R, Shengying Z (2012) Rapid analysis method of landing point error of Mars landing mission. J Harbin Inst Technol 44(7):14–20. (任高峰, 朱圣英. 火星着陆任务落点误差快速分析方法. 哈尔滨工业大学学报, Vol.44, No.7, 2012, pp. 14–20.)

Ghisu T, Parks GT, Jarrett JP, Clarkson PJ (2008) Robust design optimisation of gas turbine compression systems. In: 12th AIAA/ISSMO multidisciplinary analysis and optimization conference. Victoria, British Columbia Canada

Ghisu T, Parks GT, Jarrett JP, Clarkson PJ (2010) Adaptive polynomial chaos for gas turbine compression systems performance analysis. AIAA J 48(6):1156–1170

Hammersley JM, Handscomb DC (1964) Monte Carlo methods. Methuen, London

Harbitz A (1986) An efficient sampling method for probability of failure calculation. Struct Saf 3(1):109–115

Hasofer AM, Lind NC (1974) Exact and invariant second-moment code format. J Eng Mech Div 100(EMI):111–121

Liu BD, Iwamura K (1998) Chance constrained programming with fuzzy parameters[J]. Fuzzy Sets Syst 94(2):227–237

Liu BD (1999) Uncertain Programming[M]. New York, Wiley

Liu B (2000) Dependent-chance programming in fuzzy environments. Fuzzy Sets Syst 109:97–106

Liu B, Haftka RT, Akgun MA (2000) Two-level composite wing structural optimization using response surfaces. Struct Multi Optim 20(2):87–96

Liu D, Choi KK, Youn BD, Gorsich D (2006) Possibility-based design optimization method for design problems with both statistical and fuzzy input data. Trans ASME 128:928–935

Lurati LB (2008) Robust airfoil design under uncertain operation conditions using stochastic collocation. In: 46th AIAA aerospace sciences meeting and exhibit. Reno, Nevada

Madsen HO, Krenk S, Lind NC (1986) Methods of structural safety. NJ, Prentice-Hall, Englewood Cliffs

Manan A, Cooper JE (2008) Uncertainty of composite wing aeroelastic behaviour. In: 12th AIAA/ISSMO multidisciplinary analysis and optimization conference. Victoria, British Columbia Canada

Manan A, Cooper JE (2009) Robust design of composite wings for gust response. In: 50th AIAA/ASME/ASCE/AHS/ASC structures, structural dynamics, and materials conference < br > 17th. Palm Springs, California

Manan A, Cooper J (2009) Design of composite wings including uncertainties:a probabilistic approach. J Aircraft 46(2):601–607

Melchers RE (1989) Importance sampling in structural system. Safety 6(1):3–10
Moffitt BA (2010) A methodology for the validated design space exploration of fuel cell powered unmanned aerial vehicles. PhD thesis of Georgia Institute of Technology
Myers RH, Montgomery DC (1995) Response surface methodology. Wiley, New York
Oladyshkin S, Nowak W (2012) Data-driven uncertainty quantification using the arbitrary polynomial chaos expansion. Reliab Eng Syst Saf Elsevier 106:179–190. https://doi.org/10.1016/j.ress.2012.05.002
Palle TC, Michael JB (1982) Structural reliability theory and its applications. Springer, Berlin, Heidelberg
Qu XY, Venkataraman S, Haftka R et al. (2001) Reliability, weight, and cost tradeoffs in the design of composite laminates for cryogenic environments[C]. In: Proceedings of the AIAA/ASME/ASCE/AHS/ASC 42nd structures, structural dynamics and materials conference. Seattle, Washington
Rubinsrein RY (1981) Simulation and Monte Carlo methods. Wiley, New York
Saliby E (1990) Descriptive sampling: a better approach to monte carlo simulation. J Opl Res Soc 41(12):1133–1142
Schueller GL, Bucher CG, Bourgund U, Ouypornprasert W (1987) On efficient computational schemes to calculate failure probabilities. In: Lin YK, Schueller (eds) Stochastic structural mechanics, lecture notes in engineering, vol 31. Springer, New York, pp 388–410
Shaojuan S (2007) Research and application of uncertainty analysis method for coastal dry bulk shipping. PhD thesis of Wuhan technology university. (苏绍娟. 沿海干散货船舶运输的不确定性分析方法研究及应用. 武汉理工大学博士论文. 2007.)
Shisong M, Jinglong W, Xiaolong P (1998) Advanced mathematical statistic. Higher education press and Springer. (茆诗松,王静龙,濮晓龙. 高等数理统计. 高等教育出版社&斯普林格出版社, 1998.)
Ting P, Yunqing Z, Jinglai W (2011) Uncertainty analysis of flexible multi-body system based on polynomial chaos method. China Mech Eng 22(19):2341–2348. (皮霆, 张云清, 吴景铼. 基于多项式混沌方法的柔性多体系统不确定性分析. 中国机械工程,第22卷第19期, 2011, pp. 2341–2348.)
Wei Z, Weicheng C (1997) Direct integral method for structural reliability calculation. J. Shanghai Jiao Tong Univer 31(2):114–116. (张伟,崔维成. 结构可靠性计算的直接积分法. 上海交通大学学报, 第31卷, 第2期, 1997, pp. 114–116.)
Wiener N (1938) The homogeneous chaos. Am J Math 60(4):897–936. https://doi.org/10.2307/2371268
Xiaoyu J (2006) Network optimization under uncertain environment. PhD thesis of Tsinghua University. (计小宇. 不确定环境下的网络优化. 清华大学博士论文, 2006.)
Xiu D (2010) Numerical methods for stochastic computations: a spectral method approach. Princeton University Press. ISBN 978-0-691-14212-8
Xiu D, Karniadakis GE (2002) The Wiener-Askey polynomial chaos for stochastic differential equations. SIAM J Sci Comput 24(2):619–644. https://doi.org/10.1137/S1064827501387826
Youn BD, Choi KK (2006) Selecting probabilistic approaches for reliability-based design optimization. AIAA J 42(1)
Youn BD, Choi KK, Gu L, Yang R (2004) Reliability-based design optimization for crashworthiness of side impact. J Struct Multi Optim 27(3):272–283
Youn BD, Choi KK, Du L (2005a) Adaptive probability analysis using a enhanced hybrid mean valued method. Struct Multi Optim 29(2):134–148
Youn BD, Choi KK, Du L (2005b) Enriched performance measure approach for reliability-based design optimization. AIAA J 43(4):874–884
Youn BD, Wang P, Xi Z (2007) Complementary interaction method (CIM) for system reliability analysis. In: 48th AIAA/ASME/ASCE/AHS/ASC structures, structural dynamics, and materials conference.
Yufu N (2006) Intelligent algorithm based on simulation and its application. PhD thesis of Tianjin University. (宁玉富. 基于模拟的智能算法及其应用. 天津大学博士论文, 2006.)

Zhao LY, Dawes WN, Parks G, Jarrett JP (2009) Robust airfoil design with respect to boundary layer transition. In: 50th AIAA/ASME/ASCE/AHS/ASC structures, structural dynamics, and materials conference—17th. Palm Springs, California
Zhu P, Zhang Y, Chen GL (2011) Metamodeling development for reliability-based design optimization of automotive body structure. Comput Ind 62:729–741
Ziha K (1995) Descriptive sampling in structural safety. Struct Saf 17:33–41

Chapter 5
Reliability Based Multi-disciplinary Design Optimization Based on Reliability

The reliability based multidisciplinary design optimization is composed of reliability design and multidisciplinary design optimization. The preceding part of the paper has introduced the theory and key technology of multidisciplinary design optimization, as well as the basis of reliability design: Uncertainty theory and reliability analysis. This chapter will first introduce the reliability design method, and then introduce the multidisciplinary design optimization method based on reliability.

5.1 Reliability Design Methods

For products with complex engineering, the core of improving reliability is to reduce cognitive uncertainty and deal with the existing cognitive uncertainty and inherent uncertainty in scientific way. With the development of science, human beings have used the existing knowledge to design various complex projects. However, people's current knowledge is not enough to explain all the problems encountered in engineering design. In many cases, the knowledge they have is not sufficient, and some phenomena haven't been explained by better theories. The existing theories are based on certain simplification and assumptions, so the theories describing engineering problems inevitably contain approximation and uncertainty, namely, theoretical uncertainty. In addition to the theoretical description, the engineering design is often tested and analogized by means of model test and other means, while the test model cannot completely simulate the real engineering structure and its working environment, and there are uncertainties caused by many factors, such as environment, instrument, and many human factors in the measurement of the test process. When people try to design the real engineering structure according to the data obtained from the test, the approximation and uncertainty will be inevitably included, i.e. test

B. Pan and W. Cui, *Multidisciplinary Design Optimization and Its Application in Deep Manned Submersible Design*, Ocean Engineering & Oceanography 13,
https://doi.org/10.1007/978-981-15-6455-0_5

and measurement uncertainty. Engineering structures are generally constructed of metal materials, civil materials, and composite materials, etc. The chemical composition and microstructure of these materials cannot be completely stable, and there are inevitably various micro defects inside the materials, so the macro performance of the materials is uncertain, i.e. the material uncertainty. Even after the materials are prepared, the cutting, stamping, machining, welding and other manufacturing processes of materials will also lead to R-inch deviation, residual stress, and other uncertainties, that is to say, the manufacturing and processing of engineering structures also have uncertainties. In addition, there are various uncertainties in the working environment of engineering structures. For example, in the field of ship and offshore structure design, engineering structures need to bear various environmental loads such as water pressure, wind, wave, current, rain, snow, ice, and corrosion, etc., and the changing climate and water conditions of the marine environment are typical uncertainties of these loads.

In the face of these objective uncertainties, whether the engineering structure can ensure safety and reliability in the service life is not only related to the economy and reputation of the society, units, and individuals, but related to the safety of human life and social stability in many cases. In particular, with the continuous deterioration of the earth's environment, there are increasingly frequent extreme climates. Even the United States, with the most advanced science and technology, has no choice but to face a super hurricane like "Sandy". In the *Global Times* on October 31, 2012, there was a comment that "in front of 'Mother Nature', Americans 'kneel down'". Therefore, how to ensure the safety and reliability of large and complex engineering structures in the design life is still a major technical problem for engineers and technicians.

The systematic research of engineering reliability rose in the electronic industry of American aviation during the Second World War, and then expanded rapidly from aviation to nuclear energy, machinery, electricity, transportation, and other industries. With the development of the world economy, the number of engineering structures is increasing rapidly at an unprecedented rate, with more and more engineering accidents. In recent years, nuclear power plant leakage, aircraft crash, oil leakage of drilling platform, sudden collapse of high-rise buildings, high-speed railway train collision, and other major events have occurred at home and abroad, which often cause a large number of casualties, property losses, and even serious environmental pollution. The reliability of engineering structures is attracting unprecedented attention from the public, and governments and research institutions all over the world are aware of the importance of carrying out reliability research of engineering structures. In addition to the reliability evaluation of the existing engineering structures, it is more important to ensure the reliability of the engineering structures in the engineering design stage, so that the reliability runs through the concept of the engineering structures, and each stage from drawing to material objects. Such a design method is called Uncertainty Based Design (UBD) or Reliability Based Design (UBD).

After more than 60 years of development, the concept of reliability has now been generally accepted. Reliability has been extended from the cutting-edge aerospace, nuclear industry and other fields to almost all engineering industries. However, due to the increasing complexity of engineering products, there are increasing number of engineering projects and products, and the increasing frequency of engineering failure accidents. These accidents have brought huge life and property losses, so the importance of reliability has been emphasized repeatedly, and reliability research has always been a hot research issue.

The reliability of engineering products is closely related to the design, manufacturing, assembly and management in the process of product development, in which design plays a crucial role in the reliability of engineering products, and subsequent manufacturing, assembly and management play a role in ensuring that the products reach the reliability level of design, so the main content of reliability research is mainly the design of uncertainty information. At present, Uncertainty Based Design (UBD) can be divided into three categories below:

① Robust Design Optimization (RDO) is also called stable design optimization. The engineering system function it mainly studied is the objective function of the optimization model of engineering product design. The aim is to find the design scheme that makes the objective function (in design, the objective function is usually the performance index pursued by the design, and these indexes are usually not the necessary requirements like safety) of engineering products least sensitive to the disturbance caused by uncertain factors.

② Based on Uncertainty Based Design Optimization (UBDO), this paper mainly studies the influence of uncertainty factors on the constraint function of engineering product design optimization model (in design, the constraint function usually refers to the requirements or indicators that must be achieved, such as the stress level requirements related to safety, and key performance requirements, etc.), and the value of uncertainty variable in the objective function takes its expected value (mean value).

③ The 6-Sigma robust design, which studies the influence of uncertain factors on constraint function and objective function, is suitable for the design of engineering products with high requirements on reliability and robustness. Generally, the $\pm 3\sigma$ design of engineering products is extended to more strict $\pm 6\sigma$. The treatment of constraint function in this method is similar to that of reliability index $\beta = 6\sigma$ in UBDO, while the objective function is the weighted sum of mean and square of original objective function, and the weight coefficient need to be specified artificially. Therefore, the 6-sigma robust design can be regarded as the UBDO with higher reliability requirement of constraint function and objective function rewritten.

It can be seen that the purposes of the above three types of UBD methods are different. Robust Design Optimization (RDO) focuses on improving the anti-interference ability of the objective function; the purpose of uncertainty based design optimization (UBDO) is to ensure that the disturbance of the constraint function caused by the uncertain information does not exceed the allowable range (i.e. the reliability of the constraint conditions is met), and optimize the mean value of the objective function as much as possible. The 6-sigma robust design is not only pays attention to the reliability of constraint function, but tries to improve the anti-interference ability of objective function.

Generally speaking, the design of manned submersibles has clear requirements for structural safety and rigid functional indicators, that is to say, important indicators are taken as constraint conditions. Therefore, the reliability design of manned submersible should be classified into UBDO problem. This book will focus on UBDO problem instead of RDO and 6-sigma robust design. But the technology and research content of these three types of reliability design methods are basically the same. The research content (i.e. the common technology of uncertain design) can be divided into the following three main parts:

① Description theory of uncertain information;
② Reliability evaluation;
③ Uncertainty optimization

The uncertain information description theory, the theoretical system to describe and deal with uncertain information, is the basis of reliability design. The reliability evaluation method is used to calculate the reliability of each design scheme, and it is an essential tool for reliability design. After integrating uncertainty theory and reliability evaluation method into uncertainty optimization model, uncertainty optimization is used to solve the problem. The most common method is to transform uncertain constraints into deterministic constraints, so that deterministic optimization algorithm can be used to solve them. However, due to the existence of uncertain information, the disturbance of the optimization model is large, which may lead to the difficulty of convergence in traditional deterministic optimization algorithm. There are various ways to transform uncertain constraints into deterministic constraints, which are also classified as the research content of uncertain optimization.

When using different uncertainty information description theory, UBDO is also different. At present, the most commonly used uncertainty information description theories are probability theory, fuzzy theory, and interval theory, and UBDO methods based on these three theories are also different.

(1) UBDO—RBDO based on probability theory

When using probability theory to describe uncertain information, UBDO is called Reliability Based Design Optimization (RBDO). Probability theory is the most widely used uncertainty information description theory in engineering, so Reliability Based Design Optimization (RBDO) is the most common UBDO. In RBDO, the optimization model is usually expressed as:

$$\begin{aligned} & \min f(d, \mu_X) \\ & s.t.\ P(g_u(d, X) < 0) < P_f^t \\ & \qquad g_d(d) \le 0 \\ & \qquad d^l \le d \le d^u,\ \mu^l \le \mu_X \le \mu^u \\ & \qquad \sigma\ or\ \mathrm{cov}\ is\ known \end{aligned} \tag{5.1}$$

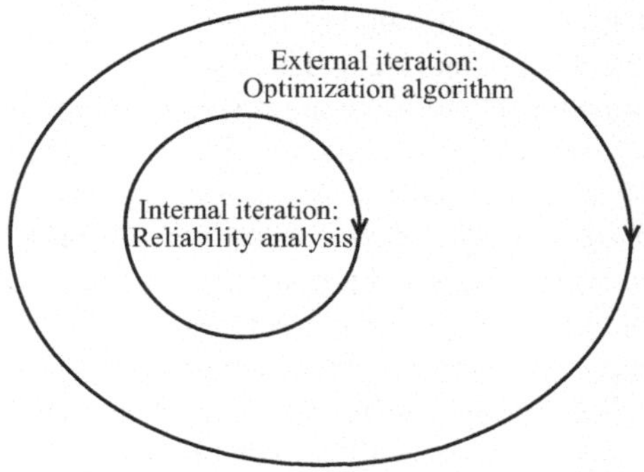

Fig. 5.1 Double loop RBDO algorithm

In above formula: d is the design variable; X is the random variable; μ_X is the mean value of X; g_u is the uncertain constraint function; g_d is the deterministic constraint function; P_f^t is the limit of failure probability; σ is the standard deviation, and cov is variable coefficient, which only needs to know one of them; $P(g_u(X) \leq 0) < P_f^t$ is called the Probabilistic Constraints, which is the biggest difference between RBDO optimization model and deterministic optimization model. It can be seen that RBDO includes the process of reliability analysis, so RBDO algorithm includes two nested iterative processes: The external iteration is the optimization algorithm, and the internal iteration is the reliability analysis. Therefore, this sort of RBDO algorithm is also called Double Loop (Allen and Maute 2004; Royset et al. 2001; Wang et al. 1995; Baabbad 2004), as shown in Fig. 5.1.

外部迭代: 优化算法 内部迭代: 可靠度分析	External iteration: Optimization algorithm Internal iteration: Reliability analysis

Because of the large amount of calculation in the double loop RBDO algorithm, we naturally think that the approximate model can be used to replace the original high-precision state function with large amount of calculation. Some approximate models can be used to approximate the value of the state function in the whole solution domain, of which some only approximate the state function near the checking point. The RBDO algorithm with the approximate model is widely used in engineering (Krishnamurthy and Romero 2002; Burton and Hajela 2002; Papadrakakis and Lagaros 2002; Gayton et al. 2003).

In addition to the use of approximate technology, another method is to use the physical meaning of the design checking point (MPP) to change the inclusion relationship between optimization and reliability analysis into a sequential relationship, and calculate the design checking point from the reliability analysis, then carry out the deterministic optimization design with the mean value as the design variable at the design checking point, and next carry out the reliability analysis at the optimized mean value… iterate until convergence. In this kind of algorithm, optimization and reliability analysis alternate, so this kind of RBDO algorithm is also called Sequential Optimization and Reliability Analysis (SORA) (DU and Chen 2002; Wu and Wang

1998). In each round of iteration in SORA algorithm, the coordinates of the design checking points are updated, so the reliability analysis algorithm in SORA algorithm must be able to solve the design checking points.

Wu et al. first combined the traditional safety factor method with the reliability analysis method in structural design (Wu et al. 2001). In this algorithm, they first assume a safety factor value and substitute it into the limit state formula, and then rewrite the original probability constraint into the deterministic constraint, thus, the optimization problem is converted into an optimization problem with only the deterministic constraint; next, they solve the deterministic optimization problem and make reliability analysis for the state function at the optimization solution, and then adjust safety factor according to the difference between the reliability of the state function and the target reliability, so as to carry out the next round of deterministic optimization design. It can be seen that the basic process of this algorithm is similar to that of SORA algorithm, except that each update is not the coordinates of the design checking point, but the safety factor. In this book, the method is extended from structural design constraints to all inequality constraints, and the importance sampling algorithm is introduced to reduce the calculated amount, so a new efficient and general RBDO algorithm and program is established. The algorithm can be regarded as a Safety Factor based Sequential Optimization and Reliability Analysis (SFSORA). This algorithm has the characteristics that SORA algorithm reduces the computation, and any reliability analysis algorithm can be used for reliability analysis. When the constraint function is in the form of $g < c$ $(c > 0)$ that is similar to structural design, the relationship between the safety factor and the actual reliability can be obtained by the method; when the constraint function is in other forms, the updated factor in each iteration of the method is not the actual safety factor, but the offset factor of the constraint function. It can be seen from the calculation example that this algorithm is a RBDO algorithm with strong practicability, and the specific algorithm is described below.

Chen et al. proposed a single loop RBDO algorithm (Chen et al. 1997) based on the physical meaning of reliability index in FORM algorithm, in which reliability analysis was replaced by target reliability index. The probability constraint condition was converted into certainty constraint condition by using the relationship between mean points and checking points. This algorithm does not need reliability analysis and its calculated amount is greatly reduced, but the gradient matrix of the last optimization iteration is used to approximate the gradient matrix of the current optimization iteration during optimization solution, the requirements of this algorithm for the optimization algorithm are higher than that of the double loop RBDO algorithm, namely, for the same problem and the same optimization algorithm, more optimization iterations may be needed to get the solution. Similar single loop RBDO algorithms are also introduced in the literature (Agarwal et al. 2004).

(2) Research status of PBDO algorithm

When fuzzy theory or evidence theory is used to describe uncertain information, UBDO has a specialized term: Possibility Based Design Optimization (PBDO).

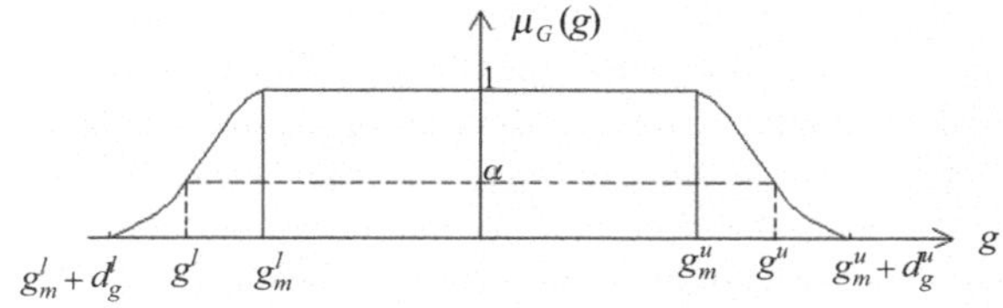

Fig. 5.2 An example of membership function with fuzzy constraint

When the uncertain variable does not have enough statistical information to establish its probability density function, fuzzy variable can be used to describe the uncertain variable, for example, cognitive uncertainty is expressed as fuzzy variable (Liu et al. 2006; Savoia 2002). At this time, the reliability design optimization of the system is based on the fuzzy theory. Generally, the reliability design optimization (fuzzy RBDO) model based on the fuzzy theory can be expressed as follows:

$$\begin{aligned} &\min f(d, u_m) \\ &s.t.\ \mu(g^l \le g(d, X) \le g^u) \ge \alpha^t \\ &\quad g^l_d \le g_d(d) \le g^u_d \\ &\quad d^l \le d \le d^u, u^l \le u_m \le u^u \end{aligned} \tag{5.2}$$

In above formula: d is the deterministic design variable; X is the fuzzy variable; u_m is the maximum membership vector of each component of X; μ is the membership function; g is the fuzzy constraint function; g^l and g^u are the fuzzy lower limit and upper limit respectively; g_d is the deterministic constraint function; and α^t is the allowable failure probability.

For the common constraint function: $g^l \le g(X) \le g^u$, suppose that its membership function have the properties shown in Fig. 5.2, where d^u_g and d^l_g is the length of transition area, i.e. the allowable deviation of upper and lower limits. It can be seen that the fuzzy constraint $\mu(g^l \le g(X) \le g^u) \ge \alpha$ is determined by α level, and there are different optimum solutions for different α levels. The final optimum solution depends on the selection of α level. The interval [gl, gu] is also called α level cut set.

The most commonly used method to solve the fuzzy RBDO is vertex analysis. In addition, there are α level cut set method and multi level cut set method. However, the calculated amount and accuracy need to be improved before they can be applied to the actual engineering design.

Du proposed the Maximal Possibility Search (MPS) to solve the fuzzy RBDO problem, and MPS can be regarded as the improvement of vertex analysis method. When the maximum point of constraint function is not in the design space vertex, the maximum point can be solved by interpolation, so as to improve the accuracy (DU 2006). The detailed steps of MPS method are not described here. The example of Nikolaidis et al. shows that the result of fuzzy RBDO is more conservative than that of probability RBDO (Maglaras et al. 1997), which is consistent with the fact that the information quantity of fuzzy variable is less than that of random variable.

As mentioned above, the axiom of fuzzy theory is not complete, and there is no unified method to select membership function and fuzzy logic function. Moreover, PBDO mainly relies on random simulation methods, such as MCM, to get more stable results when it is applied in engineering. When probability theory is extended to the generalized probability theory, it not only maintains a perfect axiom system, but expresses cognitive uncertainty. Besides, RBDO algorithm based on probability theory has developed relatively mature, and has made many achievements in engineering application. In a long time, it will also be the mainstream method of engineering reliability design, so this book will mainly introduce RBDO method, and will not discuss PBDO method in detail.

(3) Research status of IBDO algorithm

When the interval theory is used to describe uncertain information, UBDO can be called Interval Based Design Optimization (IBDO). The optimization model of IBDO algorithm can be expressed as follows:

$$\begin{aligned} & \min f(d, x) \\ & s.t.\ \min(g_u(d, x)) \geq 0 \\ & \quad g_d(d) \leq 0 \\ & \quad d^l \leq d \leq d^u, x^l - \Delta x^l \leq x \leq x^u + \Delta x^u \end{aligned} \tag{5.3}$$

In above formula: d is the deterministic variable; x is the interval number. Suppose the original lower limit of x is written as the form $\left[x^l - \Delta x^l, x^l + \Delta x^l\right](\Delta x^l > 0)$, and the upper limit is $[x^l - \Delta x^l, x^l + \Delta x^l](\Delta x^l > 0)$. According to the interval algorithm, the value range of x can be rewritten as $x^l - \Delta x^l \leq x \leq x^u + \Delta x^u$. It is required that at any point in the value range of x, the interval constraint condition $g_u(d, x) > 0$ is constantly workable, which can be equivalent to that the minimum value of the interval constraint function in the value range of x is greater than 0, i.e. $\min(g_u(d, x)) > 0$. It can be seen that IBDO algorithm can be regarded as two nested deterministic optimizations: The internal optimization is the minimum value to look for $g_u(d, x)$, and the external optimization is to look for the optimal objective function value.

In addition, when the generalized probability theory is used to describe the uncertain information, the characteristic parameters of random variables, such as mean, standard deviation (or variable coefficient) and even the event probability are no longer a certain number. These statistical characteristic parameters may be fuzzy numbers or interval numbers. At this time, UBDO can still be calculated by RBDO algorithm, but the RBDO algorithm should include algorithm processes like PBDO algorithm or IBDO algorithm.

At present, RBDO algorithm is widely used in engineering. Next, three main algorithms of RBDO algorithm are introduced in detail: Double Loop RBDO algorithm (DLRBDO, sometimes referred to as Double Loop), SORA algorithm and Single Loop RBDO algorithm (SLRBDO, sometimes referred to as Single Loop). In this book, the SFSORA algorithm based on safety factor method and SORA method is also introduced, which can establish the relationship between safety factor and system reliability intuitively. In the process of introducing these methods, this book will also introduce some improvements to the existing methods, such as improving the Single Loop RBDO algorithm, making SLRBDO algorithm more applicable, more stable, and less calculated amount. When the single parameter generalized probability theory is used to describe the uncertain information, these RBDO algorithms do not need to make great adjustments. When interval numbers are used to express uncertain parameters in generalized probability theory, only the upper and lower limits of variables need to be set as the upper and lower limits of interval numbers in the deterministic optimization, then the deterministic optimization method can be applied to reliability design optimization.

5.1.1 Double Loop

The principle of Double Loop RBDO (DLRBDO) has been introduced in Fig. 5.1. After the reliability analysis algorithm is programmed, the flow of DLRBDO is similar to that of conventional deterministic optimization. It should be noted that when RIA algorithm, PMA algorithm or Importance Sampling Algorithm (IS) are used as the reliability analysis module, the calculation may not converge or false optimization value may be obtained, so it is necessary to verify the results.

When the optimization model of engineering problems is established, the designable (changeable) variables are classified into the design variables, and the immutable (constant in deterministic optimization) parameters are classified into the auxiliary variables; there may be random variables in both the design variables and the auxiliary variables, so the following situations are often encountered in practical engineering:

① If there is no random variable in the design variable, there must be random variable in the auxiliary variable (otherwise, it is the deterministic optimization problem rather than the RBDO problem);
② There are random variables in the design variables and no random variables in the auxiliary variables;

③ Design variables contain random variables, and auxiliary variables also contain random variables.

To enable the program to handle all these situations, the DLRBDO program written in this book has made specific regulations on function call form and function writing method:

$$[DVopt, Fopt, exitflag, output] = DLRBDO(Fhobj, GhU, GhD, \ldots$$
$$R\_req, name\_DVxu, std\_DVxu, DV0, LDV, UDV, FhRA, AVxu)$$

In above formula:

Fhobj: The function handle of optimization objective function only supports single objective optimization currently, such as @ fweight.

GhU: The function handle of probability constraint condition, stored in the form of cellarray, can pass multiple objective functions, such as {@conU1, @conU2, @conU3}.

GhD: The function handle of deterministic constraints, stored in the form of cellarray, can pass multiple objective functions, such as {@conD1, @conD2, @conD3}. A cellarray is written as empty when there is no deterministic constraint (that is, {}).

R_req: Reliability requirements of probability constraints (target reliability), for example: [0.99, 0.987, 0.999].

name_DVxu: The name of the distribution type of the random design variable, stored in the form of the string cellarray, for example: {'normal', 'lognorml', 'beta'}. When there is no random variable in the design variable, assign a null cellarray (that is: {}).

std_DVxu or coe_DVxu: The standard deviation or variable coefficient of a random design variable, written in the form of a numerical vector, for example: [1, 1, 1].

DV0: Initial value of design variable. When there are both random and determinate variables in the design variables, DV0 = [DV0xu, DV0xd] is specified, that is, the random design variables are written in front of the determinate design variables.

LDV and UDV: They are the lower limit and upper limit of the design variable, respectively, with same writing method as DV0.

FhRA: The optional FhRA programs the reliability analysis program can provide are JC, CMC, RIA, PMA, and IS (Importance Sampling). It is found that JC and CMC are the most stable FhRA programs, followed by RIA and PMA, and IS is of the worst stability, because the structure of IS algorithm is complex, and includes the process of calling JC algorithm to calculate MPP. When the reliability of state function is low (R < 0.5), the error of IS algorithm is very large, even the error result of R = 1 is obtained.

AVxu: The random auxiliary variable is stored in the form of nx2 (n is the number of random auxiliary variables). The first column of AVxu is the distribution name of variable (data type is string), and the second column is the distribution parameter (numerical row vector), such as {*'norm'*, [0, 1]; *'logn'* [0.5, 1]; *'wbl'* [0.5, 2]}, which means that there are three random auxiliary variables: The first is normal random variable, i.e. the parameter $\mu = 0, \sigma = 1$ in normal density function; the second is lognormal random variable, parameter $\mu = 0.5, \sigma = 1$ in density function; the third is random variable in Weibull distribution, with parameters in density function— scale parameter: $\alpha = 0.5$, shape parameter: b = 1. The density functions of all distribution types and the meaning of their parameters are shown in the help document of Matlab statistical toolbox. To determine the auxiliary variable, it only needs to pass through the global variable instead of passing through the input parameter by the DLRBDO program.

DVopt: The optimum point.

Fopt: The value of the objective function at the optimum point.

exitflag: The value of reflecting whether the optimization converges or not. The return value of each optimization algorithm is slightly different. Please refer to the optimization toolbox of Matlab.

Output: The information returned by the optimization algorithm, including the number of function calls, the number of iterations, and the satisfaction of constraint functions, etc.

When AVxu exists (i.e. auxiliary variable contains random variable), the writing method in GhU is specified as GhU (DV, AV), that is to say, the function has two input parameters: The first one is design variable, and the second one is random auxiliary variable; when AVxu does not exist, GhU is written as GhU (DV), namely, the input parameter only has the design variable.

For DLRBDO is a double loop structure, with high requirement of calculated amount, the DLRBDO algorithm program compiled in this book adopts the optimization algorithm composed of the conventional numerical algorithm and the direct search algorithm: The program optimization first uses the conventional numerical optimization algorithm fmincon (the internal algorithm uses the Sequential Quadratic Programming (SQP)) for solution, and it calls the pattern search (PS) algorithm (open Mesh Accelerator) if it does not converge.

5.1.2 SORA Algorithm

The SORA algorithm was first proposed by Du and Chen (2002). In SORA algorithm, optimization and reliability analysis are no longer nested but sequential, and reliability analysis is no longer needed in optimization but is only carried out after optimization, see Fig. 5.3 for the flow chart. It can be seen that the core idea of SORA algorithm is to translate the probability constraint function and transform the probability constraint condition into the deterministic constraint condition at the MPP point, so that the optimization only needs to be carried out in the deterministic range, and then the MPP point is modified through reliability analysis. The SORA algorithm reduces the calculated amount greatly by removing the nest relation between optimization and reliability analysis. However, MPP points are needed in SORA algorithm, which means that the reliability analysis algorithm of MPP points cannot be used in SORA algorithm, such as Monte Carlo method.

5.1.3 SFSORA Algorithm

In engineering design, the safety factor method is the oldest and most commonly used method to ensure reliability. For example, in the design of manned submersible cabin, the 1.5 safety factor is generally used to ensure its reliability in current national codes. Based on the most commonly used safety factor method in structural design

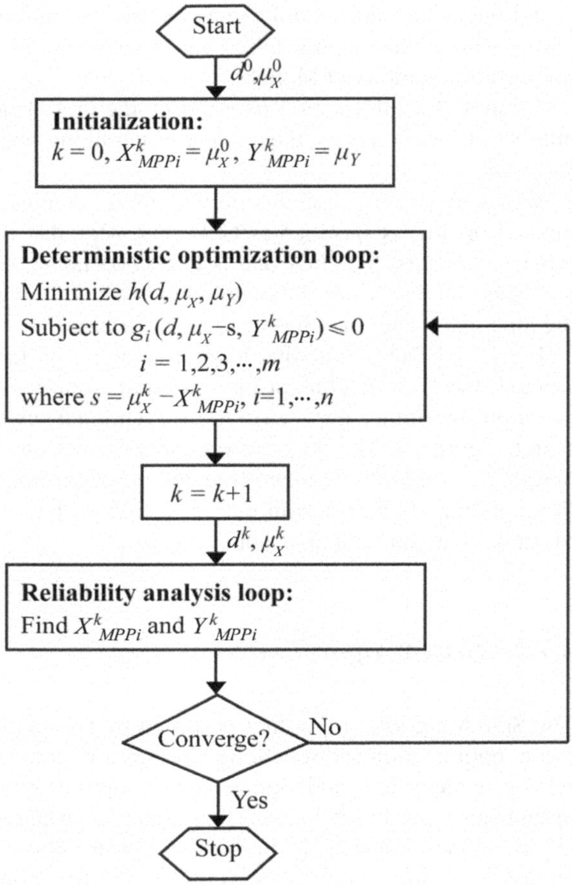

Fig. 5.3 SORA algorithm flow (Li 2008)

and combined with SORA algorithm, a SORA algorithm based on safety factor is proposed, which is called SFSORA algorithm.

In the design of engineering structure, the constraint condition is usually displacement or stress. Taking the most commonly used stress as an example, the safety domain of the safety factor method can be written as (the most intuitive writing of the safety domain is: $\sigma \leq \frac{\sigma_{0.2}}{sf}$, but due to $\sigma_{0.2}$, it is usually uncertainty, so it is rewritten):

$$g = \frac{\sigma}{\sigma_{0.2}} \leq \frac{1}{sf} \tag{5.4}$$

In above formula: σ is the calculated stress of the structure; $\sigma_{0.2}$ is the stress of material failure, usually the yield strength of the material (also the uncertainty); Sf is the safety factor (usually sf > 1). In the design of manned submersibles, the safety factor is often taken according to the existing experience and specifications, but in this way, it is not known how much the reliability of the system is after selecting the safety factor.

It is easy to analyze from formula (5.4) that: For a certain system, the larger the safety factor is obtained, the smaller the right side of formula (5.4), the greater the possibility of the left side being larger than the right side, that is, the greater the failure probability of the system. The SFSORA algorithm is to explore the relationship between safety factor and system reliability (or failure probability), to find the appropriate safety factor, so that the probability constraint function can be met (that is, to meet the reliability requirements).

The SFSORA algorithm is applicable to the problem that the probability constraint function can be expressed as formula (5.5) (the probability constraints of general engineering RBDO problems can be rewritten as this form, especially the structural design problem).

$$P(g(d, X) \le g^U) > R_f^t$$
$$Or\ equivalent\ form: P(g(d, X) > g^U) < P_f^t \tag{5.5}$$

In above formula: g is a probability constraint function (state function); d is a determinate or controllable variable; X is an uncertain variable, generally described by a random variable; g^U is the upper limit of the constraint function (it is required to be a determinate quantity greater than zero, and generally g^U has a clear physical meaning).

For the common upper limit constraint problem $g(d, X) \le \bar{g}$ in engineering, it can be discussed in the following two cases:

① If $\bar{g} > 0$, directly meet the requirements of formula (5.5);
② If $\bar{g} < 0$, it can be converted into $(g(d, X) - \bar{g} + C) - \frac{C}{sf} \le 0$, in formula (5.5), $g^U = C$, where C is greater than 0 and is any non-zero constant (the larger the C is, the greater the effect of sf is, namely, the greater the change of the limit of constraint function when the SF value changes). In this case, sf is no longer of physical significance of safety coefficient, but becomes a parameter that changes the limit of constraint function purely.

For another common lower bound restricted problem in engineering: $g(d, X) \ge \underline{g}$, it can be discussed in the following two cases:

① If $\underline{g} > 0$, it can be converted into $\left(-g(d, X) + \underline{g} + C\right) - \frac{C}{sf} \le 0$, where $C > 0$, which is similar to the case above;
② If $\underline{g} < 0$, it can be converted into $-g(d, X) - \frac{-\underline{g}}{sf} \le 0$, i.e. in formula **5.5**, $g^U = -\underline{g}$;

Therefore, the optimization in SFSORA algorithm can be expressed as a deterministic optimization problem

$$\begin{aligned} & \min f(d, \mu_X) \\ & s.t.\ g_u(d, \mu_X) - \tfrac{g^U}{sf} \le 0 \\ & \quad\ \ g_d(d) \le 0 \\ & \quad\ \ d^l \le d \le d^u, \mu^l \le \mu_X \le \mu^u \end{aligned} \tag{5.6}$$

In above formula, sf is the safety factor vector.

The steps of SFSORA algorithm are as follows:

① For each probability constraint condition, assume an initial sf_i ($i = 1, \ldots, m$, m is the number of probability constraint functions), generally, the initial value of sf_i is recommended to be 1;

② sf_i is substituted into each probability constraint g_u^i to form the deterministic optimization problem of formula (5.6), which is solved by the optimization algorithm, and d_{opt} and μ_{opt} (optimal mean point) are obtained.

③ The reliability analysis of each original probability constraint function ($g^U - g \ge 0$) is carried out at d_{opt} and μ_{opt}, and the failure probability $P_f = [P_{f_1}, \ldots, P_{f_i}, \ldots P_{f_m}]$ (m is the number of probability constraints) is obtained.

④ Compare P_f with the target failure probability P_f^t: If $P_{f_i} > P_{f_i}^t + \varepsilon$ ($P_f = [P_{f_1}, \ldots, P_{f_i}, \ldots P_{f_m}]$, ε is the convergence criterion, with the common forms of $\varepsilon = \frac{P_{f_i}^t}{10}$ or $\varepsilon = \frac{P_{f_i}^t}{100}$), then increase $sf_i : sf_i^k = sf_i^{k-1} + \Delta sf_i$ (where $\Delta sf_i > 0$ is the amount of increasing each iteration safety factor, k is the number of times to update sf); if $P_{f_i} < P_{f_i}^t - \varepsilon (i = 1, \ldots, m)$, $sf_i^k = sf_i^{k-1} - \Delta sf_i$; if $P_{f_i}^t - \varepsilon < P_{f_i} < P_{f_i}^t + \varepsilon$, then stop the iteration, otherwise convert to ②, and carry out the next iteration.

It can be seen that the SFSORA algorithm is similar to the SORA algorithm, in which the optimization and reliability analysis are carried out in order, and can reduce the calculated amount. Moreover, the SFSORA algorithm only needs the reliability (or failure probability) and does not need to find the MPP point coordinates, so all the reliability analysis methods can be used in the SFSORA algorithm. In addition, the updating of the safety factor in SFSORA has obvious rules— a simple numerical optimization algorithm can be used to complete the adjustment of the safety factor. In many cases, even a few safety factor adjustments can be made directly by hand to get the solution meeting the reliability requirements, and the safety factor and reliability of the system at the optimization design point can be known at the same time. Besides, if the deterministic optimization model and reliability analysis program of the existing system do not even need additional RBDO algorithm, only the designer needs to slightly modify the constraint function of the deterministic optimization model, and then carry out the deterministic optimization—reliability analysis—manual adjustment of the safety factor—deterministic optimization in turn according to the process of SORA algorithm. Therefore, the SORA algorithm is a RBDO algorithm with strong engineering practicability, especially in the reliability design of structures, which will be discussed in Chap. 6 of this book.

When SFSORA algorithm is used to deal with RBDO problems with random variables in auxiliary variables, it does not need to make special processing—just substitute the mean value of auxiliary random variables into the deterministic optimization model formula (5.6).

5.1.4 Single Loop

For the probability constraints of normal distribution random variables $x \sim N(\mu_x, \sigma_x)$ and state function $h(d, x)$:

$$P(h(d, x) \geq 0) > R^t \tag{5.7}$$

In above formula, R^t is the target reliability.

According to the definition of reliability index, the corresponding target reliability index can be obtained:

$$\beta^t = \Phi^{-1}(R^t) \tag{5.8}$$

Standardize x:

$$y = \frac{x - \mu_x}{\sigma_x} \tag{5.9}$$

Then, the state function can be converted into a function about y:

$$g(d, y) = h(d, x) = h(d, \mu_x + \sigma_x \times y) \tag{5.10}$$

The original probability constraints are converted into constraints on standard normal random variable $y \sim N(0, 1)$ and state function $g(d, y)$

$$P(g(d, y) \geq 0) > R^t \tag{5.11}$$

Taking the simplest 2D case of $y = [y_1, y_2]$ as an example, in the standard normal coordinate system, after drawing the $g(d, y) = 0$ and circle $\|y\| = \beta^t$, we can easily find that the two curves are tangent at the checking point (MPP point), as shown in Fig. 5.4, then the probability constraint condition formula (5.11) can be equivalent to the deterministic constraint condition at the MPP point:

$$g(d, y^*) \geq 0 \tag{5.12}$$

According to the tangent property of the curve, the coordinates of MPP points can be expressed as follows:

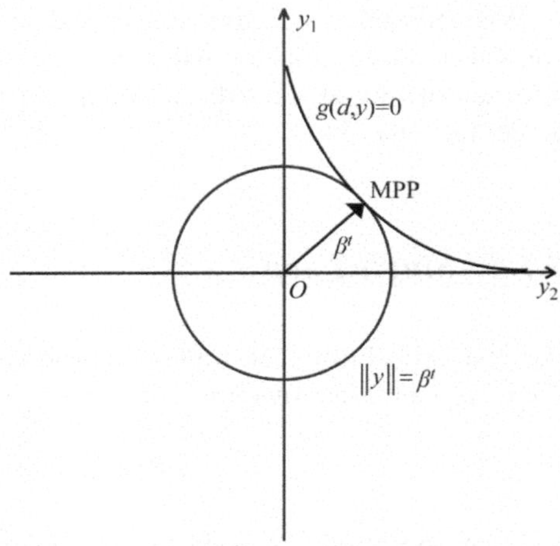

Fig. 5.4 Schematic diagram of SLRBDO algorithm

$$y^* = \pm\beta^t \frac{\nabla_y g|y^*}{\|\nabla_y g|y^*\|} \tag{5.13}$$

In above formula, $\nabla_y g|y^* = \left[\frac{\partial g}{\partial y_1}\Big|_{y^*}, \frac{\partial g}{\partial y_2}\Big|_{y^*}\right]$ is the gradient vector of state function g at MPP point ($y^* = [y_1^*, y_2^*]$), $\|\nabla_y g|y^*\| = \sqrt{\left(\frac{\partial g}{\partial y_1}\Big|_{y^*}\right)^2 + \left(\frac{\partial g}{\partial y_2}\Big|_{y^*}\right)^2}$ is the module of gradient vector. For the case of high dimension, formula (5.13) is still valid. In fact, as early as 1984, Ang and Tang (1984) deduced the relational expression.

Formula (5.13) is a system of nonlinear equations in MPP coordinates. If formula (5.13) is directly applied to RBDO problem, the internal cycle of double loop RBDO algorithm changes from reliability analysis to solving system of nonlinear Eq. (5.13). Based on the characteristics of numerical algorithm for solving system of nonlinear equations, Chen et al. used the gradient vector of the last iteration to replace the gradient vector of the current iteration, rewritten formula (5.13), put the measure of solving nonlinear formulas and searching the optimal mean point in the same optimization algorithm, so that the single loop RBDO algorithm (SLRBDO) is obtained. For the algorithm does not need to add additional design variable, it also known as Single Loop Single Vector (SLSV).

The principle of SLRBDO algorithm is to substitute the gradient vector of the previous iteration into the current iteration, and obtain the approximate value of MPP point. Finally, the coordinates and gradients of MPP point will converge to that of the true MPP point, as shown below:

$$y^k = \pm\beta^t \frac{\nabla_y g\big|_{y^{k-1}}}{\left\|\nabla_y g\big|_{y^{k-1}}\right\|} \tag{5.14}$$

In above formula, y^k means the coordinate vector of MPP point in the k-th iteration, and $\nabla_y g|y^{k-1}$ is the gradient vector obtained in the previous (K-1st) iteration [generally, the forward difference algorithm is used to obtain the approximate value of gradient vector, so the state function n + 1 (n is the number of uncertain variables) shall be called at least after each iteration].

Substitute formulas (5.9) and (5.10) into formula (5.14), we can obtain:

$$\frac{x^k - \mu_x^k}{\sigma_x} = \pm\beta^t \frac{\nabla_x h \cdot \nabla_y x\big|_{y^{k-1}}}{\left\| \nabla_x h \cdot \nabla_y x\big|_{y^{k-1}} \right\|} = \pm\beta^t \frac{\sigma_x \cdot \nabla_x h|_{x^{k-1}}}{\|\sigma_x \cdot \nabla_x h|_{x^{k-1}}\|} \tag{5.15}$$

When SLRBDO algorithm is used for RBDO problems with multiple probability constraints, the iteration of formula (5.15) is required for each probability constraint. At this time, the gradient vectors of all probability constraints are written in a gradient matrix.

The SLSV algorithm in the literature (Chen et al. 1997) has been included in the Multi-disciplinary Design Optimization platform iSIGHT. In this book, the algorithm has been improved and modified to solve more complex RBDO problems and improve the stability of the algorithm.

For the first iteration, in the literature (Chen et al. 1997), an x^0 is assumed firstly, and then the gradient matrix of the probability constraint function at x^0 (the line number of matrix = the number of probability constraint functions m, the column number of the matrix = the number of random variables n) is obtained, next, the gradient matrix along each column is summed to obtain a 1xn vector, and μ_x^0 can be inversely calculated by substituting formula (5.15), as the initial point of the optimization algorithm to start optimization. It can be seen that the initial point entered by the user is actually used as x^0, and the real optimum solution starting point is the transformed μ_x^0.

In the derivation and preparation of SLRBDO algorithm program in this book, it is found that this kind of complex processing method is not needed. In the first iteration, the μ_x^0 and $\nabla_x h|_{\mu_x^0}$ is directly taken as the initial point of optimization iteration and the initial gradient matrix of formula (5.15), so that the initial point of optimization entered by the user is the real initial point of optimization algorithm. When the user provides a better initial point, the initial point of the optimization solution will not deviate due to the transformation of the literature (Chen et al. 1997), which can quickly converge to the optimum point, and the algorithm is more stable.

Compared with the original SLSV algorithm, the SLRBDO algorithm written in this book also has made the following two improvements:

(1) The SLSV algorithm in iSIGHT only supports the conventional numerical optimization algorithm. The SLRBDO in this book provides five sorts of optimization algorithms: 0 means that the conventional numerical optimization algorithm SQP is used first, if no optimization solution is found, then compare the SQP algorithm to make the constraint function cross the border of the minimum point. If the cross-border value of the constraint function of this point is smaller than

that of the initial point, then take this point as the new initial point, otherwise, the initial point is unchanged. The process is called determining whether to update the optimized initial point and then calling the Pattern Search (PS) algorithm without search prior to the polling. If it does not converge, it will decide whether to update the initial point of optimization. Finally, we call the Latin Hypercube Search (LHS) to perform the PS algorithm with search prior to the polling, so the 0 is the hybrid optimization algorithm; 1 means that the conventional numerical optimization algorithm is used to get solution. The internal algorithm of the numerical algorithm first uses the Interior Point method. If it does not converge, it judges and decides whether to update the initial point, then uses the SQP algorithm. If it does not converge, it judges and decides whether to update the initial point, and finally uses the Active Set method to search; 2 means that the PS algorithm without search prior to the polling is used; 3 means that the Latin Hypercube Search is used as PS algorithm with search prior to the polling (LHSPS); 4 means that the Genetic Algorithm (GA) is used first to find out the better points (the points with small objective function value and small boundary value of constraint function), and then the interior point method (to find the better points) is used to solve the problem. It can be seen that SLRBDO algorithm's single loop features not only reduce the calculated amount of the Double Loop, but transfer the dual tasks of searching the optimal mean point and solving the nonlinear formula to the same optimization algorithm. The improvement of solving complexity will put forward higher requirements for the solving ability of the optimization algorithm. In many cases, the conventional numerical algorithm cannot get the convergence solution. In this book, examples like this can be seen as well. The SLRBDO algorithm in this book combines the optimization algorithm with state of the art into a hybrid optimization algorithm and then introduces SLRBDO, which can broaden the application scope of SLRBDO algorithm.

(2) For general engineering problems, the target reliability meets the requirement of $R^t > 0.5$. Before this iteration, first check whether the value of the function at the mean point meets the $h(d, \mu_x^k) \geq 0$ (that is, check whether the mean point generated by the optimization algorithm falls in the safety domain, if so, the reliability is generally impossible to be greater than 0.5), if not, do not update the gradient matrix and MPP point, so that every iteration will add a call to the state function, but it can directly skip the points generated by the optimization solver that obviously do not meet the reliability requirements, avoiding the calculation of the gradient matrix with a large amount of calculation at these points. In many cases, this will reduce the total calculated amount.

Since SLRBDO algorithm is derived from normal distribution and only normal distribution random variables are supported at present, so when it is applied to RBD0 problem with non-normal distribution random variables, the random variables should be replaced by approximate normal variables first. Moreover, since the optimization and solution of nonlinear equation system in the algorithm are completed by the same optimization algorithm, SLRBDO algorithm does not support the problem that auxiliary variables (non-design variables) contain random variables (for example, the material strength in common structural design is also random variables, but once the materials are selected in the design of structures, the strength is considered to be non-designed. For such examples, see example 2 in Sect. 6 of this chapter). At this time, the special processing is needed to substitute the auxiliary random variable as the upper limit (equal to the lower limit of the mean and the design variable of the mean of the original auxiliary random variable) of the mean value into the SLRBDO algorithm.

5.1.5 Design Optimization Example Based on Reliability

To compare the characteristics of reliability design optimization methods, this book uses Matlab to write these RBD0 algorithm programs, and carries out reliability design for three mathematical and engineering examples in the literature.

5.1.5.1 Example 1

This example is from a mathematical example in the literature (Yang and Gu 2004). Because there are only two random variables in this problem, it is convenient to draw and analyze in 2D graphics, and the state functions of this problem are all nonlinear functions, which have enough complexity and are suitable for studying and comparing RBDO algorithm. The RBDO problem of this example is as follows:

$$\begin{aligned}
&\min f = \mu_{X_1} + \mu_{X_2} \\
&s.t.\ P(g_i(X_1, X_2) \geq 0) \geq 0.9987,\ i = 1, 2, 3 \\
&\quad g_1(X_1, X_2) = \frac{X_1^2 X_2}{20} - 1 \\
&\quad g_2(X_1, X_2) = \frac{(X_1+X_2-5)^2}{30} + \frac{(X_1-X_2-12)^2}{120} - 1 \\
&\quad g_3(X_1, X_2) = \frac{80}{X_1^2+8X_2+5} - 1 \\
&\quad X_j \sim N(\mu_{X_j}, 0.3),\ j = 1, 2 \\
&\quad -10 \leq \mu_{X_j} \leq 10,\ j = 1, 2
\end{aligned} \tag{5.16}$$

The safety zone of the problem consists of two disconnected regions. As shown in Fig. 5.5, the local optimum point lies in the safety zone on the right (region of $X_1 > 0$), and the global optimum point lies in the safety zone on the left (region of $X_1 < 0$). It can be seen that the problem is of high complexity, which cannot only test the global solution ability of the optimization algorithm, but test the stability of the reliability analysis algorithm.

See Table 5.1 for the solution of this problem in literature (Li 2008), which points out that the current RBDO algorithm can only find the locally optimal solution close to the initial point in this example, but not the globally optimal solution.

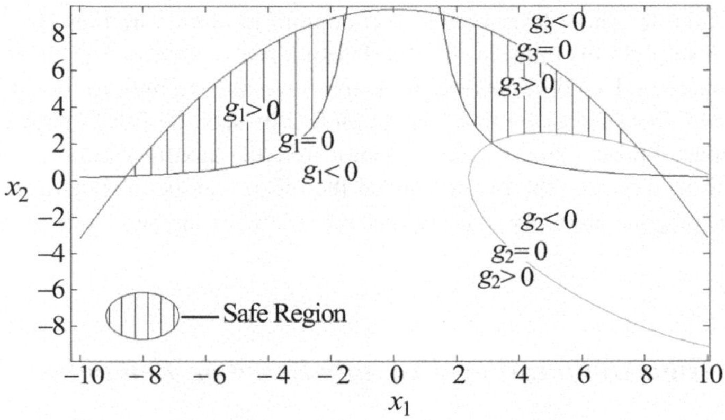

Fig. 5.5 Safety zone of RBDO example 1

Table 5.1 Results of RBDO example 1 literature

Optimization initial point	Optimization point	Optimization objective function value	Description
[4, 5]	[3.4409, 3.2909]	6.732	Locally optimum solution
[–4, 5]	[−7.031, 1.291]	−5.740	Globally optimum solution

(1) Double Loop result

First, the traditional double loop RBDO algorithm is used to solve the problem, and two different starting points are taken respectively to start optimization. Firstly, SQP optimization algorithm is used to solve the problem, PS algorithm is used to solve the problem if it is not convergent, and exit if it is not convergent either. See Table 5.2 for calculation results and time consumption of different reliability analysis algorithms.

It can be seen that the conventional numerical optimization algorithm SQP can only converge to the local optimum point close to the initial point, which is consistent with the conclusion in literature (Li 2008). In this book, however, the hybrid optimization algorithm with SQP + PS is adopted, which can converge to the local optimum point when the reliability analysis algorithm is stable.

It can also be seen that whether the Double Loop converges or not is closely related to the selected reliability analysis algorithm. The CMC method with enough sampling times is the most stable reliability analysis algorithm, followed by JC method. For the importance sampling algorithm includes JC method and importance sampling process, its stability is further reduced, and RIA algorithm that directly uses SQP to solve the formula (4.14) has the worst stability. In addition, some literatures have pointed out that PMA method is more stable than RIA algorithm (Lee et al. 2002;

Table 5.2 Calculation results of double loop in RBDO example 1

Reliability analysis algorithm	Optimization initial point	Optimization point	Optimization objective function value	Elapsed time (s)	Description
Results of literature	[4, 5]	[3.4409, 3.2909]	6.732	–	Locally optimum solution
	[−4, 5]	[−7.031, 1.291]	−5.740	–	Globally optimum solution
JC	[4, 5]	[3.4409, 3.2909]	6.7318	2.8[a]	SQP convergence
	[−4, 5]	[3.5156,3.3182]	6.8338	438.9	SQP misconvergence, PS convergence
TJT A	[4, 5]	–	–		Misconvergence
RIA	[−4, 5]	–	–		Misconvergence
CMC (76923 times of sampling)	[4, 5]	[−7.0287, 1.2964]	−5.7323	2549.2	SQP misconvergence, PS convergence
	[−4, 5]	[−7.0287, 1.2964]	−5.7323	2552.0	SQP misconvergence, PS convergence
IS (1000 times of sampling)	[4, 5]	[3.4641, 3.2607]	6.7248	8.3*	SQP convergence
	[−4, 5]	–	–		Misconvergence

Note [a]The algorithm runs 3 times, and the value is taken after the calculation time is stable

Choi and Youn2002), but the calculation in this book shows that PMA method which directly uses SQP solution has not got the optimum solution in this case.

It can also be seen that the double loop RBDO algorithm, especially the Double Loop using CMC method as the reliability analysis algorithm, needs a lot of calculation. For this numerical example with a small calculated amount, the Double Loop using CMC method as the reliability analysis algorithm also needs to run for more than 40 min to find the solution. It can be seen that when the double loop RBDO algorithm is applied to the reliability design of actual projects, the calculated amount will become a serious problem.

(2) SFSORA algorithm results

It can be seen from formula (5.16) that the probability constraint function form of the original problem does not meet the requirements of SFSORA algorithm. However, this problem can be rewritten into the form required by SFSORA algorithm according to the method discussed in Sect. 5.1.3 of this chapter. In this example, formula (5.16) can be expressed as the following deterministic optimization problem with safety factor [it can be known from the discussion in Sect. 5.1.3 of this chapter that when constant C is taken as different value, the specific value in the expression transformed from formula (5.16) will be different, but the form will not change]:

$$\begin{aligned}
&\min f = \mu_{X_1} + \mu_{X_2} \\
&s.t.\ h_i(\mu_{X_1}, \mu_{X_2}) \le 0,\ i = 1, 2, 3 \\
&where \\
&h_1 = \left(-\frac{\mu_{X_1}^2 \mu_{X_2}}{20} + 11\right) - \frac{10}{sf_1} \\
&h_2 = \left(-\frac{(\mu_{X_1} + \mu_{X_2} - 5)^2}{30} - \frac{(\mu_{X_1} - \mu_{X_2} - 12)^2}{120} + 2\right) - \frac{1}{sf_2} \\
&h_3 = \left(-\frac{80}{\mu_{X_1}^2 + 8\mu_{X_2} + 5} + 2\right) - \frac{1}{sf_3} \\
&-10 \le \mu_{X_j} \le 10,\ j = 1, 2
\end{aligned} \tag{5.17}$$

To simulate the most convenient use of SFSORA algorithm in engineering, this book does not write an independent SFSORA program, and uses the method of manually adjusting the safety factor to calculate this example:

① In the first iteration, $sf_i = 1, \quad (i = 1, 2, 3)$ is substituted, and SQP, PS or Particle Swarm Optimization (PSO) algorithm is used to solve the optimization problem of formula (5.17). When the optimization starts from [4, 5], the SQP optimization algorithm finds the locally optimal solution, while the PS and PSO algorithms can search for the global optimum point; when the optimization starts from [–4, 5], the three optimization algorithms can find the global optimum point, as shown in Table 5.3.

Take the global optimum point to get the optimal mean point: $Xopt = [-8.5324, 0.2747]$, which is the solution of the deterministic optimization problem when the original problem does not consider the randomness of variables. See Fig. 5.6 for "the optimum point of deterministic optimization". Use the CMC method with 10^5 times of sampling to analyze the reliability of the three state functions at this point, and get the reliability $R = [0.4982, 1, 0.5005]$

Table 5.3 SFSORA optimization results of the first iteration in RBDO example 1

Optimization initial point	Optimization point	Optimization function value	Description
[4, 5]	[3.1139, 2.0626]	5.1765	SQP
	[–8.5324, 0.2747]	−8.2577	PS, PSO
[–4, 5]	[–8.5324, 0.2747]	−8.2577	SQP, PS, PSO

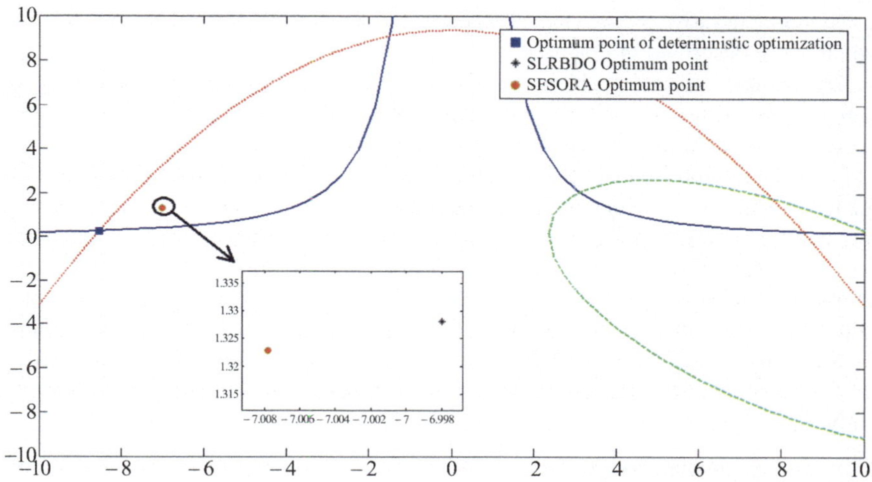

Fig. 5.6 Optimization points of different algorithms in RBDO example 1

确定性优化最优点 最优点	Optimum point of deterministic optimization Optimum point

It can be seen that the reliability of g2 is 1, that is to say, taking 1 as the safety factor can meet the requirements; while the reliability of g1 and g2 is very low. According to SFSORA algorithm, the safety factors of g1 and g3 are increased to 1.1.

② sf == [1.1, 1, 1.1] generation (5.17) is substituted into formula (5.17) and optimized in the second round to get Xopt = [–7.9703,0.6011]. The same reliability analysis method as the previous step is used to get the reliability of the state function at this time: R = [0.8283, 1, 0.8890]. It can be seen that the reliability of g1 and g3 cannot reach the target reliability (0.9987).

③ Continue to increase the safety factor to sf = [1.2, 1, 1.2], and get the optimization point = [–7.4796, 0.9533], and correspondingly get R = [0.9760, 1, 0.9850]; the reliability of g1 and g3 cannot reach the target reliability.

④ Continue to increase the safety factor to SF = [1.3, 1, 1.3], get the optimization point = [–7.0182, 1.3431], and correspondingly get R = [0.9992, 1, 0.9985]. It can be seen that R_1 > 0.9987 but R R_3 < 0.9987, and the gap between it and the target reliability is not large, the safety factor of g1 needs to be slightly reduced, while the safety factor of g3 needs to be slightly increased.

⑤ Adjust the safety factor to sf = [1.29, 1, 1.31], get the optimization point = [–7.0078, 1.3228], and correspondingly get the reliability R = [0.9989, 1, 0.9989]. It can be seen that the difference between R_1, R_3 and the target reliability is only 0.0002, so it can be considered that the solution with enough accuracy has been obtained. At this time, the target function value is –5.6850.

The same iterative process can also be used to determine the local optimum point, which is not described in this book.

Because the original problem of this example does not conform to the form of SFSORA algorithm, the sf here is not a real physical safety factor, and can only be considered as a parameter that moves the boundary of constraint function.

From the demonstration of this example, we can see that the SFSORA algorithm is a RBDO algorithm with strong practicability. Because of the obvious relationship between the safety factor and the reliability in the SFSORA algorithm, under the existing deterministic optimization model and the reliability analysis model, the reliability based design optimization can be carried out by manual safety factor adjustment. Certainly, in this example, the steps to adjust the safety factor in steps ①–⑤ can also be accomplished by simple numerical optimization algorithm, which only need a function that encapsulates deterministic optimization and reliability analysis.

(2) Single algorithm results

As described in Sect. 5.1.4, the SLRBDO algorithm program in this book provides five optimization options, which will be solved by these optimization algorithms respectively. See Table 5.4 for the comparison between the results and the literature.

It can be seen that no matter the starting point of optimization is in the left safety domain or the right safety domain, PS optimization algorithm, LHSPS optimization algorithm and genetic algorithm can find the global optimum point, but the traditional SQP cannot get the convergent solution. In this book, a variety of new optimization algorithms are introduced to replace the traditional single numerical optimization algorithm in the single loop RBDO algorithm. The use of these algorithms and their combinations greatly improves the possibility of solving the global optimum point of RBDO problem.

After comparing optimization solution obtained from three algorithms with optimum solution in the literature, the results show that the DLRBDO objective function is $\frac{5.740-5.7323}{5.740} = 0.13\%$ larger than the globally optimal solution in the literature, the SFSORA optimization objective function value is 0.96% larger than the literature solution, and the SLRBDO optimization objective function value is 1.22% larger than the solution in the literature, which shows that their errors are all very small. In addition, the CMC method with 10^5 times of sampling and the optimization scheme obtained by three algorithms are used to analyze the reliability of literature, thus, the reliability is obtained: $R_{\text{literature}} = [0.9984, 1, 0.9988]$, $R_{\text{DLRBDO}} = [0.9985, 1, 0.9988]$, $R_{\text{SFSORA}} = [0.9989, 1, 0.9989]$, and $R_{\text{SLRBDO}} = [0.9990, 1, 0.9989]$. It can be seen that the reliability R_1 at the optimization point in the literature is lower than the required reliability (0.9987), and the reliability of the optimization scheme obtained by the algorithm in this book is higher than the required reliability, so the objective function at the optimization point in this book is slightly larger than that in the literature, but it can be seen that the error is small and can be ignored.

Compare the optimization points calculated by SFSORA and SLRBDO algorithm in this example (Fig. 5.6). It can be seen that in SF-SORA algorithm, the optimization

Table 5.4 Calculation results of SLRBDO algorithm in RBDO example 1

RBDO algorithm	Optimization algorithm options	Optimization initial point	Optimization point	Optimization objective function value	Elapsed time (s)	Description
Results of literature	–	[4, 5]	[3.4409, 3.2909]	6.732	–	Local optimum point
		[−4, 5]	[−7.031, 1.291]	−5.740	–	Global optimum point
SLRBDO	0	[4, 5]	[−6.9980, 1.3281]	−5.6699	0.58[a]	SQP misconvergence, PS convergence
		[−4, 5]	[−6.9980, 1.3281]	−5.6699	0.54[a]	SQP misconvergence, PS convergence
	1	[4, 5]	–	–	–	Misconvergence
		[−4, 5]	–	–	–	Misconvergence
	2	[4, 5]	[−6.9980, 1.3281]	−5.6699	0.51[a]	
		[−4, 5]	[−6.9980, 1.3281]	−5.6699	0.50[a]	
	3	[4, 5]	[−6.9976, 1.3213]	−5.6763	0.70[a]	
		[−4, 5]	[−7.0120, 1.3266]	−5.6854	0.67[a]	
	4	[4, 5]	[−6.9101, 1.3734]	−5.5367	79[a]	
		[−4, 5]	[−6.9101, 1.3734]	−5.5367	79[a]	

Note [a]The algorithm runs six times continuously and takes the average value of the last three times

solution with small error can be obtained by manual adjustment. For the problem of normal distribution of random variables, SLRBDO algorithm uses the solution of nonlinear formulas instead of reliability analysis, which can reduce the calculated amount. The calculated amount of SFSORA algorithm mainly comes from the deterministic optimization and reliability analysis, so the optimization algorithm with faster convergence and importance sampling algorithm can be used for reliability analysis to reduce the calculated amount. However, it should be noted that the importance sampling algorithm may instead obtain an inauthentic solution with a reliability of approximately 1 when the reliability is very low (close to 0), as shown in example 2 below.

5.1.5.2 Example 2

This example is a structural design problem of a cantilever beam with rectangular section subjected to vertical and lateral random loads, as shown in Fig. 5.7. In the figure, t and w are the height and width of beam section respectively, and Y and Z are vertical and lateral loads, respectively.

The cantilever beam structure is designed to minimize the beam weight and the weight of the beam $= L \times t \times w = 100 \times t \times w$, so the objective function is as follows:

$$Fobj = tw \tag{5.18}$$

There are two constraints when the cantilever beam is stressed:

① The stress constraint:

$$\sigma = \frac{600Y}{wt^2} + \frac{600Z}{w^2t} \le \sigma_Y \tag{5.19}$$

In above formula, σ_Y is the yield strength of the material.

② The displacement constraint:

$$D = \frac{4L^3}{Ewt}\sqrt{\left(\frac{Y}{t^2}\right)^2 + \left(\frac{Z}{w^2}\right)^2} \le D0 \tag{5.20}$$

In above formula, D0 is the allowable displacement, in this case, the fixed value D0 = 2.5; E is the Young modulus of the material.

When the load and material parameters are random variables, the structural design problem becomes RBDO problem:

$$\begin{aligned}
&\min f = wt \\
&s.t.\ P(g_i(w, t, Y, Z, \sigma_Y, E) \le 0) \ge 0.9987,\ i = 1, 2 \\
&\quad g_1(w, t, Y, Z, \sigma_Y) = \frac{600Y/(wt^2) + 600Z/(w^2t)}{\sigma_Y} - 1 \\
&\quad g_2(w, t, Y, Z, E) = \frac{4L^3}{Ewt}\sqrt{\left(\frac{Y}{t^2}\right)^2 + \left(\frac{Z}{w^2}\right)^2} - 2.5 \\
&\quad Y \sim N(1000, 100),\ Z \sim N(500, 100) \\
&\quad \sigma_Y \sim N(40000, 2000),\ E \sim N(29 \times 10^6, 1.45 \times 10^6) \\
&\quad 1 \le w, t \le 5
\end{aligned} \tag{5.21}$$

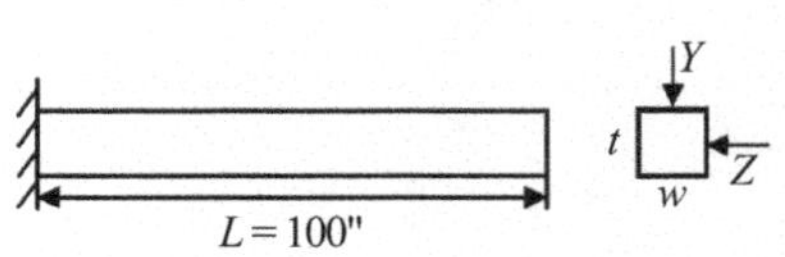

Fig. 5.7 Example 2 cantilever beam

It can be seen that the design variables (w, t) in this problem are determining variables and the auxiliary variables Y, Z, σ_Y, E are random variables, which is the feature of most structural reliability design. Next, we use Double Loop RBDO, SFSORA and Single Loop RBDO algorithm to solve them respectively. (1) The result of Double Loop RBDO algorithm.

The double loop RBDO algorithm written in this book can handle all situations:

① Only design variables contain random variables;
② Only auxiliary variables contain random variables;
③ Both design variables and auxiliary variables include random variables.

This problem belongs to category ②, and the calculation results are shown in Table 5.5.

It can be seen that for the structural design problem, when the safety domain is not as complex as the example 1 mentioned above, JC and RIA algorithms can also be used as the reliability analysis algorithm to get the convergence solution, but in this example, IS algorithm is used as the reliability analysis algorithm to get the untrue solution of w = t = 1, at this time, the real reliability R = [0, 0], while IS algorithm gets the opposite result of reliability R = [1, 1]. The IS algorithm is a reliability analysis algorithm which is derived for the case of high reliability. When the reliability is very low (close to 0), the calculation error is very large, even the result is unacceptable. In the double-layer RBDO algorithm, the mean point is generated by the optimization algorithm, which inevitably results in the mean point with low reliability, where IS algorithm will not be trusted, limiting the application of IS algorithm in RBDO.

Table 5.5 Calculation results of DLRBDO algorithm in RBDO example 2

Reliability analysis algorithm	Optimization initial point	Optimization point	Optimization objective function value	Elapsed time (s)	Description
JC	[3, 4]	[2.4961, 3.8120]	9.5149	17.8[a]	SQP misconvergence, PS convergence
RIA	[3, 4]	[2.4961, 3.8120]	9.5149	267.8[a]	SQP misconvergence, PS convergence
CMC (76923 times of sampling)	[3, 4]	[2.4985, 3.8112]	9.5223	1151.7	SQP misconvergence, PS convergence
IS (1000 times of sampling)	[3, 4]	[1, 1]	1	5.9	SQP convergence, but unreal solution is obtained

Note [a]The algorithm runs 3 times, and takes the value after the calculation time is stable

(2) SFSORA algorithm result

The SFSORA algorithm is inspired by the safety factor method of structural design, so the form of the problem directly meets the requirements of SFSORA algorithm. Combined with the characteristics that only auxiliary variables contain random variables, the optimization solution can be directly obtained by substituting the mean value of safety factor and auxiliary variables. The expression of the deterministic optimization problem with safety factor in SFSORA algorithm is as follows:

$$\begin{aligned} & \min f = wt \\ & s.t.\ P(g_i(w,t) \le 0) \ge 0.9987,\ i = 1,2 \\ & \quad g_1(w,t) = \frac{600\mu_Y/(wt^2)+600\mu_Z/(w^2t)}{\mu_{\sigma_Y}} - \frac{1}{sf_1} \\ & \quad g_2(w,t) = \frac{4L^3}{\mu_E wt}\sqrt{\left(\frac{\mu_Y}{t^2}\right)^2 + \left(\frac{\mu_Z}{w^2}\right)^2} - \frac{2.5}{sf_2} \\ & \quad \mu_Y = 1000,\ \mu_Z = 500 \\ & \quad \mu_{\sigma_Y} = 40000,\ \mu_E = 29 \times 10^6 \\ & \quad 1 \le w,t \le 5 \end{aligned} \tag{5.22}$$

The SFSORA iteration process of this problem is as follows:

① Take sf = [1, 1], and use SQP to solve the optimization problem formula (5.22) with safety factor, and then get the optimization point (w, t) = [2.0470, 3.7460]; use CMC method with10^5 times of sampling to get the reliability R = [0.4994, 0.4932], or use IS algorithm with 1000 times of sampling to get the reliability R = [0.3828, 0.4078].

② Increase the safety factor to sf = [1.1, 1], get (w, t) = [2.0207, 4.0411] as above, get reliability R = [0.7878, 0.6787] by 10^5 sampling CMC method, and get reliability R = [0.7228, 0.6275] by IS algorithm with 1000 times of sampling, and the error of two reliability analysis algorithms is about 8% when R is not high. Compared with ① and ②, R1 and R2 increased with the increase of SF.

③ Take sf = [1.1, 1.1] to get (w, t) = [2.0510, 3.9817], use the CMC method with 10^5 times of sampling to get the reliability R = [0.7893,0.7193], and use the IS algorithm with 1000 times of sampling to get the reliability R = [0.7248,0.6752]; R1 increases slightly with the increase of SF2, while R2 increases significantly with the increase of SF2.

④ Take sf = [1.2, 1.1] to get (w, t) = [2.0801, 4.1601], use CMC method with 10^5 times of sampling to get reliability R = [0.9427, 0.8880], and use IS algorithm with 1000 times of sampling to get reliability R = [0.9163, 0.8794]; R1 and R2 continue to increase with the increase of SF1.

⑤ Take sf = [1.2, 1.2] to get (w, t) = [2.0816, 4.1572], use CMC method with 10^5 times of sampling to get reliability R = [0.9428, 0.8892], and use IS algorithm with 1000 times of sampling to get reliability R = [0.9164, 0.8809]; R1 and R2 slightly increase with the increase of SF 2.

⑥ Take sf = [1.3, 1.2] to get (w, t) = [2.1363, 4.2726], use CMC method with 10^5 times of sampling to get reliability R = [0.9903, 0.9755], and use IS algorithm with 1000 times of sampling to get reliability R = [0.9843, 0.9751]; R1 and R2 continue to increase with the increase of sf1.

⑦ Take sf = [1.3, 1.3] to get (w, t) = [2.1364, 4.2726], use CMC method with 10^5 times of sampling to get reliability R = [0.9903, 0.9755], and use IS algorithm with 1000 times of sampling to get reliability R = [0.9843, 0.9751]; R1 and R2 continue to increase with the increase of sf2.

⑧ Take sf = [1.4, 1.3] to get (w, t) = [2.1898, 4.3794], use CMC method with 10^5 times of sampling to get reliability R = [0.9988, 0.9966], and use IS algorithm with 1000 times of sampling to get reliability R = [0.9981, 0.9965], it can be seen that in the case of high reliability 10^5, the errors of CMC method and IS algorithm can be ignored basically, but the difference of calculated amount is $\frac{10^5}{1000} = 100$ times; in the case of high reliability, use IS algorithm can reduce lots of calculated amount. Compared with ⑦ and ⑧, it can be seen that R1 and R2 continue to increase with the increase of sf1.

⑨ Take sf = [1.5, 1.3] to get (w, t) = [2.2407, 4.4814], use CMC method with 10^5 times of sampling to get reliability R = [0.9999, 0.9996], and use IS algorithm with 1000 times of sampling to get reliability R = [0.9998, 0.9997]; R1 and R2 are both larger than the target reliability.

⑩ Take SF = [1.45, 1.3] to get (w, t) = [2.2155, 4.4310], use CMC method with 10^5 times of sampling to get reliability R = [0.9996, 0.9990], and use IS algorithm with 1000 times of sampling to get reliability R = [0.9994, 0.9989].

⑪ Take sf = [1.44, 1.3] to get (w, t) = [2.2104, 4.4208], use CMC method with 10^5 times of sampling to get reliability R = [0.9995, 0.9986], and use IS algorithm with 1000 times of sampling to get reliability R = [0.9993, 0.9987], which is considered as convergence, at this time, the target function value is 9.7719.

It can be seen that when the safety factor sf2 of the displacement constraint is more than 1.3, the reliability of the system no longer increases with the increase of sf2. In the next step, the approximate optimum solution can be found only by increasing sf1.

As the original form of the problem can meet the form requirements of SFSORA, sf here is a safety factor with practical physical significance. From the optimization results, we can draw a conclusion: When using the safety factor method in the traditional code to design the structure, if the cantilever beam has a reliability of more than 99.87% under random load, the safety factor of more than 1.44 must be taken as the stress constraint condition, but it is enough for safety factor of displacement constraint to take 1.3. It can be seen that SFSORA is an excellent algorithm for structural reliability design, and the calculation results can reflect the reliability and safety factor of structural design scheme at the same time, which is often what the designer needs and also what other RBDO algorithms cannot provide.

(3) Single loop RBDO algorithm result

In this example, the design variable is a determinate variable, and the auxiliary variable contains random variables, which is not supported by SLRBDO algorithm by default. The random variables in the auxiliary variable need to be written as a design with equal upper and lower limits of the mean, as shown in formula (5.23).

$$
\begin{aligned}
& \min f = wt + 0 \times \mu_Y + 0 \times \mu_Z + 0 \times \mu_{\sigma_Y} + 0 \times \mu_E \\
& s.t.\ P(g_i(X_1, X_2) \le 0) \ge 0.9987,\ i = 1, 2 \\
& \quad g_1(X_1, X_2) = \frac{600Y/(wt^2) + 600Z/(w^2t)}{\sigma_Y} - 1 \\
& \quad g_2(X_1, X_2) = \frac{4L^3}{Ewt}\sqrt{\left(\frac{Y}{t^2}\right)^2 + \left(\frac{Z}{w^2}\right)^2} - 2.5 \\
& \quad Y \sim N(1000, 100),\ Z \sim N(500, 100) \\
& \quad \sigma_Y \sim N(40000, 2000),\ E \sim N(29 \times 10^6, 1.45 \times 10^6) \\
& \quad 1 \le w, t \le 5 \\
& \quad 1000 \le \mu_Y \le 1000 \\
& \quad 500 \le \mu_Z \le 500 \\
& \quad 40000 \le \mu_{\sigma_Y} \le 40000 \\
& \quad 29 \times 10^6 \le \mu_E \le 29 \times 10^6
\end{aligned}
\tag{5.23}
$$

It can be seen that the number of design variables increases from 2 ([w, t]) of the original problem to 6: $[w, t, \mu_Y, \mu_Z, \mu_{\sigma_Y}, \mu_E]$, and the upper and lower limits of the last 4 design variables are the same.

After the above processing, SLRBDO algorithm can be used for solution, and five optimization algorithms are used for solution respectively. See Table 5.6 for the results.

Compared with the optimization objective function obtained by the three algorithms, the minimum objective function value obtained by DLRBDO algorithm is 9.5149; the objective function obtained by SFSORA algorithm is 9.7719, and the

Table 5.6 Calculation result of SLRBDO algorithm in RBDO example 2

Optimization initial point	Optimization algorithm options	Optimization point [w,t] value	Optimization objective function value	Elapsed time (s)	Description
	0	[2.6809, 3.5662]	9.5607	0.3[a]	SQP misconvergence, Ps misconvergence, LHSPS convergence
[3, 4]	1	–	–	–	Misconvergence
	2	[2.5000, 3.8125]	9.5313	0.15[a]	
	3	[2.6809, 3.5662]	9.5607	0.18[a]	
	4	[2.4804, 3.8446]	9.5362	6.3[a]	

Note [a]The algorithm runs six times continuously and takes the average value of the last three times

relative error between this value and DLRBDO algorithm is $\frac{9.7719-9.5149}{9.5149} = 2.7\%$; the minimum objective function value obtained by SLRBDO algorithm is 9.5313, and the relative error with DLRBDO algorithm is 0.17%. It can be seen that SFSORA and SLRBDO algorithm can get the optimization solution with very small error by much less calculated amount than DLRBDO algorithm or even by manual optimization.

5.1.5.3 Example 3

This example comes from iSIGHT (Engineous Software, iSIGHT9.0 document), which is the design of a tension spring. The objective function is the lightest weight; there are four constraints: Minimum deformation, shear stress, oscillation frequency and spring outer diameter; the design variables are three aspects: Spring material diameter D, spring mid diameter D, and effective number of turns n, so this problem can be described as follows:

$$\begin{aligned}
&\min f = (\mu_n + 2)\mu_D \mu_d^2 \\
&s.t.\ P(g_i(d, D, n) \leq 0) \geq 0.95,\ i = 1, 2, 3, 4 \\
&\quad g_1(d, D, n) = \frac{d^4}{nD^3} - 1/71875 \\
&\quad g_2(d, D, n) = \frac{4D^2 - dD}{12566(Dd^3 - d^4)} + \frac{1}{5108d^2} - 1 \\
&\quad g_3(d, D, n) = \frac{nD^2}{d} - 140.45 \\
&\quad g_4(d, D, n) = D + d - 1.5 \\
&\quad d \sim N(\mu_d, 0.005\mu_d) \\
&\quad D \sim N(\mu_D, 0.05\mu_D) \\
&\quad n \sim N(\mu_n, 0.1\mu_n) \\
&\quad 0.04999 \leq \mu_d \leq 0.1 \\
&\quad 0.09999 \leq \mu_D \leq 1 \\
&\quad 1 \leq \mu_n \leq 50
\end{aligned} \tag{5.24}$$

The solution of this problem has been given in iSIGHT, see Table 5.7. Since the statistical parameters of random variables known in this example are not the standard deviation but the variable coefficient; in this book, the input parameters of DLRBDO, SFSORA and SLRBDO algorithm programs are modified from the standard deviation

Table 5.7 Calculation result of iSIGHT algorithm in RBDO example 3

Optimization algorithm	Optimization model	Optimization initial point	Optimization point	Optimization objective function value
Feasible direction method corrected	Deterministic optimization	[0.05, 0.1, 1.0]	[0.0515, 0.351, 11.633]	0.0127
	RBDO	[0.05, 0.1, 1.0]	[0.0554, 0.417, 12.790]	0.0190

to the variable coefficient, and the program with the variable coefficient as the input variable is obtained. The modified program is used to calculate this example, and the results are compared with those of iSIGHT.

(1) Double Loop result

In this problem, only the design variables contain random variables. See Table 5.8 for the results calculated by the DLRBDO program with the input parameter as the variable coefficient.

It can be seen that Matlab's SQP and PS optimization algorithms are more sensitive to whether the initial point is close to the boundary of the design space. When the initial point is close to the boundary of the design space, it is difficult to find the optimization solution. Even if the optimization starts from the internal point of the design space, the double loop RBDO algorithm only uses JC method or CMC method as the reliability analysis algorithm to get the optimization solution. It can be seen that the double loop RBDO algorithm has high requirement for the optimization calculation, which proves that JC and CMC are still the most stable reliability analysis algorithms.

Table 5.8 Calculation result of DLRBDO algorithm in RBDO example 3

Reliability analysis algorithm	Optimization initial point	Optimization point	Optimization objective function value	Elapsed time (s)	Description
iSIGHT	[0.05, 0.1, 1.0]	[0.0554, 0.417, 12.790]	0.0190		Feasible direction method corrected
JC	[0.05, 0.1, 1.0]	–	–	–	Misconvergence
	[0.075, 0.55, 25.5]	[0.055, 0.4071, 13.3309]	0.0189	47.2	SQP convergence
RIA	[0.05, 0.1, 1.0]	–	–	–	Misconvergence
	[0.075, 0.55, 25.5]	–	–	–	Misconvergence
	[0.05, 0.1, 1.0]	–	–	–	Misconvergence
CMC (2000 times of sampling)	[0.075, 0.55, 25.5]	[0.0553, 0.3304, 25.4968]	0.0278	95.9	SQP convergence
IS (1000 times of sampling)	[0.05, 0.1, 1.0]	[0.05, 0.1, 1.0]	7.4963e-4	8.4	SQP convergence, but the unreal solution is obtained
	[0.075, 0.55, 25.5]	[0.05, 0.1, 1.0]	7.4963e-4	36.6	SQP convergence, but the unreal solution is obtained

By comparing the results of double loop RBDO algorithm and single loop SLSV algorithm of iSIGHT, it can be found that in this example, the approximate solution can be obtained by using JC method as the reliability analysis algorithm, and the solution obtained by CMC method deviates greatly. It can be seen that the traditional Double Loop is not only low in calculation efficiency, even if the calculation efficiency is not considered, the result obtained by using CMC method as the reliability analysis algorithm is not necessarily stable.

In addition, in this example, we can see that the importance sampling algorithm gets an untrue solution again, so it is necessary to demonstrate and confirm the calculation results when using IS algorithm in RBDO (by calculating the constraint function value of the mean point, we can judge whether the mean point falls in the reliability domain, so as to preliminarily judge whether the reliability calculation of IS algorithm is right or wrong).

(2) SFSORA algorithm result

It can be seen that to meet the requirements of SFSORA, the original problem formula (5.24) is directly substituted into the safety factor to obtain the optimization model with safety factor in SFSORA algorithm, as shown in formula (5.25).

$$\begin{aligned}
&\min f = (\mu_n + 2)\mu_D\mu_d^2 \\
&s.t.\ P(g_i(\mu_d, \mu_D, \mu_n) \le 0) \ge 0.95,\ i = 1, 2, 3, 4 \\
&\quad g_1(\mu_d, \mu_D, \mu_n) = \frac{\mu_d^4}{\mu_n\mu_D^3} - \frac{1/71875}{sf_1} \\
&\quad g_2(\mu_d, \mu_D, \mu_n) = \frac{4\mu_D^2 - \mu_d\mu_D}{12566(\mu_D\mu_d^3 - \mu_d^4)} + \frac{1}{5108\mu_d^2} - \frac{1}{sf_2} \\
&\quad g_3(\mu_d, \mu_D, \mu_n) = \frac{\mu_n\mu_D^2}{\mu_d} - \frac{140.45}{sf_3} \\
&\quad g_4(\mu_d, \mu_D, \mu_n) = \mu_D + \mu_d - \frac{1.5}{sf_4} \\
&\quad 0.04999 \le \mu_d \le 0.1 \\
&\quad 0.09999 \le \mu_D \le 1 \\
&\quad 1 \le \mu_n \le 50
\end{aligned} \tag{5.25}$$

The iteration process of SFSORA algorithm is as follows:

① sf = [1, 1, 1, 1], get the optimum point Xopt = [0.0517, 0.3569, 11.2935], where reliability R = [0.5060, 0.4845, 1, 1]; it can be seen that g3 and g4 do not work at the optimization point, and sf3 and sf4 take 1.

② First, adjust sfi, take sf = [1.1, 1, 1, 1], and get the optimum point Xopt = [0.0522, 0.3699, 11.6283], where reliability R = [0.6965,0.4630,1,1], it can be seen that sf2 also needs to be increased.

③ Take sf = [1.2, 1.1, 1,1] to get the optimum point Xopt = [0.0562,0.4250,11.2467], where the reliability R = [0.8330, 0.9840, 1, 1]; it can be seen that g2 is very sensitive to the increase of safety coefficient. When sf2 increases to 1.1, R2 has exceeded the target reliability 0.95, and sf2 needs to be reduced slightly.

Table 5.9 Comparison of calculation results between SFSORA algorithm and iSIGHT algorithm in RBDO example 3

RBDO algorithm	Optimization initial point	Optimization point	Optimization objective function value	Reliability R (2000 times of sampling CMC)
iSIGHT	[0.05, 0.1, 1.0]	[0.0554, 0.417, 12.790]	0.0190	[0.95, 0.9465, 1, 1]
SFSORA	[0.05, 0.1, 1.0]	[0.0557, 0.4224, 12.5819]	0.0191	[0.9505, 0.9555, 1, 1]

④ Take sf = [1.3, 1.05, 1, 1], and the optimum point Xopt = [0.0567, 0.4634, 9.7393], where the reliability R = [0.9190, 0.85, 1, 1].
⑤ Take sf = [1.4, 1.075,1,1], and get the optimum point Xopt = [0.0577,0.4757,10.3370], where the reliability R = [0.96,0.9555,1,1].
⑥ Take sf = [1.375,1.0725,1,1], and the optimum point Xopt = [0.0556,0.4222,12.5809], where reliability R = [0.9530,0.9465,1,1].
⑦ Take sf = [1.375,1.073,1,1], and the optimum point Xopt = [0.0557,0.4224,12.5819], where the reliability R = [0.9505,0.9555,1,1] and which is considered as convergence, at this time, the target function value is 0.0191.

Compared with the results of iSIGHT (Table 5.9), it can be seen that the target function value of SFSORA algorithm is increased by about 0.5% compared with that of iSIGHT, but the reliability of SFSORA algorithm is higher than that of iSIGHT at the optimum point.

(3) Single Loop result

All optimization algorithms are selected for solution, and the results obtained by SLRBDO algorithm are shown in Table 5.10. It can be seen that except that the PS optimization algorithm and the conventional numerical algorithm with the initial point on the boundary of the design space have not been optimized, other optimization algorithms all have been optimized.

It can be seen that when SLRBDO algorithm is used in this example, the minimum error between the optimal value of objective function and the optimum solution of iSIGHT is 0 (PS optimization algorithm is used), but when genetic algorithm is used, the error reaches $\frac{0.0306-0.019}{0.019} = 61\%$. Combining the results of the previous examples 1 and 2, it is found that when the genetic algorithm is used in SLRBDO algorithm, the calculation results are greatly affected by the precocity of GA algorithm.

Comparing the results of DLRBDO, SFSORA and SLRBDO in this example, it can be seen that DLRBDO algorithm has a large amount of calculated amount and is difficult to get convergence results, while SFSORA and SLRBDO algorithm in this book can get optimization solutions with a small calculated amount. In addition, in use, it is better not to take the initial point of optimization at the edge of the design space.

Table 5.10 Calculation result of SLRBDO algorithm in RBDO example 3

RBDO algorithm	Optimization algorithm options	Optimization initial point	Optimization point	Optimization objective function value	Elapsed time (second)	Description
iSIGHT	–	[0.05, 0.1, 1.0]	[0.0554, 0.417, 12.790]	0.0190	–	Feasible direction method corrected
	0	[0.05, 0.1, 1.0]	[0.0506, 0.3025, 24.6308]	0.0206	1.5[a]	SQP misconvergence, PS misconvergence, LHSPS convergence
		[0.075, 0 55, 25.5]	[0.0500, 0.2844, 26.5921]	0.0203	0.81[a]	SQP convergence
	1	[0.05, 0.1, 1.0]	–	–	–	Misconvergence
		[0.075, 0.55, 25.5]	–	–	–	Misconvergence
SLRBDO	2	[0.05, 0.1, 1.0]	–	–	–	Misconvergence
		[0.075, 0.55, 25.5]	[0.0541, 0.3828, 14.9531]	0.0190	1.5[a]	
	3	[0.05, 0.1, 1.0]	[0.0506, 0.3025, 24.6308]	0.0206	1.4[a]	
		[0.075, 0.55, 25.5]	[0.0581, 0.4863, 9.7513]	0.0193	1.4[a]	
	4	[0.05, 0.1, 1.0]	[0.0674, 0.7322, 7.1957]	0.0306	11.0[a]	GA is independent of the initial point
		[0.075, 0.55, 25.5]	[0.0674, 0.7322, 7.1957]	0.0306	11.0[a]	GA is independent of the initial point

Note [a]The algorithm runs 3 times, and takes the value after the calculation time is stable

5.2 Multidisciplinary Design Optimization (MDO) Based on Reliability

In the multidisciplinary design optimization (MDO) on basis of reliability and in combination with the reliability design and multidisciplinary design optimization (MDO), the constraining condition of reliability design are not just indicators within the single discipline or a single system any longer, but the hard design index of the complex total system that consists of multiple coupled subsystems. When analyzing the reliability of the whole system, it is necessary to adjust and balance the internal subsystem, which will make the calculation more complicated. Therefore, the first problem to be solved in multidisciplinary design optimization (MDO) based on reliability is to improve computing efficiency, which needs starting with both reliability design theory and MDO theory. Since the beginning of the 21st century, there have been some studies on reliability based multidisciplinary design optimization, for example, Monte Carlo method is used as a reliability analysis algorithm to deal with the transmission of uncertain information between coupled subsystems according to the worst condition method. That is, a multidisciplinary reliability design framework based on C0 algorithm is established by ignoring the variation of coupling parameters and directly substituting the coupling parameters which makes the system performance the worst. This is the easiest way to handle with. However, it is difficult to find the coupling parameter values that make the performance of complex system worst, especially nonlinear systems which exist widely in practical engineering. In order to reduce the computation complexity of system analysis, an approximate model to replace the real system analysis is developed, for example, Kriger model is the global approximation of multidisciplinary systems, which can effectively reduce the computational complexity. (Simpson et al. 2008). Since the approximate model brings about the reduction of computation, the double-cycle approach can be adopted in the reliability design. The implementation of this approach relies on the approximation technique with high accuracy and good stability, but it is difficult to simulate the performance of the whole system with an approximation technique when the system complexity is high. Some scholars systematically analyze and compare the combination of different reliability design optimization methods, reliability analysis algorithms and different multidisciplinary design optimization by algorithm cases. (Smith 2007). From these cases it can be found that different approaches are applicable to different kinds of problems, and the approaches applicable to all problems have not yet been derived. The method of reliability multidisciplinary design optimization (MDF), IDF, CSSO, Co, etc. in combination with the alternative method of double-cycling, optimization reliability analysis and the application (Yuan et al. 2005; Huang et al. 2009; Xia Qing et al. 2010; Meng et al. 2011; He et al. 2011; Yu et al. 2012; Yunping et al. 2013) have been studied domestically; However, it can be seen that the domestic research focuses on the application, the theoretical research is a little insufficient.

Reliability design and multidisciplinary design optimization (MDO) themselves are complicated design techniques, in which the amount of computation is very large.

The computation of RBMDO problem composing of both will further increase. So the current reliability design is mostly limited to the simple systems of single disciplines, of which applications are not many in multi-disciplinary complicated system. Compared with RBDO algorithm in single disciplines, it will result in three-nested cycles when dual-cycle RBDO algorithm is directly applied to undecoupled multidisciplinary design optimization (MDO) system,: system balance iteration, reliability analysis, optimization iteration In many cases, the amount of computation is far beyond the acceptable range of the project. The research literature on multidisciplinary system RBDO mostly focuses on the field of aeronautics and astronautics. Yao reviewed the research status of reliability design applied in aircraft design. The reader may also refer to this paper for the overview of its development and technology.

5.2.1 RBMDO (MDF-RBMDO)

In Chap. 2, the MDF method of certainty multidisciplinary design optimization is introduced, which is suitable for the problem of less subsystems and the close coupling. The subsystem balance iteration is completed in the MDA cycling. The objective function, constraints and design variables of the system layer are the same as the single-disciplinary RBDO problem, so it can be treated directly as the single-disciplinary RBDO problem:

(1) The MDF model is solved by the dual-cycling RBDO algorithm, and the MDF-RBMDO algorithm is adopted to solve the three-level nested iteration: subsystem balance, reliability analysis, optimization solution search. Therefore, the calculation is always large that are not applicable anytime in the actual engineering problems.
(2) The MDF model is solved by the SFSORA algorithm, at this time when the optimization and reliability analysis of MDF-RBMDO problems are performed sequentially: Optimization solving is a certainty optimization process in which the safety factor is treated as a known parameter, and the optimization process in which the objective function and the system constraint function are analyzed also needs the subsystems balance, that is, it includes the system analysis (MDA) iteration. Therefore, it includes two nested iterations; Similarly, the reliability analysis process also needs subsystem balancing. Therefore, it includes nested iterations of reliability analysis and MDA. Therefore, when the SFSORA is used in MDF-RBMDO problem, the three nested iterations are transformed into two alternative two-nested iterations and the computational complexity will be reduced. However, the complexity of calculation is often beyond the actual engineering problems acceptable.
(3) MDF model is solved by the single-cycle RBDO algorithm, in which it is necessary to convert all non-normal random design variables into approximate normal random variables. Then the MDF problem can be treated as a single-disciplinary

problem and solved directly by the single-cycle method. The single-cycle MDF-RBMDO problem includes the optimal solution and two-level nested iterations of MDA, so the computational complexity is lower compared with the Sfsora algorithm. It is important to note that the process of the non-normal distribution random variables transformed into approximately normal distribution variables will introduces the extraordinary errors. Therefore, in terms of the problems with few subsystems and close coupling, the solution efficiency by MDF model is acceptable. When the random variables are close to normal distribution, it is recommended to adopt the single cycle method for MDF-MDF-RBMDO solution.

5.2.2 RBMDO (CO-RBMDO)

C0 method of certainty multidisciplinary design optimization decouples subsystems by introducing auxiliary design variables at the system level, which causes the necessity of dealing with these auxiliary design variables when modeling with C0 method in RBMD0 problem. As the system-level design variables and subsystem-level design variables in C0-RBMD0 are random variables, the coupling variables of each subsystem will become random variables, and the corresponding auxiliary design variables should also be random variables. The random distribution type and feature parameters of the auxiliary design variables are unknown, and even the range of mean value is given artificially, just like the certainty C0 method. The most common approach is to assume that the auxiliary design variables are normal distribution random variables, and after the confidence is obtained based on engineering experience, it is converted into the value of variation coefficient. But for complex systems, there are often no data or experience of these coupling parameters, so compatibility constraints will be transformed into reliability constraints, and the target reliability needs to be given artificially.

Assuming that after the auxiliary design variables are randomly distributed, and the optimization of the C0 model at the system level and that of each subsystem level are nested, the optimization process of subsystem RBD0 still needs multiple system optimization and subsystem optimization nesting though the optimization process of each subsystem is a simple unconstrained problem. Therefore, the computational complexity of C0-RBMD0 is not necessarily lower than that of MDF-RBMD0, and the C0 method doesn't satisfy the KKT condition which leads to the unavailability of efficient numerical optimization search algorithms. Gradient-independent optimization algorithms can only be used, which makes it possible that C0-RBMD0's computational load will increase rather than decrease compared to MDF-RBMD0. In addition, the artificial assumptions of random distribution of auxiliary design variables and objective reliability and compatibility constraints will introduce additional errors. Therefore, it is not recommended to adopt C0-RBMD0 method for reliability design optimization of complex practical project problems when no new solutions appear in this article.

5.2.3 RBMDO (AP-RBMDO)

From the previous introduction of C0-RBMD0, it can be found that the method of decoupling by introducing auxiliary variables, which is widely used in certainty MD0 method, will bring a series of problems and errors in RBMD0 problem. The other decoupling approach in certainty MD0 is the approximation model. When the approximate technology is applied to RBMD0 problems two types can be divided as follows according to the approximate object:

(1) The approximate object is the subsystem's coupling variable. First, the sufficient data points of each coupling variable and design variable are obtained through multiple MDA analyses. Then, approximate technology in response to the surface, Kriging or neural network are adopted to establish the approximate relationship between the coupling variables and the design variables and to perform the error tests. Then, the approximate model is adopted to model each subsystem replacing the actual parameter transfer, and the natural decoupling between each subsystem is realized. The system level can directly call the necessary state parameters attained by each subsystem and approximate model, that is, MD0 model can be transformed into single discipline optimization model. The dual-cycle algorithm, SFS0RA algorithm or single-cycle algorithm can be directly applied to the equivalent single-disciplinary optimization model.
(2) The approximate object is the subsystem's state variable. It is also necessary to obtain sufficient data points of each subsystem's state variables and design variables through multiple MDA analysis, and then the approximate model between each state variable and design variable will be established. Then, the approximate model of state variables can be directly used to replace the analysis of each subsystem. The MD0 problem can be transformed into a single-discipline optimization problem just including system-level analysis and many approximate models. After this, it can be treated and solved according to the single subject RBD0 problem. Compared with the method of establishing approximate model for coupling variables in (1), the method of directly establishing approximate model for state variables allows the optimization process completely separate from the subsystem analysis, which the calculation efficiency is higher.

The AP-RBMD0 optimization solution and computation is much more efficient than other RBMD0 methods with the introduction of approximation technique. But the precision of AP-RBMD0 depends on that of approximation model, so high precision approximation technique is the foundation of AP-RBMD0. In addition to the approximation technique itself, sufficient data points are needed to establish the approximate model, which requires multiple system analysis, (that is, multi-wheel system equilibrium iteration). Complexity of RBMDO computation is usually large. In contrast, the computational complexity of the system analysis needed to obtain these data points is often much smaller, and usually one round of system analysis can generate a data point of approximate variables (multiple coupling variable or state variable).Therefore, it is not necessary to make a separate system analysis for each approximate variable. The actual project data is always more and more by accumulation, and the precision of the approximate model depending on the data will be higher and higher. When the data sample is enough, the approximate model can reach the precision of the real model. In addition, in many cases, for the engineering research, especially when frontier engineering research, is ahead of theoretical research, there may be only sample data without theoretical support, or only a simple theory without perfect calculation and analysis methods. At this time, the approximation model will play an important role in the establishment of these unknown functional relationships, which is the same idea as the popular theory of big data in the field of the social science. Therefore, AP-RBMDO will become a practical method for RBMDO applied to practical engineering product design. In this book, it's recommended to use this method in priority for RBMDO problems in which approximate models can be established.

5.2.4 Computation Case of Multidisciplinary Design Optimization Based on Reliability

Take example 1 at the right side in Sect. 5.1 in this chapter for example to introduce RBMDO:

$$\begin{aligned}
&\min f = \mu_{X_1} + \mu_{X_2} \\
&s.t.\ P(g_i(X_1, X_2) \geq 0) \geq 0.9987,\ i = 1, 2, 3 \\
&\quad g_3(X_1, X_2) = \tfrac{80}{X_1^2+8X_2+5} - 1 \\
&sub_1: \\
&\quad g_1(X_1, X_2, b_1, b_2) = \tfrac{b_1 X_2}{20} - 1 \\
&\quad b_1 = X_1(b_2 - X_1 + 5) \\
&sub_2: \\
&\quad g_2(X_1, X_2, b_1, b_2) = \tfrac{b_2}{30} + \tfrac{(X_1-X_2-12)^2}{120} - 1 \\
&\quad b_2 = \tfrac{b_1}{X_2} + X_2 - 5 \\
&where: \\
&\quad X_j \sim N(\mu_{X_j}, 0.3),\ j = 1, 2 \\
&\quad 0 \leq \mu_{X_1} \leq 10 \\
&\quad 0 \leq \mu_{X_2} \leq 10 \\
&\quad X_1 \neq X_2
\end{aligned} \tag{5.26}$$

It is easy to see that this problem is derived from the upper right part of Example 1 in this chapter, ($0 \leq \mu_{X_1}, \mu_{X_2} \leq 10$ see in Fig. 5.5) after g3 is omitted and overwritten. Subsystem 1 and Subsystem 2 of the problem are coupled by the Parameters b1

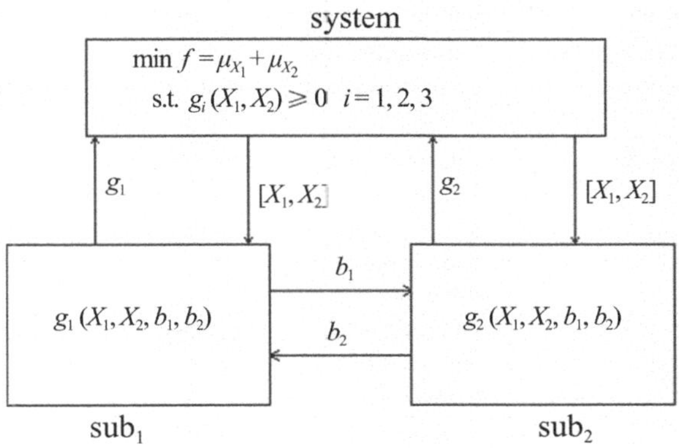

Fig. 5.8 RBMDO example parameter delivery relation

Table 5.11 Solution of equivalent RBDO problem for RBMDO example

Initial point	Optimization point	Optimization function value
[4, 5]	[2.3475, 4.0608]	6.4083

and b2 and are two-leveled MDO problems containing two subsystems, as shown in Fig. 5.8.

When the expressions of B1 and B2 are simultaneous and the equations are solved, the following can be easily solved:

$$\begin{cases} b_1 = X_1 X_2 \\ b_2 = X_1 + X_2 - 5 \end{cases} \tag{5.27}$$

The problem can be reduced to an equivalent ordinary RBDO problem with no subsystem coupling by substituting formula (5.27) into (5.26). The solution of this RBDO problem can be easily obtained by using the SLRBDO algorithm program in the earlier text, as shown in Table 5.11.

In this paper, the multidisciplinary design optimization (RBMDO) algorithm based on reliability is performed to solve the RBMDO problem with different MDO algorithms, which is compared with the equivalent RBMDO solution.

5.2.4.1 MDF-RBMDO

If not decouple, MDF algorithm is used for compiling,, the MDO problem runs as follows:

(1) Optimization algorithm of the system layer will generate the design scheme X = [Xi, X2]. X is passed to the subsystems Sub1 and Sub2.
(2) Balance iteration of subsystem layer. Sub1 assumes a B2 value first, calculates bi and magic value. Use sub2 with bi as input to get a new b2. Compared with

Table 5.12 MDF-RBMDO examples and optimization results of the original problem

Optimal initial point	Sub-equilibrium iteration method	Optimization point	Objective function value	Reliability check (10^6 CMC method)
[4, 5]	Literature results	[3.4409, 3.2909]	6.732	[0.9968, 0.9977, 1]
[4, 5]	b2initial value = 0, and it's direct iteration	[2.3475, 4.0608]	6.4083	[0.9984, 1, 1]
	b2 initial value = 1 and it's direct iteration	[4.2672, 7.37]	11.6371	[1,1,1]
	b2 initial value = 4 and it's direct iteration	The balance iteration of subsystem is not convergent	–	–
	fzero	[2.3475, 4.0608]	6.4083	[0.9984, 1, 1]

previous round, the difference between the b2 and the new b2 sees if it is less than the specified value, (take value here 10^{-5}. The next iteration of subi takes the new b2 as the input variable yields a new subi... until the b2 difference value between two iterations is less than the specified value. The magic and magic values of the balance iteration in the subsystem are passed back to the system level.

(3) In the system level, judge if the constraint function can meet it according to the magic and magic value, and the target function is evaluated. Thus, it is first to assume that the initial value of the coupling variable b2 starts the equilibrium iteration of the subsystem, or that the initial value of bi starts the equilibrium iteration of the subsystem. The MDF-RBMDO problem is solved by a single-cycle SLRBDO algorithm. The results are shown in Table 5.12.

It can be seen that in actual running, when assuming that the initial value of b2 performs the equilibrium iteration of the subsystem, it is found that when the initial value of b 2 has a great influence on the result, and the improper initial value needs many iterations to obtain the equilibrium solution of the subsystem. Even the equilibrium solution of the subsystem cannot be obtained. According to the characteristic that the subsystem equilibrium iteration has only one parameter in this example, an optimization algorithm "fzero" is used to replace the direct iteration method for the subsystem equilibrium iteration, and the equivalent RBDO problem can be obtained quickly.

From this example, in the MDF model, each operation of the system layer needs a round of subsystem balanced iteration. When adopting double-cycle Rbdo algorithm, MDF-RBMDO needs three nested cycles: optimization, reliability analysis and subsystem balance iteration, which will greatly increase the computation. And it is not always possible to get a balanced subsystem at every design point optimized at the system level. The subsystem balancing is actually a process of being similar to solving equations, and the corresponding optimization algorithm is usually more effective than the direct iterative cycle method.

In addition, it can be seen that in MDF-RBMDO model, the probability constraint is located at the system level, and the subsystem level has no probability constraint. So the coupling parameters between subsystems can be treated as certainty variables when the subsystems iterate.

5.2.4.2 CO-RBMDO

Decouple Subsystem 1 and Subsystem 2 of the problem. The model is reconstructed by CO method, and the CO-RBMDO problem is obtained. See the formula (5.28). So, the system-level design variables increases by 4: μ_{X_1}, μ_{X_2}, b_1^{sys}, b_2^{sys}, the constraints are increased by2 compatibility constraints, where the artificially assume that the compatibility constraints satisfy the reliability requirements; Since the two subsystems do not have internal design variables(local design variables), the two subsystems need optimizing at the subsystem level. Subsystem i just needs to perform a system analysis according to the parameters passed by the system level, and then pass $b_i(i = 1, 2)$ and $g_i(i = 1, 2)$ back to the system level.

$$\begin{aligned}
&\text{sys}: \\
&\min f = \mu_{X_1} + \mu_{X_2} \\
&s.t.\ \ P(|b_1^{sys} - b_1| \le \varepsilon) \ge 0.9987 \\
&\qquad P(|b_2^{sys} - b_2| \le \varepsilon) \ge 0.9987 \\
&\qquad P(g_1 \ge 0) \ge 0.9987 \\
&\qquad P(g_2 \ge 0) \ge 0.9987 \\
&\qquad \text{设计变量}: (\mu_{X_1}, \mu_{X_2}, b_1^{sys}, b_2^{sys}) \\
&sub_1: \\
&\qquad g_1(X_1, X_2, b_1) = \frac{b_1 X_2}{20} - 1 \\
&\qquad b_1 = X_1(b_2^{sys} - X_1 + 5) \\
&sub_2: \\
&\qquad g_2(X_1, X_2, b_2) = \frac{b_2}{30} + \frac{(X_1 - X_2 - 12)^2}{120} - 1 \\
&\qquad b_2 = \frac{b_1^{sys}}{X_2} + X_2 - 5
\end{aligned} \tag{5.28}$$

In which,

$$X_j \sim N(\mu_{X_j}, 0.3),\ \ j = 1, 2$$

b_1^{sys}, b_2^{sys} random distribution is unknown.

$$1 \le \mu_{X_1} \le 10$$
$$1 \le \mu_{X_2} \le 10$$

$\varepsilon =$ Small positive number, here take 10^{-4}.

After modeling with CO-RBMDO, it is found that it is difficult to give reasonable assumptions on random distribution parameters(mean and variance) as the newly introduced auxiliary design variable b_1^{sys} and variable b_2^{sys} have no practical engineering significance.

5.2.4.3 AP-RBMDO

As pointed out earlier in this book and in the literature (Pan and Cui 2008,2009), the core idea of MDO algorithm is decoupling and approximation. However, it can be seen for the previous text that the MDO algorithm that is decoupled with auxiliary design variables in RBMDO will face with new problems. The AP-RBMDO based on approximate decoupling technique has many advantages in practical engineering.

The functional relationship between the coupling parameters b1 and b2 and the design variables [X1, X2] is determined by the Formula (5.26) and (5.27). Formula (5.27) is an explicit function expression which can be obtained by directly solving the equation, but it is not easy to get the function expression for complex engineering problems. At this point, the approximation technique can be used to obtain the approximate expression of the function.

(1) Data point acquisition

The design matrix is produced by DOE in the design space. For this problem, there are only two design variables, the full factorial design method can be directly adopted to establish design matrix (11 levels for each variable). The design matrix is shown in Table 5.13.

Table 5.13 Design variable 11 leveled full factorial design matrix

(0, 0)	(1, 0)	(2, 0)	(3, 0)	(4, 0)	(5, 0)	(6, 0)	(7, 0)	(8, 0)	(9, 0)	(10, 0)
(0, 1)	(1, 1)	(2, 1)	(3, 1)	(4, 1)	(5, 1)	(6, 1)	(7, 1)	(8, 1)	(9, 1)	(10, 1)
(0, 2)	(1, 2)	(2, 2)	(3, 2)	(4, 2)	(5, 2)	(6, 2)	(7, 2)	(8, 2)	(9, 2)	(10, 2)
(0, 3)	(1, 3)	(2, 3)	(3, 3)	(4, 3)	(5, 3)	(6, 3)	(7, 3)	(8, 3)	(9, 3)	(10, 3)
(0, 4)	(1, 4)	(2, 4)	(3, 4)	(4, 4)	(5, 4)	(6, 4)	(7, 4)	(8, 4)	(9, 4)	(10, 4)
(0, 5)	(1, 5)	(2, 5)	(3, 5)	(4, 5)	(5, 5)	(6, 5)	(7, 5)	(8, 5)	(9, 5)	(10, 5)
(0, 6)	(1, 6)	(2, 6)	(3, 6)	(4, 6)	(5, 6)	(6, 6)	(7, 6)	(8, 6)	(9, 6)	(10, 6)
(0, 7)	(1, 7)	(2, 7)	(3, 7)	(4, 7)	(5, 7)	(6, 7)	(7, 7)	(8, 7)	(9, 7)	(10, 7)
(0, 8)	(1, 8)	(2, 8)	(3, 8)	(4, 8)	(5, 6)	(6, 8)	(7, 8)	(8, 3)	(9. 8)	(10, 8)
(0, 9)	(1, 9)	(2, 9)	(3, 9)	(4, 9)	(5, 9)	(6, 9)	(7, 9)	(8, 9)	(9, 9)	(10, 9)
(0, 10)	(1, 10)	(2, 10)	(3, 10)	(4, 10)	(5, 10)	(6, 10)	(7, 10)	(8, 10)	(9. 10)	(10, 10)

The data points corresponding to the coupling variables b1 and b2 are obtained by sub-system balance iteration according to the design matrix, as shown in Tables 5.14 and 5.15, respectively.

(2) Establishment of metamodel

Based on the known data points, the quadratic response surface (SSR) can be adopted to obtain the approximate model with small error:

$$\begin{aligned} b_1 &= f_1(X_1, X_2) \\ &= p_0 + p_1 X_1 + p_2 X_2 + p_3 X_1^2 + p_4 X_1 X_2 + p_5 X_2^2 \\ p_0 &= 6.569 \times 10^{-16},\ p_1 = -1.656 \times 10^{-16},\ p_2 = 1.634 \times 10^{-15} \\ p_3 &= -2.286 \times 10^{-16},\ p_4 = 1,\ p_5 = -1.554 \times 10^{-16} \end{aligned} \quad (5.29)$$

Table 5.14 b 1 data point

0	0	0	0	0	0	0	0	0	0	0
0	1	2	3	4	5	6	7	8	9	10
0	2	4	6	8	10	12	14	16	18	20
0	3	6	9	12	15	18	21	24	27	30
0	4	8	12	16	20	24	28	32	36	40
0	5	10	15	20	25	30	35	40	45	50
0	6	12	18	24	30	36	42	48	54	60
0	7	14	21	28	35	42	49	56	63	70
0	8	16	24	32	40	48	56	64	72	80
0	9	18	27	36	45	54	63	72	81	90
0	10	20	30	40	50	60	70	80	90	100

Table 5.15 b 2 data point

5	6	7	8	9	10	11	12	13	14	15
6	7	8	9	10	11	12	13	14	15	16
7	8	9	10	11	12	13	14	15	16	17
8	9	10	11	12	13	14	15	16	17	18
9	10	11	12	13	14	15	16	17	18	19
10	11	12	13	14	15	16	17	18	19	20
11	12	13	14	15	16	17	18	19	20	21
12	13	14	15	16	17	18	19	20	21	22
13	14	15	16	17	18	19	20	21	22	23
14	15	16	17	18	19	20	21	22	23	24
15	16	17	18	19	20	21	22	23	24	25

Similarly, it was found that a single response surface can be used to construct a high precision approximation of B2:

$$\begin{aligned} b_2 &= f_2(X_1, X_2) \\ &= p_0 + p_1X_1 + p_2X_2 \\ p_0 &= -5,\ p_1 = 1,\ p_2 = 1 \end{aligned} \tag{5.30}$$

(3) Optimization model.

The approximate model (functions f1 and f2)to replace the relation emulation of b1 and b2 in subsystems 1 and 2, the optimization model is transformed into the traditional single disciplinary RBDO model:

$$\begin{aligned} &\min f = \mu_{X_1} + \mu_{X_2} \\ &s.t.\ P(g_i(X_1, X_2) \geq 0) \geq 0.9987,\ i = 1, 2 \\ &\quad g_1(X_1, X_2, b_1, b_2) = \tfrac{b_1 X_1}{20} - 1 \\ &\quad b_1 = f_1(X_1, X_2) \\ &\quad g_2(X_1, X_2, b_1, b_2) = \tfrac{b_2}{30} + \tfrac{(X_1 - X_2 - 12)^2}{120} - 1 \\ &\quad b_2 = f_2(X_1, X_2) \end{aligned} \tag{5.31}$$

Among them,

$$\begin{aligned} &X_j \sim N(\mu_{X_j}, 0.3),\ \ j = 1, 2 \\ &0 \leq \mu_{X_1} \leq 10 \\ &0 \leq \mu_{X_2} \leq 10 \\ &X_1 \neq X_2 \end{aligned}$$

(4) Calculation result.

The model is solved by the single-cycle RBDO method. The results are shown in Table 5.16. As the high precision of the approximation model is higher, the solution

Table 5.16 AP-RBMDO algorithm results

Method	Initial point	Optimization point	Objective function value	Reliability check (10^6 CMC method)
Leiterature	[4, 5]	[3.4409, 3.2909]	6.732	[0.9968,0.9977, 1]
RBDOproblem solution	[4, 5]	[2.3475, 4.0608]	6.4083	[0.9984, 1, 1]
AP-RBMDO	[4, 5]	[2.3444, 4.0641]	6.4085	[0.9985, 1, 1]

of AP-RBMDO is very close to the solution of the equivalent RBDO problem, and the error of the objective function is just: $\frac{|6.4085-6.4083|}{6.4083} = 0.0031\%$.

From this example, it can be seen that in addition to the large computation, the initial design value of MDF-RBMDO has a great influence on the result, and convergence cannot be guaranteed. Co-RBMDO needs to introduce auxiliary design variables, and it is difficult to determine the random distribution type and parameters of these new variables. The complexity of AP-RBMDO implementation lays in obtaining enough data points by systematic analysis and establish an approximate model with satisfactory accuracy while it's much easier for the establishment of the optimization model and solving.

References

Agarwal H (2004) Reliability based design optimization: formulations and methodologies. PhD thesis, Notre Dame

Allen M, Maute K (2004) Reliability-based design optimization of aeroelastic structures. Struct Multi O 27:228–242

Ang A, Tang WH (1984) Risk and reliability. Wiley, New York

Baabbad MA (2004) Reliability-based design optimization of a nonlinear elastic plastic thin-walled t-section beam. PhD Dissertation, Dept. of Aerospace and Ocean Engineering, Virginia Polytechnic Inst. and State Univ., Blacksburg, VA

Burton SA, Hajela P (2002) Variable complexity reliability-based optimization with neural network augmentation. In: AIAA 2002-1474

Chen XG, Hasselman TK, Neill DJ (1997) Reliability based structural design optimization for practical applications. In: AIAA-97-1403, pp 2724–2732

Choi KK, Youn BD (2002) On probabilistic approaches for reliability-based design optimization (RBDO). In: 9th AIAA/ISSMO symposium on multidisciplinary analysis and optimization. Atlanta, Georgia, AIAA-2002-5472

Du L (2006) Reliability-based and possibility-based design optimization using inverse analysis methods. PhD thesis of graduate college of the university of Iowa

Du X, Chen W (2002) Sequential optimization and reliability assessment method for efficient probabilistic design. In: Proceedings of the ASME design engineering technical conference, vol 2. American Societyof Mechanical Engineers, pp 871–880

Gayton N, BourinetJ M, Lamaire M (2003) CQ2RS: a new statistical approach to the response surface method for reliability analysis. Struct Saf 25(1):99–121

He Q, LI Y, AO L, Wen Z, Yue Q (2011) Reliability and multidisciplinary design optimization of turbine blades based on single cycle method. Propulsion Technology (5), 658–663. (贺谦, 李元生, 敖良波, 温志勋, 岳珠峰. (2011). 基于单循环方法的涡轮叶片可靠性及多学科设计优化. 推进技术(5), 658-663.)

Huang H, Yu H, Yuan Y, Zhang X, Li Y (2009) Optimization of multidisciplinary reliability design based on single discipline feasible method. Acta Aerophenica Sinica 30(010):1871–1876. (黄洪钟, 余辉, 袁亚辉, 张小玲, & 李彦锋. (2009). 基于单学科可行法的多学科可靠性设计优化. 航空学报, 30(010), 1871-1876.)

Krishnamurthy T, Romero VJ (2002) Construction of response surface with higher order continuity and its application to reliability engineering. In: AIAA 2002-1466

Lee J, Yang Y, Ruy W (2002) A comparative study on reliability-index and target-performance-based probabilistic structural design optimization. Comput Sand Struct 80(3–4):257–269

Li F (2008) Collaborative product design under uncertainty. PhD thesis of Arizona State University

Liu D, Choi KK, Youn BD, Gorsich D (2006) Possibility-based design optimization method for design problems with both statistical and fuzzy input data. Trans ASME 128:928–935

Maglaras G, Nikolaidis E, Haftka RT et al (1997) Analytical-experimental comparison of probabilitic and fuzzy set based methods for designing under uncertainty. Struct Optim 13:69–80

Meng D, Huang H-W, Xu H, Zhang X, Zhang X (2011) A method for uncertainty analysis of multidisciplinary systems – Improvement of collaborative uncertainty analysis. Chinese J Mech Eng (19):133–139. (孟德彪, 黄洪钟, 许焕卫, 张小玲, & 张旭东. (2011). 一种多学科系统不确定性分析方法——协同不确定性分析法的改进. 机械工程学报(19), 133-139.)

Pan BB, Cui WC (2008) Multidisciplinary design optimization methods for ship design. J Ship Mech 12(6):914–931

Pan BB, Cui WC, He L (2009) A ship hull transform program for multidisciplinary design optimization. J Ship Mech 13(6):886–894

Papadrakakis M, Lagaros ND (2002) Reliability-based structural optimization using neural networks and montecarlo simulation. Comput Methods Appl Mech Eng 191(32):3491–3507

Qing X, Hong C, Shifeng Z (2010) Application of reliability optimization method in multidisciplinary design optimization of airborne missiles. J Missile Guidance 30(1):40–42. (夏青, 蔡洪, 张士峰. 可靠性优化方法在飞航导弹多学科设计优化中的应用. 弹箭与制导学报, 2010, 30(1), pp. 40-42.)

Royset JO, DerKiureghian A, Polak E (2001) Reliability-based optimal design of series structural systems. J Eng Mech 127(6):607–614

Savoia M (2002) Structural reliability analysis through fuzzy number approach, with application to stability. Comput Struct 80:1087–1102

Simpson TW, Toropov V, Balabanov V, Viana FAC (2008) Design and analysis of computer experiments in multidisciplinary design optimization: a review of how far we have come-or not. In: 12th AIAA/ISSMO multidisciplinary analysis and optimization conference. Victoria, British Columbia Canada. AIAA 2008-5802

Smith N (2007) Probabilistic design of multidisciplinary systems. PhD thesis of Vanderbilt University

Wang L, Grandhi RV, Hopkins DA (1995) Structural reliability optimization using an efficient safety index calculation procedure. Int J Numer Meth Eng 38:1721–1738

Wu YT, Wang W (1998) Efficient probabilistic design by converting reliability constraints to approximately equivalent deterministic constraints. Trans SDPS, J Integr. Des Process Sci 2:13–21

Wu YT, Shin Y, Sues R, Cesare M (2001) Safety-factor based approach for probability-based design optimization. In: Proceedings of the 42rd AIAA/ASME/ASCE/AHS/ASC structures, structural dynamics, and materials Conference,AIAA2001–1522, Seattle, WA

Yang RJ, Gu L (2004) Experience with approximate reliability-based optimization. Struct Multi Optim 26:152–159

Yu M, Liu Y, Li L, Li Y, Yue Q (2012) Reliability-based multidisciplinary design optimization of centrifugal impeller with double-loop strategy. Acta Aeronautica et Astronautica Sinica (4). (于明, 刘永寿, 李磊, 李元生, & 岳珠峰. (2012). 基于双循环的离心叶轮多学科可靠性优化设计. 航空学报, 33(004), 650–657.)

Yuan W, Huang H, Wu J (2005) A new multi-discipline design optimization method based on reliability. In: Proceedings of Vehicle and engineering equipment quality and reliability forum, 2005 national conference on mechanical reliability. (原薇, 黄洪钟, & 吴宝贵. (2005). 一种新的基于可靠性的多学科设计优化方法初探. 2005年全国机械可靠性学术交流会暨"车辆与工程装备质量与可靠性论坛"论文集.)

Yunping L, Jun Z, Bing Z, Jin S (2013) Optimization of multidisciplinary reliability design based on Bliss and PMA. Acta Aeronautica sinica, 34(010):2349–2356. (刘云平, 张俊, 张冰, & 孙瑾. (2013). 基于bliss和pma的多学科可靠性设计优化. 航空学报, 34(010), 2349–2356.)

Chapter 6
Design of Manned Submersible

The success of the submersible "Jiaolong" has allowed China to be one of the few countries that have mastered deep-diving technology, and achieved the record on the maximum depth of the G-2 manned submersible. More importantly, this has brought about a series of social effects, so that the country and the people realized the importance of developing the ocean and deep-sea equipment. While we are sure of our achievements, we should also see that the submersible "Jiaolong" is only the beginning of our research on deeper manned submersibles. Firstly, the high-strength titanium-alloy manned module, most of the sonar and navigation equipment, buoyancy materials, robots and etc. for submersible "Jiaolong". are imported from abroad. Secondly, the design of the submersible "Jiaolong" is still type on basis of experience. Although the method of multidisciplinary design optimization (MDO) has been explored in the process of development and production, a certainty evaluation method based on MDO theory has been initially established, and many uncertainties in practice have not been considered. When the assurance of reliability only depends on the management of the strengthening of safety and quality according to design specification, the qualitative analysis of reliability outline, the reliability design based on uncertainty theory has not been really carried out. In addition, the operational cost of the submersible "Jiaolong" is very expensive, constraining its application scope. In order to solve these problems and further improve the level of research and development of China's marine equipment, the Office of Ocean Affairs in Nation's 863 Program has also approved a project of developing a 4500 m manned submersible. The localization rate of the 4500 m manned submersible should be 85% or above in which to improve the reliability and lower the operation cost. As the need of abyssal science is becoming more and more urgent, the development of deep-sea manned submersible is on the agenda. The huge pressure of reaching 11,000 m is beyond

B. Pan and W. Cui, *Multidisciplinary Design Optimization and Its Application in Deep Manned Submersible Design*, Ocean Engineering & Oceanography 13,
https://doi.org/10.1007/978-981-15-6455-0_6

the limits of most existing commercial deepwater equipment, allowing many equipment to be customized, redesigned or even redesigned. This makes the reliability of deep-sea manned submersible exceed other deep-sea manned submersible design.

This chapter will begin with the main structure of manned submersible, and introduce the key technology of manned submersible design step by step.

6.1 Main Components of Manned Submersible

6.1.1 Manned Cabin

The main task of manned submersible is to carry human to the deep sea for observation and operation and safely back, so the core component of manned submersible is a huge- pressure structure against the marine pressure, which is called manned cabin. The manned cabin is a cylindrical or spherical hollow pressure-resistant structure composed of high-strength steel or titanium alloy. The typical manned cabin is a sphere with the inner diameter of about 2 m and can contain 2–3 persons. There are access hatches for personnel and transparent windows for deep-sea viewing. After the personnel enter the manned cabin, the hatch cover is closed, and the manned cabin is isolated from the outside world. In order to maintain the environment suitable for human survival in the manned cabin, the cabin is equipped with a life-supporting system including an oxygen supply system, CO_2 absorption system, temperature and humidity control system, food, etc. The entire manned submersible is operated by the pilot inside the manned cabin, and it has equipment such as computers, control systems, display screens and other equipment. These cabin equipment is connected to extravehicular propulsion, manipulator, camera system and other equipment outside the cabin through cable, so manned cabin is also equipped with a water-sealed patch bay.

6.1.2 Buoyancy Material

The weight of the manned cabin and other equipment makes the gravity of the submersible below the water far greater than the buoyancy, so it is necessary for a buoyancy system to balance the weight. Devices or materials that can provide the buoyancy has a lot, from wood, foam plastic to balloons, empty barrels. In the design of submersible, they are named as buoyancy devices or buoyancy materials. Ordinary buoyancy materials can hardly be used for the deep manned submersibles, since manned submersibles descend thousands or even tens of kilometers into the sea, the ordinary buoyant materials are compressed into buoyancy-losing slabs or collapsed directly under the pressure of thousands' atmospheres of water. Therefore, the buoyancy material for the manned submersible must be able to withstand high pressure. Not only the ultimate strength of the buoyancy material is higher than the working pressure of the manned submersible, but also the deformation of the buoyancy material under high pressure cannot cause loss of the large buoyancy. In addition, there are requirements of durability, water absorption. In the later chapters of this book, a variety of commonly used buoyancy materials will be introduced.

6.1.3 Propulsion and Ballast

Once deployed from the surface support platforms like the mother ship, the manned submersibles become autonomous underwater carrier. The motion of manned submersible is mainly divided into horizontal motion (mainly searching motion along the sea floor), vertical motion and vertical motion (floating upward and diving downward motion). The horizontal motion is mainly driven by thrusters to drive the vehicle forward, backward and turn. Therefore, a propulsion and control system is needed on the submersible, and the system's power cannot be too large due to the energy limit. Usually manned submersibles move much slower horizontally than submarines and have a limited horizontal range. Manned submersibles need run for long distances vertically, compared to shorter horizontal distances. For example, the submersible "Jiaolong", which is designed to reach a depth of 7000 m, means that it has to dive at a depth of 7000 m each time. After finishing the task, 7000 m's floating motion will be carried out. It would take a lot of energy to complete such a long vertical motion if the manned submersible were driven by thrusters. Since the Earth's gravity and sea water buoyancy exist, people have come up with a method to realize the vertical movement of the manned submersible by adjusting the ballast of the manned submersible and making use of the gravity buoyancy difference., the submersible using this principle does not need consume energy in the process of floating upward and diving downward, which is called non-power motion. Therefore, the manned submersible will be equipped with a set or sets of ballast control system, the specific implementation solution has the adjustable water tank type, solid ballast jettisoned type, adjustable hydraulic oil tank type, etc. These content will be introduced in details in the following chapter of this book.

6.1.4 Deep-Sea Observation Operation

After the crew in the manned submersible can observe the environment and life in the deep sea through the transparent windows of the manned cabin after they have been safely carried to the deep sea. Since the light cannot reach the deep sea hundreds of meters below, a lighting system is needed to provide lighting on manned submersibles. As photography technology advances, people have installed the camera system on the manned submersible to record the strange environment and biological Figures at the deep sea. In addition to viewing, it is more desirable to for human being to measure quantitatively the parameters of the marine environment, such as basic temperature, salinity and depth. Therefore, the manned submersibles carry a variety of sensors and marine scientific instruments. for measuring the marine environment. In addition to on-site observations, samples of seafloor sediments, deep-sea organisms and deep-sea water have begun to extracted.. After manned submersibles bring the samples back to the surface, scientists can use large laboratories and instruments for deeper analysis, so manned submersibles usually have a variety of deep-sea sampling equipment. Some deep-sea survey and sampling equipment needs robotic personnel to operate. Sometimes manned submersibles also perform underwater tasks such as deep-sea rescue, bottom salvage, underwater engineering construction and maintenance, which allows the multiple freedom manipulator gradually to become the standard configuration for manned submersibles. As the requirements of deep-sea scientific exploration and operations is higher, more and more deep-sea equipment will be installed on manned submersibles. Therefore, it is necessary to reserve some energy control interface and allowance of the design load for the new equipment in the future at the beginning of design.

6.1.5 Communication

A manned submersible becomes an autonomous carrying vehicle after leaving the master ship, which needs to know its position and condition in the ocean for support anytime. This will need installing the underwater positioning and communication systems on both the master ship and manned submersibles. With this set of system, the master ship can track the depth and relative position of the manned submersible anytime, the crews of the master ship can communicate with those on the manned submersible in real time through an underwater communication system. The master ship can give a mission to the manned submersible or provide surface support to the manned submersible as requested by crews. Nowadays, the underwater communication system not only realizes the real-time communication between the master ship and the submersibles, but also the two-way transmission of the data. For example, the scientist on a submersible can send measurements or underwater Figures back to the master ship, which can then give the submersible new instructions or requirements based on the returned undersea parameters or Figures. When a manned submersible emerges from the water after completing the underwater task, the underwater positioning communication system of the submersible may be disturbed by the sea surface waves, which may cause the master ship to lose the coordinates and communication. Therefore, a surface positioning communication system is needed on the manned submersible, and the development of satellite positioning and communication technology has made global positioning possible. High-frequency radio communication technology can also be used for positioning and communication when the distance between the vehicle and the mother ship is not too far. Different positioning and communication technologies have their own advantages and disadvantages. In order to ensure communication security, manned submersibles are usually equipped with multiple communication positioning technologies.

6.1.6 Energy Control

On manned submersibles, energy supplies are needed for all from life support systems to propulsion system, communications system to viewing and sampling equipment. With the limits of today's underwater energy technology, the size of manned submersibles and the difficulty of withstanding pressure in the deep sea, the current manned submersibles are basically powered by pure electric energy. With equipment carried more and more, more and more complex tasks to perform, and more and more long hours for operation, the battery system of the manned submersible vehicle has become the largest component outside the manned cabin. Therefore, the energy density (per unit mass or per unit volume of energy) is an important indicator for the power supply to manned submersible. In addition to the energy density, such a large power supply would require a large and heavy pressure structure if placed in a pressure-resistant structure, so pressure-compensated batteries are usually used, in which the battery is put into a battery box and then filled with insulating oil. The pressure inside and outside the box is balanced by the volume change on the insulating oil, so that the battery box wall does not need to bear the huge pressure of the sea water. The driving systems of manned submersibles (including the propulsion system and the ballast system) are also controlled entirely by electrical signals. The operation tools on manned submersibles and a large number of deep-sea research equipment also need performing the control and data storage. Therefore, the manned submersible needs control system with the complete to configurations (the complete hardware and software system).

6.1.7 Structure (Pressure Resistance and Non Pressure Resistance Structure)

In order to fix the systems such as manned cabin, buoyancy device, propulsion system, power system, etc. with the equipment, a large frame is needed for the manned submersible. The frame is like a fish bone that links all the equipment together, so it is necessary to have sufficient strength to withstand the load. Meanwhile, the frame should take into account the overall layout of manned vehicles and the requirements of equipment installation. In addition to the large frame, manned vehicles also have a large number of accessories for installing equipment, most of these parts are the rods or plates, collectively known as non pressure resistance structure. In addition, there are a lot of equipment on the manned submersible that is not pressure-resistant or needs a dry environment. The equipment cannot be installed in the manned cabin due to safety, volume limit and operational needs. At this time, the small pressure tanks (compared to the manned cabin) are need for these equipment, either cylindrical or spherical pressure-resistant tanks are collectively known as the pressure structures of manned submersibles. As the core component of the manned submersibles, the manned cabin is a large pressure-resistant structure with greater processing difficulty. Although it is also divided in the structural system and designed according to the pressure-resistant structure, material research and structural design are usually initiated during the critical technology research phase to ensure the safety, reliability and weight optimization for the manned cabin.

6.2 Design Overview of Manned Submersibles

Life support, buoyancy device and structure (including manned cabin), propulsion, ballast, communication, observation and operation device, energy and control are the main subsystems of manned submersible. The most important task in submersible design is to integrate these subsystems and their components into an one body under the premise of satisfying the safety and functional requirements, so that the volume and weight of the submersible meet the support conditions of the master ship while reducing the cost. The process in which to realize the technical route and overall arrangement of the safety and overall index of the manned submersible is called the overall design or the overall system design of the manned submersible. At the beginning of the overall design, the equipment of each sub-system composing of the manned submersible often needs to be selected, modified and adjusted according to specific requirements, actual working condition and general layout requirements. It is also necessary to consider the connection and matching with the other subsystems' parts or equipment. The design within these subsystems is often named as sub-system design or branch system design. In addition, in the early phase of the manned submersibles development, it is necessary to review the current level of design, processing and construction, to assess whether key technologies and major equipment can be mastered, and to find out the technical difficulties and key components to be resolved as early as possible for positioning. Then, the key technology research of manned submersible is carried out by building the research om sub project to solve the technical problems and develop the equipment. This phase is called as key technology research of manned submersible.

When the key technology research has made a breakthrough and the main technical route is feasible, the design of manned submersible can be conducted according to the requirements of the design task. As mentioned above, the design of manned submersible vehicle is composed of the overall design and subsystem design. The overall design decomposes and transforms the performance index of manned submersible into the performance index of each subsystem through analysis Therefore, the overall design is the target of subsystem design; The subsystem design will also directly determine whether the criteria of overall design can be met. The most typical example is that after the steel material of U.S. manned cabin was replaced with titanium alloy, so that the working depth increases from 2000 m to 4000 m and reduces the weight of the entire vehicle, so the overall design relies on the subsystem design. Furthermore, the design of each subsystem is not independent to each other. For example, the capacity of the energy system has constrained the power of the other subsystem equipment, the buoyancy material system has constrained the weight of the other subsystems equipment, and the pressure structure have constrained the hardware volume of the electronic control system. Design of each subsystem is sometimes mutually beneficial, but more it is contradictory, so the subsystem design should consider the relation and conflict with other subsystems while satisfying the design index of this subsystem. Therefore, the design of manned submersible is that of the complex engineering associated with several subsystems.

As mentioned earlier, the classical theory for designing complex projects like manned submersibles is the serial design method. In the serial design method, the subsystems are sorted according to their importance and the main pre-and post-dependent relation. Then each subsystem is designed according to the order. The relationship between this subsystem and other subsystems is ignored when designing a subsystem. After all the subsystems have been designed, the analysis of overall system is carried out and sees if it can meet the requirements of design index. This process is called as a design cycle. If the target cannot be met, the next round of design cycle will begin… After the continuous improvement in multiple design cycle, finally achieve the design index, the cycle diagram for serial design of manned submersible is seen in Fig. 6.1.

As the theory of multidisciplinary design optimization (MDO) develops, the traditional serial design method has been replaced by multi-disciplinary design optimization (MDO) method in aviation fields. Multidisciplinary Design Optimization (MDO) has been gradually introduced into the design of manned submersibles. The design of manned submersible vehicle can be divided into overall design and subsystem design. The design of each subsystem can be further refined to the design of each equipment A schematic diagram of the sub-layer design of a manned submersible is shown in Fig. 6.2. In the framework of sub-layer design, the overall design is responsible for the sub-system design indicators, but also collecting the feedback results of sub-system design that is used for evaluating the overall performance indicators. When the design of sub-systems conflicts, the overall design will also coordinate sub-systems with the overall design of manned submersible as the starting point. This ensures that the purpose of each design adjustment is to optimize the overall system rather than the single subsystems.

总体设计	General design
生命支持	Life support
能源	Energy source
载人舱	Manned cabin
浮力材料	Buoyancy material
观测作业	Observation operation

(continued)

(continued)

结构	Structure
通信	Communication
推进	Propulsion
压载	Ballast

载人潜水器总体设计	General design of manned submersibles
结构设计	Structural design
生命支持系统设计	Life support system design
能源与配电设计	Energy and distribution design
控制系统设计	Control system design
水动力系统设计	Hydrodynamic system design
通信定位系统设计	Communication and positioning system design
灯光摄像设计	Lighting and photography design
作业系统设计	Operation system design
浮力材与舾装设计	Buoyancy material and outfitting design
耐压结构	Pressure resistant structure
非耐压结构	Non-pressure resistant structure
供氧与 CO2 吸收	Oxygen supply and CO2 absorption
电源	Power supply
配电	Power distribution
硬件集成	Hardware collection
软件开发	Software development
潜水器外形设计	Outer design of submersibles
快速性、操作性	Rapidity and operation feasibility
水下通信定位	Underwater communication positioning
灯光	Lighting
拍照摄像与显示	Photograph and display
液压系统	Hydraulic system
机械手	Manipulator
其他作业取样设备	Sampling equipment for other operations
浮力材料布置	Buoyancy material arrangement
轻外壳与其他附件	Light shell with other accessories

Compare serial design method with hierarchical multidisciplinary design optimization method:

(1) Final design scheme.

In order to achieve the purpose, the serial method cannot answer whether the final design scheme is optimal or not, and does not know whether all subsystems of the

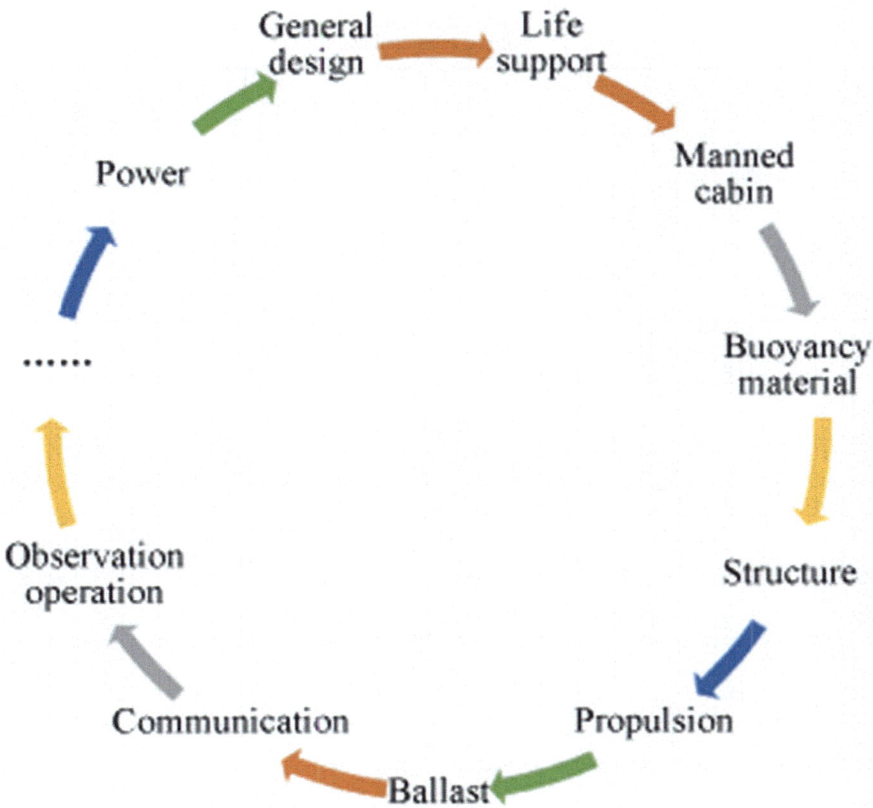

Fig. 6.1 Serial design method of manned submersible

final design scheme are in the best balance state. Just find a feasible design scheme that can have each subsystem not conflict. The hierarchical multidisciplinary design optimization (MDO) method allocates the design indexes of each subsystem by the overall design layer, evaluates the system performance according to the feedback results of the subsystem design, and then adjusts the indexes. In the overall design level, according to the system performance, the contradictions of the various subsystems are coordinated. So, multidisciplinary design optimization (MDO) is to take obtaining better system performance index as the beginning point. The final design is to have considered the scheme of the subsystem balance and overall system performance optimization. In addition, the influence weight of each subsystem and parameters on the system performance index and the interaction of each subsystem can be obtained by data exploring technology.

(2) Design efficiency.

In the serial design method, the design cycle of the sub-system in the back must await the design of the previous sub-system and get the information before starting.

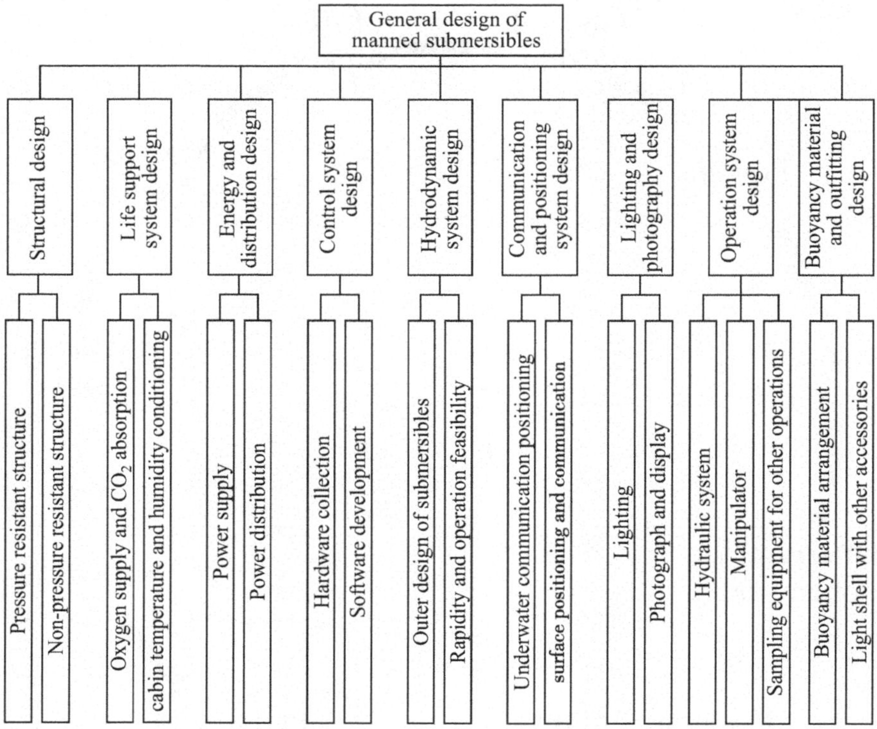

Fig. 6.2 Hierarchical design method for manned submersible

The time required for the entire design cycle is basically the overlapping of the time required for each subsystem design. In multidisciplinary design optimization (MDO), each subsystem is parallel, which means that the independent subsystems can be designed at one time. The subsystems with contradictory or dependency relation can be designed in parallel after decoupled with appropriate method. This greatly improves the design efficiency and shortens the design cycle.

(3) Personnel management.

In the serial method, the personnel of each subsystem are responsible only for the former and the latter subsystems, on basis of processing the design within this subsystem. Throughout the design cycle, the designers of each subsystem need to await the end of the previous subsystem design before they start analyzing and designing. The task of subsystem design is completed and enters the waiting state again after the data and results are sent to the next subsystem, the utilization rate of personnel is very low. In multidisciplinary design optimization (MDO), each subsystem only communicates with the overall design, but it can give indexes and design tasks to multiple or even all subsystems at one time. The designers of each subsystem design and back the results to the overall design at one time, then the overall design allocate the coordination and task to the subsystem, which reduces the idle time of the designers within each subsystem. Meanwhile, the overall design is directly communicated with each subsystem, enhancing the control of the overall design to the designers within each subsystem.

(4) Data management.

In serial design method, data is usually transmitted in one way. That is, after the design of the subsystem in the front of the design cycle is completed, all the parameters are packaged together and transmitted to the subsequent subsystem, so that the data packets received by each subsystem not only contain the necessary parameters of the subsystem, but also includes the necessary parameters for the subsequent subsystem, which has the data security reduced, and also raised the cost of data transmission. In multidisciplinary design optimization (MDO), each subsystem only communicates with the overall design, that is, each subsystem only receives the parameters and indexes of the overall design, and only submits the design results and parameters to the overall design after designed. The parameters and other data transmitted from each subsystem are managed and stored centrally by the overall design. The overall design only allocates the necessary parameters to the subsystem, ensuring the security and completeness of the data. Only those data that are effectively transmitted also means the improvement in efficiency of data transmission and the reduction in cost of data transmission.

(5) Design change.

In the serial design method, any change of subsystem design needs to carry on a round or even several rounds of complete design cycle for checking, and correct the effect of subsystem design change on the whole system. In multidisciplinary design optimization (MDO), after a subsystem design changes, the overall design first evaluates the overall performance changes resulted from the changes, and then drives other related subsystems to make corresponding design changes according to the changes of overall performance. The design of the irrelevant subsystem does not need to change, which can effectively shorten the design change cycle and reduce the cost of change.

It can be seen that compared with serial design method, the multidisciplinary design optimization method has a great advantage and this idea of hierarchical and parallel design is very suitable for the design of complex engineering products. However, the current implementation of multidisciplinary design optimization (MDO) is still faced with the problem of high computational complexity, increasing dimension of design space brought by subsystem's decoupling.

Besides, the multidisciplinary design optimization (MDO), deriving from the aviation field, is applied to the design of manned submersible, which will face many engineering problems. Such problem will be further discussed in the following chapter in this book.

6.3 Design Basis of Manned Submersible

6.3.1 Task and Mission Analysis

Before developing the manned submersible, it is necessary to investigate the technology status of the similar submersible at home and abroad, and combine the functional requirements for scientific investigation, underwater operation, mobile position. Considering the technical risk and manufacturing capability, the main technical indexes for manned submersible are determined, and the initial draft for the design mission of the submersible is prepared. It is usually necessary to revise and adjust the initial draft for the design task statement by consulting the opinions from potential target users, collecting the expert opinions of the submersible, calling meeting between the competent department and the expert demonstration and so on, in order to determine the final draft. Of course, this process may not be completed by the design unit, but by the competent authority or project sponsors and publish the design task. In the design task and mission statement, the main parameters of the submersible will be defined:

6.3.1.1 Main Task and Mission

The mission and task is the motive of manned submersible development. Mission can be divided into military, scientific exploration, underwater sightseeing and exploration, underwater operation etc. Different mission purpose will determine design parameters such as the depth of the submersible operation, underwater working time, the configuration of the main operating tool, staffing.

6.3.1.2 Weight and Range of the Main Dimension

According to the weight and main dimensions of the existing manned submersible with the same model, the reasonable range of the weight and main dimensions for the submersible is given combining the current technical situation and the situation of the master ship. On the one hand, the overall design of the submersible is constrained. On the other hand, it also provides design and layout reference for the supporting facilities of the submersible, such as lifting equipment, the testing equipment related to deck space of master ship, etc.

6.3.1.3 Staff Allocation

According to the typical working conditions, the personnel allocation can be arranged providing reference for the size and space layout design of manned cabin, and also provide basis for life support system design.

6.3.1.4 Main Performance Indicators

According to the analysis of mission and typical working condition, the main performance indexes such as working depth, underwater working time, time of upward float and down diving, maximum time of life support, speed and valid loads are determined.

6.3.1.5 Major Equipment and Components

The main equipment is determined with the basic composition and working condition of the submersibles. On the manned submersibles nowadays, manned cabin, life support systems, energy system, communication and positioning system, framework, buoyancy material, light photography and operation manipulators have become the standard configuration. In the special equipment, what equipment to be specially specified is mainly the special operation equipment, such as biological trap, special sampling device, drilling equipment, special operation tool, special sensor, etc. In order to control the development risk and reduce the design space, sometimes the more detailed requirements for main components of the submersible will be given in the design mission statement. For example, it specifies what materials to use for the manned cabin, the number of viewing windows, what absorption system for C02 for life support systems, what batteries or energy sources to use, which communication and positioning equipment to use, the range and resolution requirements for light photography, the freedom of movement of the manipulator and the grip, etc.

6.3.1.6 Working Condition Requirement

In addition to specifying the operation area of the submersible, deployment and recovery of sea, it is necessary to clarify the requirements for submersible storage on the road, transport way, lifting way, storage and maintenance at sea. In the term of some manned submersible with the special purpose, it is also necessary to propose the corresponding requirements according to the specific working condition. For example, for manned submersibles that need operate frequently in high-temperature seawater, (such as near hydrothermal vents), it is necessary to specify the allowable range of seawater temperature and the permissible operating time for the submersibles at the high temperature; However, the allowable radiation strength and working time must be specified for the underwater submersibles working in the radiation-polluted sea area.

Design mission statement is the cornerstone of the submersible design. When the draft of design mission is finalized, that means that the design work of the manned submersible can be launched. Just as the design of any complex engineering system, the design of manned submersible is a process from the shallow to the deep, constantly adding content, detailed analysis, improving details. Therefore, if the design of the submersible is divided according to the time axis, the design of the manned submersible can be divided into the stage of scheme design, preliminary design, detailed design, construction design, etc. More detailed descriptions on these design stages will be given in subsequent sections. It is necessary to note that the actual design process is sometimes not strictly divided by this process. Yet, whatever design procedure is followed, the design is always continuous deepening course, and the analysis model will gradually transition from simplified calculation to numerical simulation and simulation test, which needs to adjust the overall design optimization model. It is necessary to analyze the performance of each subsystem accurately as much as possible and to control the computation within an allowable range.

6.3.2 Design Flow

Generally, the design of manned submersible can be divided into three phases: scheme design, preliminary design and detailed design.

In the phase of scheme design, the work to be conducted is as follows:

(1) According to the requirements of product stated in the contract, on the basis of the comprehensive analysis of the product functions, indicators, key technologies, development cycle and funds, select the best product solution after the analysis, comparison and full demonstration of several schemes and solutions.
(2) After determine the product scheme, the realization schemes for products are carried out, the development process and the development cycle are determined. The network chart for product development scheme is drawn, and the technical requirements for the product component system subsystem and equipment are proposed after the demonstration.
(3) When it is necessary, the product characteristics and development process is analyzed, the outline for reliability assurance is prepared for determining the reliability working goad and compulsory work items. According to the relevant standards, make use of the reliability, maintainability, safety, supportability and optimization design techniques to demonstrate the reliability indexes of the systems and equipment. It is necessary to predict and analyze various factors affecting reliability and their influence degree, and find out main factors, to study various potential countermeasures, and to put forward scientifically the distribution and acceptance methods for reliability index.
(4) Propose key technologies and relevant research topics, and requirements for new materials and components development projects and key outsourcing projects. Propose the requirements for imported technology and equipment and materials.
(5) The scheme for system (subsystem) and equipment should pass the primary verification of the principle test and the performance simulation test.
(6) When selecting components, should pay special attention to seriation, standardization, generalization, and reduce variety and specification as much as possible, make a list for the key purchase parts, and develop special control measures.

(7) The key technology and major research projects should be basically solved in the phase of product scheme design.

In the phase of scheme design, the following and produce technical documentation should be completed:

① Product composition, main technical index, overall technical scheme and description of use requirements;
② Planning document of product realization includes quality planning and development planning of product application software;
③ Quality and reliability control measures (such as first draft of product quality assurance outline, product reliability, first draft of maintainability outline, etc.)
④ First draft of product standardization outline;
⑤ Product testing and measurement measures;
⑥ The feasibility demonstration of the new technology and equipment used;
⑦ If necessary, analysis on product development risk and evaluation should be performed
⑧ Evaluation of product cost and price;
⑨ The implementation of the products' main supporting components of;
⑩ Network map for product development;
⑪ Review report on design input, review report on project design, etc.

In the preliminary design phase, the following aspects should be carried out:

(1) According to the requirements of product planning, the initial design, analysis, calculation, test and verification of the product will be conducted;. The initial design, trial-manufacture and test of the system (subsystem) and equipment should be conducted.
(2) Do a good job in interface design and technical coordination between the system (sub-systems) and equipment, do a good job in comprehensive coordination and interface design between the equipment within the system. The technical documents shall be formed for all important technical coordination.
(3) In reliability design, mature technology should be used as far as possible, components with high reliability and long life should be adopted, such as redundancy design, functional reduction design, EMC design, etc. Vibration-proof design and protective design for the climate and environment should be considered in structural design.
(4) In terms of safety design, the possibility of damage to equipment and injury to human body resulted from operation should be considered. In product design, safety protective measures should be taken to prevent the failure of parts and components in certain parts of the product, causing the damage of major components or key parts of the product, or even damaging people.
(5) For maintainability design, the position of the inspection point, the external operability, and the method of structural assembly should be considered mainly. For design, implement the principle of modularization, and improve the degree of standardization as far as possible, which makes product's maintenance easy.

(6) The reliability test items of system (subsystem) and equipment are proposed, and the test outline is formulated.
(7) In the phase of initial design, all key technical problems during product development should be solved, basically meeting the requirements of the contract. The positive conclusion is reached after design verification.

In the phase of initial design, the following and produce technical documentation should be completed:

① Design and development planning document;
② Distribution of product's main index, product external interface relationship, etc.
③ Formal texts such as product quality assurance outline, product reliability and maintainability outline, product standardization outline, etc.
④ Technical documents such as normal regulations of product design, general plan for product process, standard of application software of product system and interface specification, detailed list for product's key and important items and analysis report (initial draft) of key characteristics;
⑤ The review report at the design and development, (including the review report of overall process plan, review report of initial design), etc.
⑥ Complete set of design documents, etc.

In the phase of the detailed design, the following tasks should be carried out:

(1) According to the contract and the result of the initial design, the technical design of the system (subsystem) and equipment should be carried out, providing the necessary drawing data for production.
(2) For the design problems left from the initial design review, measures must be taken for solving and reflected in the design drawings.
(3) For the detailed design, the technologies and measures shall be adopted as far as possible which have been appraised or tested by the principle prototype, and the technical state shall be frozen if the practice proves that the technologies are correct, reliable and meet the requirements of the technical indexes.
(4) In terms of projects required for reliability, the collection and arrangement of original reliability information should be done well in the phase of detailed design. According to the information obtained, the reliability assessment should be carried out preliminarily.

In the phase of detailed design, the following should be completed and technical documentation should be formed:

① The necessary complete set of drawings and data for production;
② Design verification data;
③ Review report of detailed design.

In each phase of the design process, the designer can use the serial design method shown in Fig. 2.1 or the hierarchical and parallel multidisciplinary design optimization method as shown in Fig. 2.2. Multidisciplinary design optimization (MDO) has become a trend of development. The hierarchical and parallel design for manned submersible can be divided into general design and subsystem design naturally. The main purpose of this book is to introduce multidisciplinary design optimization (MDO) and its application to the design of manned submersibles. In Chapter 7, we'll introduce in details with examples.

6.3.3 System Division

As manned submersible vehicle is a complex system, involving many subjects. Therefore, it should be divided into sub-system or sub-system to implement the development work. As the chief designer, it is necessary to divide combined with the specific working conditions. If the administrative system has provided the chief designer with a design team, the system division at this time is divided according to the team's existing personnel and the feature of the units involved. The submersible "Jiaolong" is the case. At that time, 11 sub-systems were divided: overall performance and general layout, carrier structure, outfitting, ballast and trim regulation system, propulsion system, power and distribution system, guidance and control system, acoustic system, hydraulic and working tool system, life support system, emergency and submarine jettisoning system. But if the chief designer can organize his own design team, it can be divided into several major. For example, when designing the 11,000 m manned submersible" Rainbow Fish," we divide the manned submersible into seven sub systems (Cm et al. 2014): overall Performance and general layout, carrier structure, mechanical system, electrical system, control system, acoustic system, and life support system.

6.4 Key Technologies in the Design of Manned Submersible

From technical viewpoint, the most core issue is the overall design and integration issue. what parts are configured to the sub, and what components are like. How all of these parts outline together become a fully functional submersible, and how well it performs during navigation and operations. Second, is the processing and manufacturing of the equipment, such as manned balls, buoyancy material, power source, electric motors, pumps, valves and so on. Here is a brief description on the current situation of these technologies.

6.4.1 Pressure-Resistant Structure

The pressure-resistant structure has an airtight atmospheric chamber, the key component of which is to provide an atmospheric space manned sphere for the occupant and the instrument equipment. Besides, the pressure-resistant structure also contains a plurality with small diameter one pressure-resistant tank or more, one adjustable ballast tanks or several and several high pressure gas tanks.

The design and construction of the manned sphere are difficult points. In terms of design, although many classification societies have design standards, they vary greatly from one another. There is no uniform prediction method for the ultimate bearing capacity of manned balls among ship classification societies. A new prediction formula for the ultimate bearing capacity of manned balls is proposed based on multiple finite element analysis. Pan et al. (2010) It is well consistent with those results of finite element analysis. Based on this, we proposed a new set of design standards which have been adopted by the China Classification Society for the Edition 2013 specification. In the design of manned ball, technical problems such as the deformation coordination of viewing window, the determination of fatigue load spectrum, the reliability analysis of fatigue life and the multi-objective optimization design are to be solved. On the premise of satisfying the safety, the evaluation on the design of the manned ball is mainly the quantity of viewing windows and the coverage of view between them. The most advanced manned ball in design is the American "New Alvin". It has 5 viewing windows, scientists and the main driver have more vision coverage, which is conducive for scientists to command the main driver for operations.

There are three ways to make the manned ball: no welding, hemispherical welding and clack welding. No welding process: adopt casting into two hemispheres, and then machining for mold, and then connect with bolts. Russia's two manned submersibles "Peace", adopts this process. What concerns mainly is the poor quality of the welding. Hemispherical molding process: use a large-sized thick plate for direct-stamping molding, and then the use electron beam to weld two hemispherical equatorial ring seams. Such as Japan's deep-sea 6500 and the titanium alloy spherical shell made by United States are adopting this process. Clack muddling process: Each hemisphere is divided into 7 ball clacks, after each sphere is separately molded, weld 7 ball clacks into hemisphere with narrow beam, and then weld two hemispheres equatorial girth. Such as the titanium alloy spherical shell that is made by Russian adopted this process, including the "Russia" "Consular", and "Jiaolong". The third process has lower requirements for rolling ability and stamping capacity for heavy titanium alloy plate, but higher requirements for welding.

If the welding quality is good, then so is the safety of manned ball. China has just completed three 4500 m titanium manned cabins made by hemispherical welding and ball clack welding, which indicates that the manufacturing capacity of manned cabin in China has reached the advanced level in the world.

6.4.2 Pressure Compensation

One of the biggest differences between manned submersibles and submarines is the extensive application of the principle of pressure compensation, which makes the pressure-resistant cabin as small and few as far as possible. All the personnel and equipment of the submarines are in the pressure-resistant cabin, so the pressure-resistant cabin is very large. Then as the diving depth increases, the wall thickness of the pressure-resistant cabin increases rapidly, reaching the limit of industrial manufacture at the depth of 500–600 m. The so-called pressure compensation is to fill the container with liquid like oil or sea water, so that the pressure in the cabin increases with the depth of submergence. The cabin wall can be designed to be very thin with little deformation as there is the same pressure inside and outside the cabin. This greatly reduced the weight of the box. Other hydraulic pipeline also applies the principle of pressure compensation by filling oil.

6.4.3 Buoyancy Device

According to the Archimedes' principle, whether the substance can provide positive buoyancy into water depends on the density of the substance itself and the volume of water drainage. Therefore, it is required that Buoyant materials are either lighter in density than water, or having a larger volume of drained water, or both. Sixty years ago, when it reached the deepest ocean floor, the "Crest" took the lighter gasoline as its buoyancy material, while various metal, glass and ceramic buoyancy balls adopted a hollow ball to drain water and provide buoyancy.

The balance of Generation 2 manned submersibles are realized in seawater with mainly solid buoyancy materials. The advancement of the solid buoyancy material is expressed by its density and water absorption under the given bearing capacity. The lower the density and water absorption is, the better it is. There are two types of buoyancy materials currently applied in submersibles. One is a processable buoyancy material made of glass beads doped with epoxy. The density of this 7,000 m high pressure buoyancy material reaching the most advanced level is 481kgfm^3, but the intensity of buoyancy materials that is allowed to be exported from the United States is 561 kgfm^3. The other is a ceramic sphere, a lighter buoyancy material with a density of just 340 kgfm^3 at sea level, but it has only been applied in unmanned submersibles. Besides, some buoyancy devices are used to adjust the buoyancy and weight of the submersible. The most common buoyancy device is the ballast tank, which uses high-pressured gas to discharge the sea water from the tank to reduce the weight of the submersible. Generally it is suitable to use near the sea area; Adjustable ballast system, change the weight of the submersible by discharging the seawater with high-pressure seawater pumps, or reducing the weight of the submersible with discarding ballast.

Take the submersible "Jiaolong" for example. The basis of adjusting weight and buoyancy is the calculation of the weight and attitude of the submersible, considering the hydrodynamic profile of the submersible. The adjustments of weight and buoyancy of the vehicle are described in the following sections, according to the functional role:

The submersible's buoyancy is mostly provided by low density buoyancy materials, which has about 12 m^3 in volume, and most are placed on the top of the submersible. The advantage of this arrangement lies in that it can increase the height of the buoyancy center of the submersible.

The function of the ballast tank is to provide the freeboard on the water surface of the submersible. The tank is located near the center of the submersible's gravity. The high-pressure air for blowing out seawater from the tank is stored in the high-pressure air tank. The air tank is located close to the tank. This reduces the arrangement difficulty of high-pressure gas path and the valve fitting.

The submersible relies on its own buoyancy to dive downward. After diving to a certain depth, the density of the sea water will increase. It needs a certain weight to balance itself. It needs a certain positive buoyancy when it floats upward. Adjustment of these weight and buoyancy is achieved by disposing ballast. The weight and position of the disposable ballast are determined by the submersible's diving downward and buoyancy calculations.

The weight of the disposable ballast is calculated before each diving, but the conditions on the sea floor will inevitably differ slightly from the estimated results. Meanwhile, the operation of the submersible will also bring about the change in the weight of the submersible. The regulation of these two parts is completed by the adjustable ballast system. The adjustable ballast tank is arranged near the gravity of the submersible, and the attitude of the submersible is not affected in the course of adjustment.

The pitch adjustment of the submersible is realized by moving the mercury in the fore-and-aft regulator. The fore-and-aft regulator shall be arranged to obtain the maximum pitch as far as possible when the hydrodynamic profile is allowable.

The main factors to be considered in selecting buoyancy materials are strength, deformation, water absorption, durability, influence of cyclic load and machinability.

6.4.4 Energy

The battery box is heavy which need to be removed from the submersible. So the two battery boxes are arranged in the rear, which makes the forward and backward bending moments of the submersible balanced and easy to maintain. The trend in the future is lithium-ion battery, Japan's "Deep sea 6500" has been realized. The 1,200 m manned submersible ICTINEU3, also adopts the Lithium polymer battery which is developed by Spain.

6.4.5 Propulsion and Control

In the design of manned submersible, users have clear requirements for the mission and overall performance indicators, of which the operation feasibility is a very important indicator. It is realized by the arrangement and control of thrusters and the coordination of auxiliary devices. Take the 7000 m manned submersible "Jiaolong" as an example. According to its operation mission and overall performance index, the requirements are as follows for its operational feasibility:

(1) With a good stability of direct navigation motion.
(2) With 6D freedom space mobility.
(3) With in situ rotation ability and hovering ability in operation.
(4) The emergent brake pitch at 1.0 knot is not greater than 6 times the length of the submersible.

According to the above requirements and the constraints of the general layout, the designers carried out the operational design for the manned submersible "Jiaolong".

According to the operational mission and the requirements of overall performance, and on the basis of the work in the phase of the scheme demonstration, the hydrodynamic layout of the 7000 m manned submersible scheme is determined as follows:

(1) The main stern propeller is composed of four ducted propellers arranged in a cross-shaped vector.
(2) The auxiliary thruster is provided with a transverse channel thruster at the bow and a vertical 90 thrust thruster at the front, left and right direction. Rotary rotatable thruster.
(3) Match the arrangement of stern thruster, the stabilizer wing adopts the X-shaped arrangement.

The feature of layout scheme is to mutually match with 7 thrusters, which enables the manned submersible to obtain operational feasibility of 6D freedom space including longitudinal, lateral, vertical, yaw, pitch and roll. It has good maneuverability. On the basis of this, we have designed the main body line and stabilizer wing appendages.

6.4.6 Control System

The control system of manned submersible is the brain and nerve of manned submersible. It collects the operation command and information of various sensors on the operation panel of manned submersibles. Through the brain's analysis and judgment, the output command controls various executive organs, which enables the submersible to complete various movements. Meanwhile, the collected information will be stored and displayed on the man-machine interface, which will provide useful reference for the submarine and the commander, and makes it easier to drive the submersible and command the operation.

The control system of manned submersible mainly includes navigation control subsystem, navigation and positioning subsystem and comprehensive information display and control subsystem. The navigation control subsystem mainly completes the bottom operation command of the submersible, the collection of sensor information and the control of various executing structures, and achieves the motion control of the submersible with 6 degrees of freedom. Meanwhile, the collected information is transmitted to the comprehensive information display and control subsystem via the network for storage, processing and display. The navigation and positioning subsystem obtains the high precision position and attitude information of the manned submersible mainly based on the underwater vehicle and various navigations and positioning sensors on the support ship. Thus this enables the sub and the surface command personnel to grasp the underwater position and the sub can quickly and accurately reach the operation point, and complete operational tasks. The comprehensive information display and control subsystem further processes and processes the information and instructions collected by the navigation control subsystem. Finally display them in the form of man-machine interface, and stores all the information collected post analysis through data analysis platform.

6.4.7 Manipulator

Manned submersible is a platform, of which operating capacity depends on the operating capacity of the manipulator to a large extent. From the open information, the current international configuration of manned deep-diving submersible (manned deep-sea vehicle) manipulators and tools are as follows:

(1) Submersible “Nautilus”.

Anchor operator: 5 freedom, solenoid valve control.

Operator: 7 freedom, electro-hydraulic servo control.

(2) ”Deep-sea 6,500”

Anchor operator: 7 freedom, use driving arm mode to control, weight is 55 kg.

Operator: 7 freedom, master-slave mode for the end. the body parts feel force feedback for control, weight is 72 kg.

(3) Alvin

Anchor operator: 6 freedom (arm: pitch and swing; Elbow: pitch; Wrist: pitch and rotation; Gripper: open and close,) electromagnetic valve control, maximum extension distance is 1752.6 m, downward lift force at full extension is 45.36 kg, the weight into the water is 53 kg.

Operator: 7 freedom (arm: pitch and swing; Elbow: pitch; Wrist: pitch, swing and rotation; Gripper: open and close), master-slave electro-hydraulic servo control, the maximum extension distance is 1879.6 m, down lifting force at the full extension is 60.04 kg, wrist torque is 4.15 kg.m, rotation speed is 65 rpm. Install a second wrist: torque is 2.075 kg. The rotation speed is 13.0 rpm and the weight into water is 53 kg.

(4) Cyana

Anchor operator: 6 freedom, electromagnetic switch control.

Operator: 6 freedom, electro-hydraulic servo control.

(5) Submersible "Jiaolong"

According to this situation, in the design of the 7000 m manned submersible "Jiaolong", we decided to place two manipulator, the main manipulator is installed on the lower right side of the manned spherical shell in the front, while the sub-manipulator is installed on the left side.

① The main manipulator is 7 freedom (Arms: pitch and swing; Elbow: pitch; Wrist: pitch, swing and rotation; Gripper: open and close,) adopt master-driven electro-hydraulic servo control, the end part for holding the object is controlled in the way of the force sense feedback bilateral control.

a. Dimensions.
Maximum arm extension: about 1850 mm.
Arm width: 380 mm.
Maximum height of the arm when curling: 950 mm.

b. Bearing capacity.
Rotation: 360. Continuous.
Wrist torque: around 4.5 kg.m (oil pressure is 14 MPa.)
Lifting capacity under full extension: about 75 kg (oil pressure 14 MPa).
Maximum operating pressure: 21 MPa.
Arm swing angle: 45. (Inner) f90. (Outer).
Weight: 58 kg or less (below water).

② Sub-manipulator is 7 freedom arm: pitch and swing; Elbow: pitch; Wrist: pitch, swing and rotation; Gripper: open and close,) using solenoid valve control.

a. Dimensions.
Maximum arm span: about 1800 mm.
Arm width: 380 mm.
Maximum height of the arm when curling: 850 mm.

b. Bearing capacity.
Rotation: 360. Continuous
Wrist torque: around 3.8 kg.m (oil pressure 14 MPa.)
Full lifting capacity under full extension: about 55 kg (oil pressure 14 MPa.)
Maximum operating pressure: 21 MPa.
Arm swing angle: 45. (Inner) F90. (Outer).
Weight: 55 kg or less (below water.)
The whole frame of manipulator is made of high-strength aluminum alloy, and the lower part of wrist joint is made of high strength titanium alloy.

6.4.8 *Life Support*

The Life support system of manned submersible is an essential system to ensure the crew to live safely and work effectively in deep sea. At present, the crew members of all manned submersibles in the world are placed in pressure-resistant cabin. Whatever depth is tens of meters, hundreds or even thousands of meters, maintain normal pressure inside pressure-resistant tank, the content of oxygen and carbon dioxide are kept within the atmospheric concentration range. Even in the deep sea, the crew's living condition is similar to that on land. All the external water pressure that increases with the water depth acts on the pressure-resistant shell.

The life support system is designed to supply oxygen to the cabin continuously in order to supplement the oxygen consumed by the crew during breathing, and to maintain the oxygen concentration in the cabin of the normal standard(ranging from17%~23%). Meanwhile, the carbon dioxide generated by the crew is continuously removed with absorbent, and the content of carbon dioxide in the ballast tank is reduced to the allowable range (< 0.5%). Besides as the hyperbaric oxygen tank is installed in the pressure-resistant chamber, consideration should be given to reducing the risk of oxygen tank leakage to the crew. Also it ensures that the crew can focus themselves on completing the deep-sea operations during the normal work. Therefore, the life support system usually adopts automatic oxygen supply mainly in normal work, in combination with manual operation. oxygen is generated by oxygen candle in emergency situation rather than high pressure oxygen tank.

(1) Normal work and life support.

In order to ensure that the crew is fully immersed in underwater operations, life support adopts the automatic oxygen supply, and the valve of oxygen supply will work according to signals of oxygen concentration: When the oxygen concentration in the pressure-resistant chamber is below 17%, the valve of the oxygen supply automatically opens; When the oxygen concentration is higher than 23%, the valve of the oxygen supply automatically closes.

According to the specification requirement, a manual device should be set up to adjust the concentration of oxygen in the pressure-resistant chamber artificially. Normal work and life support system devices include:

① Life support console: 3 life support modes can be selected and activated: normal working, emergency oxygen supply and mask life support. Pressure, temperature, humidity, oxygen concentration, oxygen partial pressure, carbon dioxide concentration, carbon dioxide partial pressure can be shown; The following parameters can be alarm: oxygen concentration, carbon dioxide concentration. ② Oxygen supply device: one XL high- pressured oxygen tank with the pressure of 15 MPa, pressure-reduction valve, flow meter, valve of oxygen supply, oxygen supply pipeline, oxygen concentration meter, partial pressure indicator and alarm. ① Carbon dioxide absorbing device: Xkg lithium hydroxide tank (packed by the special sealing box): sealing package, fan, inner tube, inner cover, machine box body, machine box cover, meter for carbon dioxide concentration, sub-pressure indicator and alarm. ④ Dehumidifier: the silicon gel is filled inside, and the rate of water absorption is 0.325 kg of water for each kg of silicon gel. ⑤ Deodorizer and excreta storage device ⑥ Air circulation fans and silencers inside champers.

(2) Emergent oxygen life support.

Since the time of the emergent oxygen supply is 36H, a 20L 15 MPa high-pressured oxygen tank is needed. Once leakage occurs, 2700L oxygen will be omitted, and the total volume of the pressure-resistant chamber is only $5m^3$ in which a large number of instruments and equipments are installed. Therefore, for submersible, oxygen candle is used to generate the O2 instead of the high-pressure oxygen tank, which greatly improves the safety performance. The device is the same as the normal life support device except that the oxygen candle is used to replace the high-pressure oxygen tank and pipeline.

Main body of oxygen candle: Ignition tool, candle body, chemical filter layer, physical filter layer, shell body, shell cover and the sealing cover.

When an Oxygen candle has finished producing oxygen, the old oxygen candle body can be removed, the seal cover of the new oxygen candle body can be twisted off, the conductor can be pulled out, its terminal plug can be inserted into the socket of the power box. The switch of the power can be opened, after 3 mm, the new oxygen candle will give off oxygen gas.

Carbon dioxide absorption device can also absorb carbon dioxide effectively by replacing lithium hydroxide tanks timely.

(3) Face mask life support.

When hazardous gases are generated in a pressure chamber, the crew can wear a face mask breathing apparatus, which provides oxygen via a pipeline from a 13L high-pressured oxygen tank, and the carbon dioxide emitted is absorbed by lithium hydroxide.

Reference

Pan BB, Cui WC, Shen YS, Liu T (2010) Further study on the ultimate strength analysis of spherical pressure hulls. Marine Struct 23(6):1–18

Chapter 7
Application of Multi-disciplinary Design Optimization in Manned Submersible Design

Under the thousands of meters deep ocean, manned deep-ocean submersible will not only encounter extreme pressure in deep ocean and other unknown work environment, but experience wave slap and wobble caused by wind and wave when launching and retrieving submersible, so the changing ocean work environment requires high reliability on manned submersible. With the fact that more than one disciplines and subsystems are involved in submersible, the uncertainty will not only exist in the integration of all of devices and subsystems but also in themselves of devices and subsystems. According to the design experience of Jiaolong manned deep-ocean submersible, the uncertainty of several main systems of manned submersible in system overall design could be simply analyzed:

(1) Shape and resistance system. The design goes this way: estimate submersible's overall dimension according to systems with large volume, like structure and energy source system, then determine submersible's shape prototype, produce models as per the prototype and conduct towing tank test, finally obtain resistance performance curve from all directions. According to this design idea and Jiaolong's design experience, the following uncertainty could be perceived in design process:

① Difference exists between submersible's actual shape and its prototype shape, namely dimension deviation. According to design experience of Jiaolong, many more devices are exposed into water than the prototype. The shape information of these devices is not available at the initial design stage, they could be counted as the uncertain information.

② Difference exists between shape prototype and model of towing tank test, namely scale effect. There are many "air containing space" in material objects allowing water to move in and out, these air containing space can affect flow field and model's shape completeness. Usually the model

B. Pan and W. Cui, *Multidisciplinary Design Optimization and Its Application in Deep Manned Submersible Design*, Ocean Engineering & Oceanography 13,
https://doi.org/10.1007/978-981-15-6455-0_7

cannot be totally processed as prototype, many simplified procession can be adopted.

③ Uncertainty exists in towing tank test, for example, experiment error. The deviation in submersible shape and resistance system could be counted as uncertainty in submersible shape and resistance calculation. However, due to lack of enough statistic data, these uncertainty information could not establish its probability density function, they should be hypothesized according to project experience, they could also be described by interval number and fuzzy number.

(2) Propulsion system. There is deviation of propeller installation on installation position and angle; the procession of screw propeller is uncertain and so the deformation; the efficiency coefficient of propulsion system cannot be completely and precisely described; domestic propulsion system has caught up with or even surpass imported systems, that is why 4500 m manned submersible adopts domestic propulsion system, but the pure domestic propulsion system is short of practical application and complete verification as well as statistic information of domestic propulsion system.
(3) Payload system. The payload carried by submersible is usually all kinds of work tools, most of time these tools is self-closed system, is not designed according to loading-carrying ability of submersible, thus the allowable load of submersible is frequently exceeded a little. To simplify design, the regular method is reserve part of load when design, stuff it by fixed ballast, and discharge it when necessary. Besides the sample basket holding ballast is generally designed as easy-to-discharge, namely there is only one explosive bolt at fixed point. The area covered by ballast will changes as ballast changes. The attack speed combined by wave speed, submersible's moving speed of up and down, submersible's launching and retrieving speed is not assured certainly. Once during the sea trial of "Jiaolong", the explosive bolt broke due to heavy attack on sample basket with terrible sea condition, and the sample basket and sample's tool is lost.
(4) Control, observation and communication and navigation system. This part is involved with a large amount of software and hardware of control, video recording and navigation. The function of these software and hardware inevitably has deviation to some extent, so the function of combined system by them will also has some uncertainty, besides, the work environment of these devices is always changing.
(5) Structure system. Structure system is composed of pressure structure and non pressure structure, while non pressure structure includes frame structure and exterior structure. According to experience of traditional structure reliability analysis, the uncertainty information of structure could be divided into loading uncertainty, material uncertainty and geometry uncertainty. For pressure structure, loading uncertainty refers to computed pressure is not strictly smaller than the relevant pressure of designed depth, which is caused mainly by different sea environment (density, etc.), submersible moving into deep ocean, etc.; material uncertainty refers to the uncertainty of pressure material's strength,

toughness, attack-proof performance, fatigue-proof performance, corrosion-proof performance, etc.; geometry uncertainty refers to thickness deviation, out-of-roundness. For frame structure, loading uncertainty refers to uncertainty of launching and retrieving sea condition, wave slapping and wind load and collision load; material uncertainty refers to the uncertainty of pressure material's strength, toughness, attack-proof performance, fatigue-proof performance, corrosion-proof performance, etc. In addition, frame structure has many weld, the performance deviation of weld could also be counted as material uncertainty; geometry uncertainty refers to procession deviation, weld deformation and part modification. Exterior structure mainly refers to buoyant material, light shell and stabilizer fin. Buoyant material has geometry uncertainty, water absorption uncertainty and density uncertainty, etc.; light shell has loading uncertainty, strength uncertainty and thickness deviation uncertainty; stabilizer fin is light as for weight, no requirement is clarified about its strength, so this book does not consider the structure uncertainty of stabilizer fin temporarily.

(6) Acoustic system. Acoustic communication will not only be influenced by sea noise, but by its own algorithm, this is similar with observation and navigation system.

(7) Life support system. Occupants have large individual difference, their oxygen consumption and discharge of carbon dioxide is uncertain. Deviation exists between oxygen tank's real oxygen storage and carbon dioxide absorber's real absorption and their designed volume.

Beside the common uncertain factors, other uncertain factors are involved in the recommended example in this book, namely the 4500 m domestic manned submersible. Its important character is that its major equipment adopts domestic underwater equipment, while domestication and reliability is correlative, namely that increase domestication means need to face bigger reliability challenge. This is the first time for our country to start so large scale research on manned submersible's relative technology and equipment, research and production of many homemade spares and equipment is also the first time. Many technology and product we don't have is produced and put into practical application, for example the manned submersible is involved in material preparation, rolling and pressing into half-ball of homemade high strength titanium alloy big thick plate, narrow gap MIG welding (electron beam welding) and machine tool procession and assembly. Underwater pressure low density buoyant material has been researched these years and developed well, but pressure ability, absorption ability and compression amount have not been verified, many hydraulic and electronic equipment are researched for the first time and applied in practical project products. We could get experience from project experience and scientific rules, the first applied technology and researched products is more risky than maturer technology and products. While the newly researched technology and products adopted by 4500 m manned submersible are waiting to be verified and solved about their reliability, which is a critical challenge for the reliability design of 4500 m manned submersible.

The uncertainty in the systems and work environment of manned submersible forces relevant person to consider submersible's reliability from design, building to operation management. Reliability means that the systematic project should start to plan and lay out at the stage of design, to conduct at the stage of building, to guarantee through management, to proceed carefully at each stage. Negligence in any stage will increase failure rate. Amount the three stages of design, building and operation, design stage will play the decisive role for product's reliability. A good design will leave enough space of reliability for manufacture and subsequent management. Thus, the reliability design of 4500 m manned submersible starts from the stage of scheme design and will exist throughout the service period of submersible.

As for the project system, like manned submersible, including more than one subsystems, traditional system reliability design method is reliability quota distribution, namely according to the relationship of subsystems and subsystem's influence on the whole system, subsystem is divided into series connection and load connection artificially, then as per the relationship and subsystem's extent of importance, the reliability quota of general system is distributed to subsystems. The process is affected heavily by people's action and contents many much deviation, so this method is heavily disputed. While for this kind of problem involving more than one subsystem and more than one discipline, multidisciplinary design optimization should be adopted. When proceeding reliability design with the background of multidisciplinary design optimization, the relationship of subsystems will not be divided int series connection or load connection artificially, but conforms to their inherent relation to transmit parameters and uncertainty information, namely "the uncertainty in one discipline may be transmitted to other disciplines by coupling variables, finally the uncertainty output by multidisciplinary system is the uncertainty accumulation of each discipline" (Jianguo et al. 2008). The defect of traditional method enables project team to research multidisciplinary design optimization based on reliability when proceeding general design of 4500 m manned submersible and apply it from the stage of scheme design.

The recommended application of reliability based multidisciplinary design optimization (RBMDO) in 4500 m manned submersible design in this chapter could be regarded as the example of RBMDO'S application in practical project. The first example is design of submersible's key part—manned cabin, whose research object is manned cabin's structure design, which belong to single disciplinary reliability based design, but has high requirement of safety and reliability duo to importance of manned submersible. This book will provide the complete reliability based design procedures, including modeling of high accuracy computation, modeling of uncertainty parameters, reliability analysis and implementation of reliability based design, they can provide direct reference for reliability based design of other high reliability required project products, especially for large scale project structure. The second example is the general design in the conception design stage of 4500 m manned submersible, presents the application of RBMDO in design of complete complex project system.

7.1 Reliability Based Design of Manned Cabin

Manned cabin is the crucial part of manned submersible ("Jiaolong" manned cabin see Fig. 7.1), it bears huge seawater pressure, provides proper living, observation and work environment for worker, and it provides much buoyancy by volume of displacement, is the heaviest too, even accounts for one quarter or one third weight of submersible, so its design should consider safety and reliability and minimum weight.

The first step of reliability based design is to establish state function with pretty good computation accuracy. As for manned cabin design, the major state function is the ultimate bearing capacity function. The current manned submersible design regulation provides design equation of manned cabin, but after comparison of computation, the computed results are very much different, and there is big difference from the manned cabin design of serving manned submersible, because the equation has a long history since established, and the technology of manned cabin procession and measurement is totally different from the modern. So when proceeding manned cabin reliability based design, the first step is to establish ultimate bearing capacity equation

Fig. 7.1 Figure of Jiaolong manned cabin

with higher accuracy. But establishing new equation to replace the old one is not easy, the process needs strict computation and verification: first analyze and compare the character of current regular equation and analyze the basic rule of bearing capacity of manned cabin crate; analyze large amount of experiment design of manned cabin based on the widely applied and accepted the nonlinear finite element method to obtain enough computation result of manned cabin design numerical value; analyze computed result and establish new equation of high accuracy and verify the accuracy of equation by breaking manned cabin ball model. After establishing pretty accurate state function, the performance statistic data of titanium alloy used in manned cabin should be obtained through many sample experiments, statistic analysis of uncertainty parameter modeling for some uncertainty design parameter of manned cabin's design should be done according to available data and experience, such as material performance, dimension and initial defects, etc., after all of the above work, the reliability design can be proceeded. This book will also adopts traditional safety factor method to design manned cabin and compares it with reliability design result. The reader could identify the quantitative relation of this kind of structure between safety factor and reliability.

7.1.1 Establishment and Verification of New Equation for Manned Cabin's Bearing Capacity

Currently, the manned cabin of serving manned deep-ocean submersible, including "Aerwen", "Yingwuluo", "Deep-ocean 6500" and "Wenlong", adopts ball-shape pressure structure. Ball-shape pressure structure has many advantages, such as equal force, lightest weight when bear same pressure, largest volume, so the manned cabin of 4500 m manned submersible will also adopt ball-shape structure. There are mainly two kinds of methods to compute the bearing capacity of manned cabin: computation based on experience equation and value computation based on the nonlinear finite element method. Both methods have advantages and disadvantages, see Table 7.1. Given that a large amount of computation required for nonlinear finite element method, this method cannot be applied directly for reliability design which has strict requirement on amount of computation, while the experience equation usually provides low accuracy. To meet the requirement on computation accuracy

Table 7.1 Computation comparison of manned cabin bearing capacity

Computation method of manned bearing capacity	Experience equation equation	Numerical value method
Accuracy	Bad to middle	Good
Recognized degree	Good	Middle
Computation amount	Small	Large
RBDO application	Good	Bad

and amount, this book will first proceed deep research and comparison for available regular equation to master theoretical basis of manned cabin's ultimate bearing capacity, then proceed nonlinear finite element method computation for hundreds of manned cabin design schemes with different sizes by design of experiment (DOE) technology, and establish more accurate computation experience equation of manned cabin ultimate bearing capacity by data analysis and fitting, verify this equation through pressure barrel break experiment of four manned cabin models, finally provide solid foundation for manned cabin's reliability design. The following will introduce in detail the concrete establishment process of this new equation, and provide reference for structure reliability design researcher.

7.1.1.1 Analysis and Comparison of Available Equation

Currently the rules of submersible we have collected includes: Norway Det Norske Veritas (DNV) publish in 1988 *Rules for Certification/Classification of Submersibles* (for short DNV 1988, others are similar with this), France Bureau Veritas (BV) publish in 1989 *Rules and Regulations for the Classification of Submersibles*, England Lloyd's Register (LR) Published in 1989 *Rules and Regulations for the Construction and Classification of Submersibles and Underwater Systems*, Russia Russian Maritime Register of Shipping (RS) publish in 2004 *Rules for the Classification and Construction of Manned Submersibles, Ship's Diving Systems and Passenger Submersibles*, China Classification Society (CCS) publish in 1996 *Introduction and Building Rules of Underwater system and submersible*, America American Bureau of Shipping (ABS) publish in 2004 *Rules for Building and Classing Underwater Vehicles, Systems and Hyperbaric Facilities 2010*, Germany Germanischer Lloyd Aktiengesellschaft (GL) publish in 2009 *Rules for Classification and Construction, 1-Ship Technology, 5-Underwater Technology, 2-Manned Submersibles*, Japan Nippon Kaiji Kyokai (NK) (date of publish unknown) publish *Rules for the Survey and Construction of Steel Ships Contents.*

Beside these rules, with finite element method's high accurate structure analysis ability and more mature scope, the general finite element software is regularly used to analyze structure of deep submersible's ball-shape manned cabin. This book will compare and analyze the computation method of each classification society in the following words.

(1) DNV (1988).

$$P \le \frac{P_{cr} \cdot \psi}{\gamma \cdot \gamma_m \cdot \kappa} \tag{7.1}$$

$$\gamma = \begin{cases} 1.3 & P \le 2 \\ 0.33(4.3 - 0.2P) & 2 < P < 5 \\ 1.1 & P \ge 5 \end{cases} \tag{7.2}$$

Note,

P is design pressure (MPa);
γ_m is material coefficient, for steel, $\gamma_m = 1.15$, for other materials value of γ_m is not given;
ψ is the coefficient reflecting post bulking behavior, for ball-shape shell $\psi = 0.75$;
K is structure coefficient, value see below equation:

$$\kappa = \begin{cases} 1.0 & \lambda < 0.5 \\ 0.7 + 0.6\lambda & 0.5 \le \lambda \le 1.0 \\ 1.3 & \lambda > 1 \end{cases} \tag{7.3}$$

$$\lambda = \sqrt{\frac{\sigma_F}{\sigma_e}} \tag{7.4}$$

Note,

σ_F is material's yield stress (MPa);
σ_e is material's elastic yield stress, value see below equation:

$$\sigma_e = 0.605 \cdot \delta \cdot \frac{t}{R} \cdot E \tag{7.5}$$

$$\delta = \frac{0.5}{\sqrt{1 + \frac{R}{100 \cdot t}}} \tag{7.6}$$

Note,

T is thickness of ball-shape shell (mm);
R is nominal radius, namely the average distance between center of ball and center of cross section of shell plate (mm);
E is elasticity modulus (MPa);
P_{cr} is anti-pressure critical pressure (MPa), value see below equation:

$$P_{cr} = 2 \cdot \Phi \cdot \frac{t}{R} \cdot \sigma_F \tag{7.7}$$

$$\Phi = \frac{1}{\sqrt{1 + \lambda^4}} \tag{7.8}$$

Λ is defined as Eq. (7.4), other parameters are defined as the former.

DNV (1988)'s rules and requirement about production deviation: the deviation (DNV defines as overall unroundness) between middle surface of actual thickness of ball-shape shell and real spherical surface of nominal radius is not bigger than 0.5%

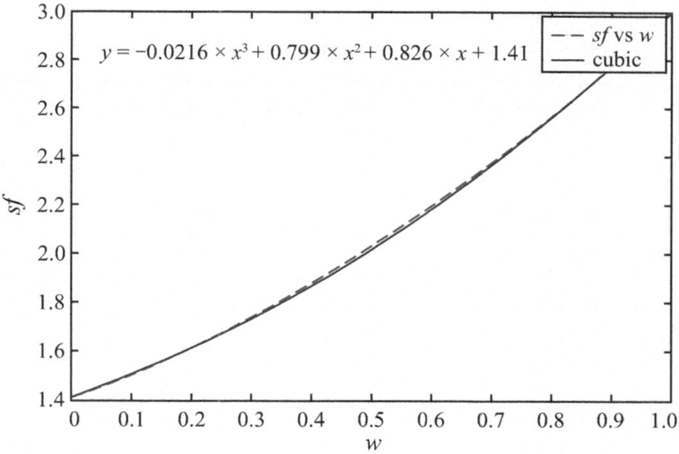

Fig. 7.2 BV (1989)'s relation of safety coefficient sf and out-of-roundness w

of nominal radius, and the local deviation measured by arc mould (DNV defines as overall unroundness) is not bigger than

$$\delta = \frac{0.04\sqrt{Rt}}{1 + 4\sqrt{t/R}}.$$

$$P \le P_b/sf \tag{7.9}$$

Note,

P is design pressure;
Sf is safety coefficient, see Fig. 7.2 (in order to make programme computation easy, this equation adopts approximation of 3 polynomial, and its approximation equation is provided in Figure

The value of out-of-roundness W adopts bigger value of the two equations:

$$w = 200\frac{D_{\max} - D_{\min}}{D_{\max} + D_{\min}} \tag{7.10}$$

$$w = 400\frac{Y_0}{D_0} \tag{7.11}$$

Note,

$D_{\max}$ is the biggest internal diameter when measuring ball-shape shell;
$D_{\max}$ is the smallest internal diameter when measuring ball-shape shell;

D_o is external diameter of ball-shape shell;
Y_0 is the depth of the flat (between fibers located at mean thickness;
P_b is the smallest buckling pressure, the smaller one will be taken between elastic buckling pressure P_e and thin film buckling P_f :

$$P_e = \frac{9.6E}{9 + 0.003\left(\frac{D_o}{t}\right)}\left(\frac{t}{D_o}\right)^2 \tag{7.12}$$

E is elasticity modulus;
D_o is external diameter of ball-shape shell;
T is thickness of shell

– P_f is calculated by Fig. 7.3:

See equation in Figure:

$$P_{th} = \frac{8E}{\sqrt{3(1-\nu^2)}}\left(\frac{t}{D_o}\right)^2 \tag{7.13}$$

$$P_p = \frac{4t\sigma_e}{D_o} \tag{7.14}$$

Note,

σ_e is material's yield stress;
D_o is external diameter of ball-shape shell;

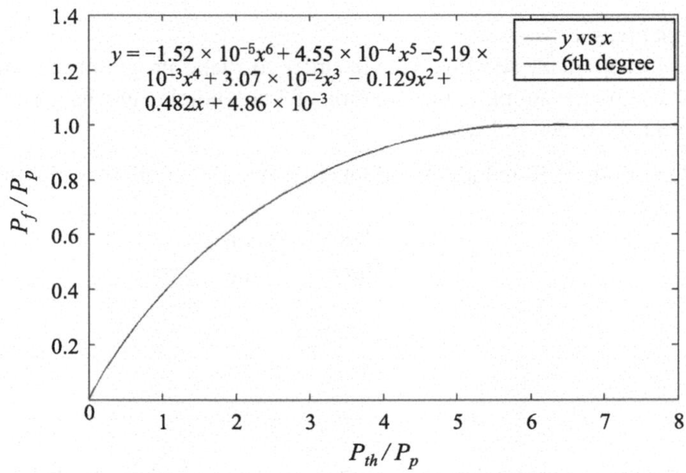

Fig. 7.3 BV (1989) relation of thin film buckling pressure and theoretical buckling pressure

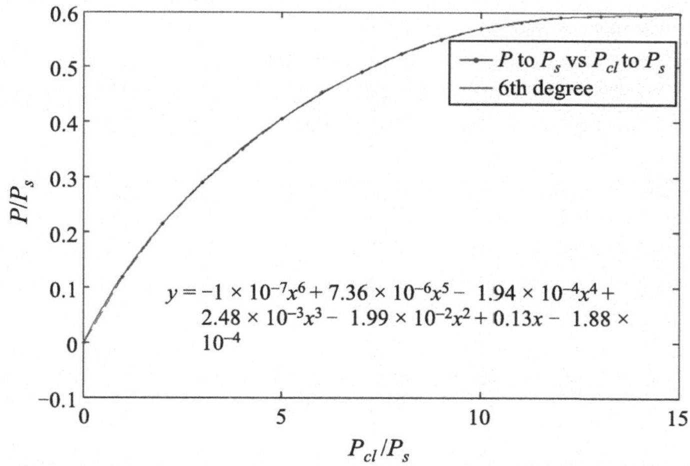

Fig. 7.4 LR (1989) relation of design pressure, elastic head lost pressure and thin film buckling pressure

V is poisson ratio

(3) LR (1989).

Design pressure P is obtained from Fig. 7.4 (to make programme computation easy, this equation adopts approximation of 6 polynomial, and its approximation equation is provided in Figure).

In Fig. 7.4, fitting formula's "x" represents value of horizontal axis, "y" represents value of vertical axis.

Ps is thin film buckling pressure, namely the pressure when film stress reaches yield stress, value see formula (7.15):

$$P_s = \frac{2t\gamma_m \sigma}{R} \tag{7.15}$$

Note,

R is nominal diameter of ball;
T is thickness of shell;
γ_m is material factor, 1.4 for alloy steel of carbon, manganese and iron; 1.1 for austenitic steel and aluminium alloy;
σ is computation yield stress. Value see Table 7.2 (only for conditions under 50 °C):
σ_y is yield limit of material at normal temperature
σ_u is ultimate tensile strength at normal temperature
$\sigma_{0.2}$ is stress when 0.2% strain, namely nominal yield limit

Table 7.2 L1989 value of computing yield stress

Material	Carbon steel, manganese steel and low alloy steel	Austenitic stainless steel	Aluminium, aluminium alloy
σ	Smaller between $\frac{\sigma_y}{1.5}$ and $\frac{\sigma_u}{2.35}$	Smaller between $\frac{\sigma_y}{1.5}$ and $\frac{\sigma_u}{2.5}$	$\frac{\sigma_{0.2}}{1.5}$

P_{cl} is elastic buckling pressure of ball-shape shell, value see formula (7.16).

$$P_{cl} = \frac{2E}{\sqrt{3(1-\nu^2)}} \cdot \left(\frac{t}{R}\right)^2 \tag{7.16}$$

Remark the original: $P_{cl} = \frac{2E}{\sqrt{3(1-\nu^2)}} \cdot (\frac{t^2}{R})$ This formula's dimension is not balanced, this book modifies it in reference with BV and ABS.

Note:

E is material's elasticity modulus, v is poisson ratio. Other parameters as defined in former formula

LR (1989) makes equation applied by limiting production deviation. The concrete rule of production deviation: the deviation between middle surface of actual shell's thickness and spherical surface of nominal radius should not be bigger than 0.5% nominal radius.

(4) RS (2004)

Verification of RS (2004) should include yield stress and buckling.

① Yield stress verification:

$$\sigma^0 = \frac{PR}{2t} \le [\sigma^0] \tag{7.17}$$

Note:

σ^0 is thin film stress

P is design pressure (MPa) $= (H_{op} + \Delta H)/100$. H_{op} is work depth(m), ΔH is hyper depth (≥ 50 m);

R is shell nominal radius (refers to middle surface of shell thickness)

$[\sigma^0]$ is allowed stress, value see Eq. (7.18):

$$[\sigma^0]=\begin{cases}\sigma_s/1.5 & 受压应力处\\ \sigma_s/2.0 & 受拉应力处\end{cases} \tag{7.18}$$

For stress-strain
For tensile stress-strain

Note:

σ_s is buckling limit

② buckling verification:

$$P \leq \frac{P_{sc}}{sf} \tag{7.19}$$

Note:

Sf is safety coefficient
P_{sc} is computation buckling pressure, value see Eq. (7.20).

$$P_{sc} = \eta P_s \tag{7.20}$$

Note:

Ps is theoretical buckling pressure, value see formula (7.21).

$$P_s = 1.21E\left(\frac{t}{R}\right)^2 \tag{7.21}$$

The definition in formula (7.21) is same as formula (7.15) and formula (7.16)

η is is correction factor, value see formula (7.22).

$$\eta = \frac{\eta_s}{\sqrt{1 + [(1 + f_s)\eta_s\delta]^2}} \tag{7.22}$$

In formula (7.22):

f_s is dimensionless max manufacturing deviation, value see formula (7.23)

$$f_s = \frac{f}{t} \tag{7.23}$$

In formula (7.23):

F is max manufacturing deviation, namely the max radial deviation between actual shell surface and real shell's surface of nominal radius.
T is defined as formula (7.15)
δ stress ratio, value see formula (7.24)

$$\delta = \frac{P_s R}{2t\sigma_s} \tag{7.24}$$

The definition of parameter in formula (7.24) is same as formula (7.15), (7.18), (7.21).

η_s is correction factor given out-of-roundness, value see formula (7.25)

$$\eta_s = \frac{1}{1 + (2.8 + f_s) f_s^{\frac{2}{3}}} \tag{7.25}$$

Note: the formula of $\eta_s = \frac{1}{1+2.8 f_s) f_s^{\frac{2}{3}}}$ in rules is apparently wrong, has been modified according to formula (Paliy 1991).

Parameters in formula (7.25) is same as formula (7.23).

RS (2004) describes manufacturing deviation by parameter f in computation formula, it could be viewed in formula (7.23)

(5) CCS (1996)

CCS (1996) will also proceed verification of stress and buckling.

③ Stress verification:

$$\sigma = \frac{PR}{2t} \leq 0.5667\sigma_s \tag{7.26}$$

When computation stress meets $\sigma_{\text{cal}} = \frac{1.5PR}{2t} \leq 0.85\sigma_s$ and computation stress in original text is computed 1.5 times as big as work pressure, the equation could be seen as the form of formula (7.26).

Note:

σ is average circumferential stress of ball-shape shell under work pressure;
P is work pressure;
R is nominal radius of shell;
T is thickness of shell;
σ_s is material yield stress

④ Buckling verification:

$$P \leq \frac{P_{cr}}{1.5} = \frac{C_s \cdot C_z \cdot P_e}{1.5} \tag{7.27}$$

Note:

P is work pressure;
Pcr is buckling pressure;
Pe is elastic buckling pressure, value see formula (7.28)

$$P_e = 0.84EC^2 \tag{7.28}$$

Note in formula (7.28)

E is elastic elasticity modulus;
C is correction factor of shell thickness and radius, value is determined by ratio of shell thickness t and shell radius R (Fig. 7.5)

The other two parameters in formula (7.27).

a. Cs is nonlinear coefficient of material physics, value is from Fig. 7.6.

Note in Fig. 7.6

σ_s is material yield limit;
σ_e is the corrected thin film stress, value see formula (7.29):

$$\sigma_e = \frac{1.5P}{2C} \tag{7.29}$$

Note in formula (7.29):

P is work pressure;

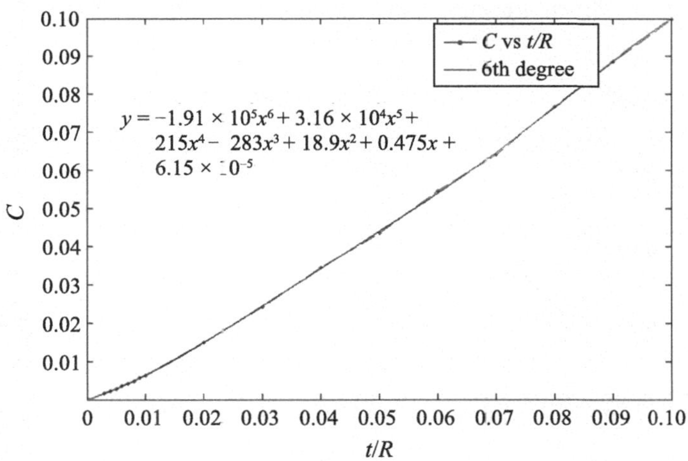

Fig. 7.5 CCS (1996) relation of parameter C and t/R

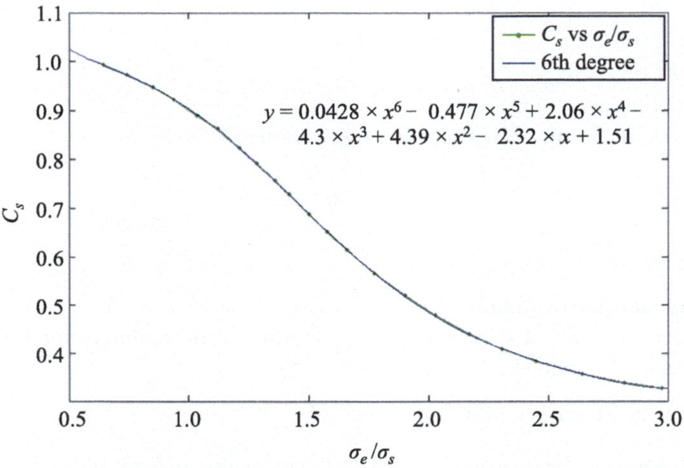

Fig. 7.6 CCS (1996) relation C s and σ_e/σ_s

C is determined by the ratio of shell thickness t and shell radius R.
b. Cz is manufacturing effect coefficient, determined by value in Fig. 7.7

The definition of parameters σ_e and σ_s as in Fig. 7.6.

One thing that deserves attention is that the scope of σ_e/σ_s in Figs. 7.6 and 7.7 is limited and so is the $0.003 \leq t/R \leq 0.1$ in atlas of coefficient in rules (CCS 1996). when given material parameters and shell radius, the shell thickness shall make the

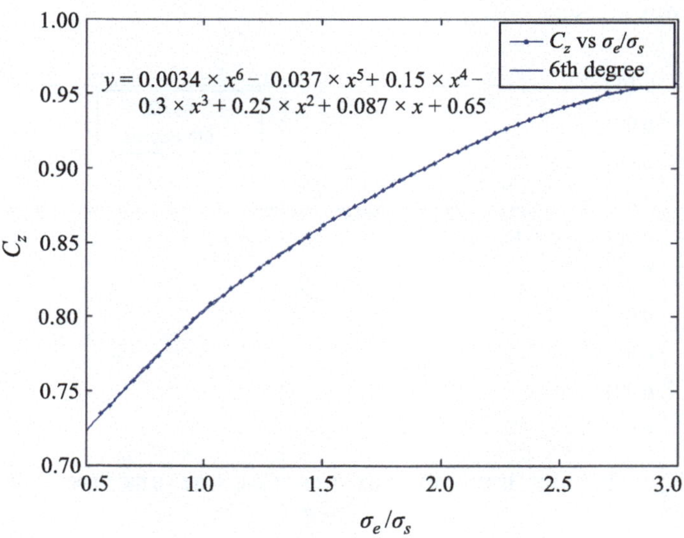

Fig. 7.7 CCS (1996) relation of Cz and σ_e/σ_s

value of σ_e/σ_s amount the scope of these Figure and table, or the verification cannot be conducted.

CCS (1996) will consider manufacturing deviation by coefficient Cz, see Fig. 7.7.

(6) ABS (2010).

$$P = \frac{P_{cs}}{sf} \tag{7.30}$$

Note:

P is design pressure (work pressure);
Sf is safety coefficient, valued as 1.5 (in rules $\frac{1}{sf} = 0.67 \approx \frac{1}{1.5}$)
P_{cs} is shell's ultimate pressure, value see formula (7.31):

$$P_{cs} = \begin{cases} P_{ys} \cdot 0.7391[1 + \left(\frac{P_{ys}}{0.3P_{es}}\right)^2]^{-\frac{1}{2}} & for \frac{P_{es}}{P_{ys}} > 1 \\ 0.2124P_{es} & for \frac{P_{es}}{P_{ys}} \le 1 \end{cases} \tag{7.31}$$

Note that this formula is different from formula in original rules, it is finally confirmed after communication with ABS: there is printing error in original rules, formula (7.30) is correct.

Note in formula (7.31):

P_{ys} is the pressure when thin film reaches yield limit, value see formula (7.32):

$$P_{ys} = \frac{2\sigma_y t}{R_o} \tag{7.32}$$

Note in formula (7.32):

R_o is shell's average exterior diameter;
σ_y is material yield stress;
T is shell thickness. P_{es}

Another parameter in formula (7.31): P_{es} is shell's elastic buckling pressure, value see formula (7.33):

$$P_{es} = \frac{2E}{\sqrt{3(1-\nu^2)}}\left(\frac{t}{R_o}\right)^2 \tag{7.33}$$

Note in formula (7.33):

E is material's elasticity modulus;
V is poisson ratio

Other parameters are defined as formula (7.32).

ABS (2010) regulates that the manufacturing deviation of manned cabin ball-shape shell shall be: the deviation between actual shell's internal radius and designed internal radius's real spherical surface shall not be bigger than 1% designed internal radius; and adopted radius equals designed internal radius, arc length shall be determined by the arc-shaped tool in Fig. 7.8, and the arc deviation between actual spherical surface and arc-shaped tool shall not be bigger than 0.5% designed internal radius.

(7) GL (2009)

GL (2009) requires verification of manned shell of stress verification and buckling verification.

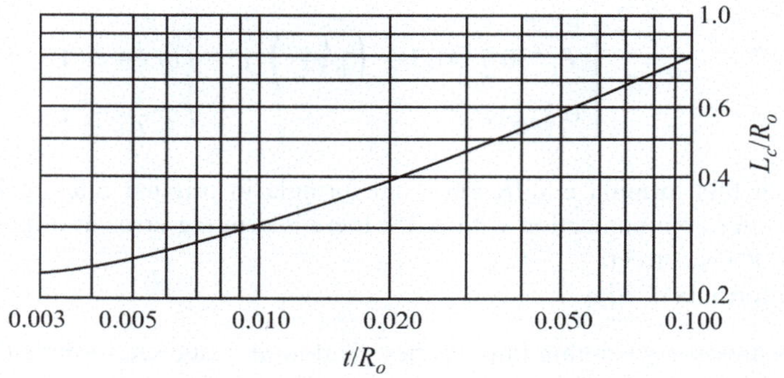

Fig. 7.8 Relation of arc length of measuring tool (Lc) and shell thickness radius ratio (t/R_o)

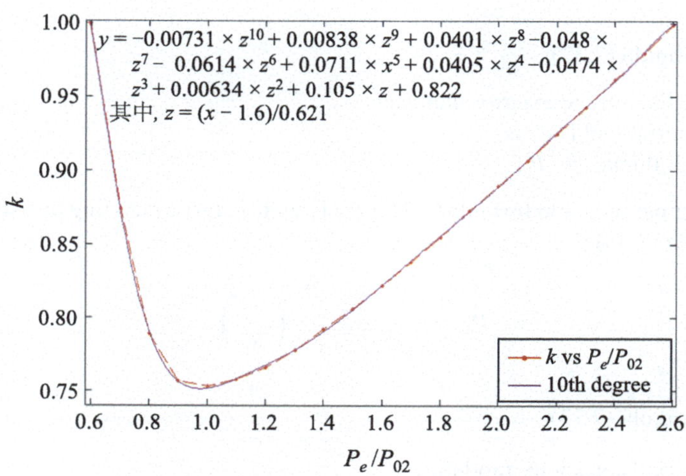

Fig. 7.9 Reduction coefficient and fitting formula of non low alloy steel material

① yield verification:

Yield verification includes three kinds of pressure: work pressure, experiment pressure and collapsing pressure.

$$\sigma = \frac{R_{ol}^2 \cdot P_{cal}}{2R_{ml}t} \leq [\sigma] \tag{7.34}$$

Note:

- σ is computation stress;
- T is shell thickness (average thickness around computation point);
- R_{ol} is the biggest local shell exterior radius including shell's out-of-roundness, the measuring tool and diagram form of shell local out-of-roundness U (also called local flattening) could be seed in appendix B-E-4, the relation formula for measuring circle's diameter (critical arc length) and local exterior radius see formula (7.35), shell local flattening U see formula (7.36). The rules points out: R_{ol} could be 1.3 times as big as nominal radius when designing (the relevant local deviation shall be 0.218t), for machined ball, the value shall be 1.05 times as big as designed exterior radius when meeting measuring local deviation is smaller than 0.035t.
- R_{ml} is biggest local shell middle surface radius given shell out-of-roundness, it could be value as $R_{ol} - t1/2$; t1 is local shell thickness average value of measured circle.
- P_{cal} is computation pressure (MPa), includes work pressure, experiment pressure and critical pressure, all of the three conditions needs verification, see formulas (7.39), (7.40), (7.41).
- P_{cal} is allowed stress. The safety coefficient in allowed press is divided into three conditions, see formula (7.42)

$$L_{c1} = \frac{2.2}{\sqrt[4]{\frac{3}{4}(1 - \nu^2)}} \sqrt{R_{ol} \cdot t1} \tag{7.35}$$

Note in formula (7.35):

- V is material's poisson ratio;
- t1 is shell thickness average value around measuring point, it could be nominal thickness when design

Other parameters are defined as formula (7.34).

$$U = x - \bar{x} \tag{7.36}$$

Note in formula (7.36):

$$\bar{x} = R_{ol} - \sqrt{R_{ol}^2 - \frac{L_{c1}^2}{4}} \tag{7.37}$$

The parameters in formula (7.37) shall be defined as the above formulas.

$$x = R_0 - \sqrt{R_0^2 - \frac{L_{c1}^2}{4}} \tag{7.38}$$

Note in formula (7.38):

R_0 is shell's nominal exterior radius. Other parameters shall be defined as the above formulas

Computation pressure:

a. Under work pressure:

$$P_{cal} = P = 0.0101 \cdot H \tag{7.39}$$

Note in formula (7.39):

P_{cal} is computation pressure (MPa)
P is work pressure (MPa);
H is design depth (m)

b. Under experiment pressure:

$$P_{cal} = \begin{cases} (3 + 11 \times P)/10 \; 0.5 \leq P < 3 \\ (12P)/10 \qquad\qquad P \geq 3 \end{cases} \tag{7.40}$$

Note in formula (7.40):

P_{cal} is computation pressure (MPa)
P is work pressure (MPa)

b. Under critical pressure:

$$P_{cal} = \begin{cases} \frac{(8+16+P)}{10} \; 0.5 \leqslant P \leq 6 \\ \frac{(17.3\times P)}{10} \quad P \geqslant 6 \end{cases} \tag{7.41}$$

Note in formula (7.41):

P_{cal} is computation pressure (MPa);
P is work pressure

Allowed stress:

$$[\sigma]=\begin{cases}\frac{\sigma_s}{1.7} & \text{工作压力下}\\ \frac{\sigma_s}{1.1} & \text{试验压力下}\\ \frac{\sigma_s}{1.0} & \text{临界压力下}\end{cases} \tag{7.42}$$

Under work pressure
Under experiment pressure
Under critical pressure

Note in formula (7.42):

σ_u is material's measured ultimate tensile strength;
σ_s is material's buckling stress

The allowed stress under work pressure in rules is $[\sigma] = \min(\frac{\sigma_u}{2.7}, \frac{\sigma_s}{1.7})$, and the rules points out that if external pressure makes shell produce compress stress, the tensile stress shall be neglected, namely $[\sigma] = \frac{\sigma_s}{1.7}$, both theoretical analysis and finite element calculation could figure out that the stress in manned cabin is compress stress (membrane stress calculated by finite element method is always negative in the whole shell thickness), so the value is $[\sigma] = \frac{\sigma_s}{1.7}$.

② Buckling verification:

The manufacturing of manned submersible's manned cabin will remove remaining stress, so this book will only adopt computation formula in rules after stress removed.

$$P \leq \begin{cases}(10P_{cdp} - 8)/16 & 1.6 \leq P_{cdp} < 10.4\\ P_{cdp}/1.73 & P_{cdp} \geq 10.4\end{cases} \tag{7.43}$$

Note in formula (7.43):

P is ultimate work pressure;
P_{cdp} is shell's critical pressure, see formula (7.44):

$$P_{cdp} = k \cdot P_{cr} \tag{7.44}$$

Note:

P_{cr} is critical pressure in low alloy steel, see formula (7.45).
K is reduction coefficient. Reduction coefficient of low alloy steel is K-1, others see Table 7.3 (as mentioned earlier, only consider conditions of removing residual stress for manned cabin). (The rules does not provide k curve of titanium, but it clarifies that when use other materials it should be negotiated with GL, and

Table 7.3 Reduction coefficient of non low alloy steel materials

P_e/P_{02}	$0.470 \le P_e/P_{02} < 0.595$	$0.595 \le P_e/P_{02} \le 2.6$	$P_e/P_{02} > 2.6$
k	1	See Fig. 7.9	1

determined through model experiment.) for titanium alloy manned cabin, due to $\frac{P_e}{P_{02}}$ is bigger than 2.6, so the factor will be counted as 1 in this book.

$$P_{cr} = \begin{cases} P_e & (P_e/P_{02} \le 0.595\text{时}) \\ P_{02}(0.475 + 0.195\frac{P_e}{P_{02}}) & (0.595 \le \frac{P_e}{P_{02}} \le 2.7\text{时}) \\ P_{02} & (\frac{P_e}{P_{02}} > 2.7\text{时}) \end{cases} \tag{7.45}$$

Note in formula (7.45):

P_{02} is theoretical elastoplasticity buckling stress, value see (7.46)

$$P_{02} = \frac{2\sigma_s t1 R_{ml}}{R_{ol}^2} \tag{7.46}$$

The parameters in formula (7.46) are defined as formula (7.34) (7.42).
Pe is elastic buckling stress, value see formula (7.47).

$$P_e = \frac{1.4}{\sqrt{3(1-\nu^2)}} \cdot E \cdot \left(\frac{t1}{R_{ol}}\right)^2 \tag{7.47}$$

Note in formula (7.47): E is material's Young's modulus.
Other parameters are defined as formula (7.35).

GL (2009) regulates:

For overall out-of-roundness: the deviation between actual shell exterior radius and real spherical surface of nominal exterior radius shall not bigger than 1% nominal exterior radius. For local out-of-roundness: with the overall out-of-roundness, measuring method of local out-of-roundness see appendix B in GL (2009) rules, the result of measured flattening U (namely local out-of-roundness) shall not bigger than 21.8% of shell nominal thickness, or the computed shell critical pressure needs correction. The rules also regulates that during design the manufacturing deviation could be valued: the biggest shell local exterior radius shall be valued as 1.3 times of nominal exterior radius; local shell thickness as nominal shell thickness.[original text: For the lay out a local radius of 1.3 times the nominal radius and a nominal thickness of the shell (eventually reduced by the corrosion addition) is to be assumed.]. For machined shell, when measured local out-of-roundness is smaller than $u = 0.035s_1/Ro$, the value could be $R_{ol} = 1.05 \cdot R_o$. (Original text: For mechanically machined spherical shells local radii less than $1.05 \cdot R_o$ are reachable from point of manufacturing. The more favorable geometrical condition of the shell can be introduced in the calculation with at minimum $R_{ol} = 1.05 \cdot R_o$ under the assumption that the measurement procedure, as described in Annex B, has proven a maximum permissible local flattening of $u = 0.035s_1/Ro$ with an accuracy of at least $0.001s_1$).

Compare the computation methods in these rules: (1) input parameter comparison.

This book will compare the input parameters in these rules and the input parameters are divided into geometry dimension, material parameters, manufacturing deviation and compensation coefficient (safety coefficient), after comparison it could be found that the parameters in these rules is not unified, see Table 7.4.

Table 7.4 Comparison of input parameters of computation methods of all classification society

Parameter category	DNV (1988)	BV (1989)	LR (1989)	RS (2004)	CCS (1996)	ABS (2010)	GL (2009)
Geometry dimension of manned cabin	Shell radius and thickness	Shell radius and thickness	Shell radius and thickness	Shell radius and thickness	Shell radius and thickness	Shell radius and thickness	Shell radius and thickness
Material parameter	Young's modulus, buckling stress	Young's modulus, buckling stress, poisson ration	Young's modulus, buckling stress, poisson ration	Young's modulus, buckling stress	Young's modulus, buckling stress	Young's modulus, buckling stress, poisson ration	Young's modulus, buckling stress, poisson ration
Manufacturing deviation		Shell local max and min diameter, deviation between actual shell exterior surface and designed real ball exterior surface		Deviation between actual shell exterior surface and designed real ball exterior surface			Local radius deviation and shell thickness procession deviation
Compensation coefficient (safety coefficient)				Safety coefficient	Safety coefficient	Safety coefficient	

As for the same necessary computation parameters, the requirement and regulation of each rules is different, for example, only ABS (2010) clearly points out the computation methods could be applied to titanium alloy, while other rules is established based on steel. So whether the computation method of rules could be applied to titanium is not inspected, and the material curves given in each rules is only for steel; besides, there are different requirement on allowed scope, measuring methods and measuring tools of manufacturing deviation in each rules.

(2) Computation methods

It can be noted that the common basis of all of these rules is circumferential stress computation formula of shell middle surface (when the circumferential stress of shell middle surface reaches material's yield stress, the relevant exterior pressure is yield pressure, its formula is called theoretical yield pressure formula and elastic critical buckling pressure, but there is different key points and correction coefficient in rules of various classification society). The previous rules, like DNV (1988) and LR (1989), are established mainly based on theoretical yield pressure, structure defects (including out-of-roundness, thickness deviation and material defect, etc.) and buckling in theoretical buckling formula is corrected by reduction coefficient. In 1990s, the rules, like RS (1994) and CCS (1996) corrected yield pressure and buckling pressure at the same time, the smaller value is used as design pressure, these rules thought the main effect for structure defects comes from buckling pressure, so the reduction coefficient representing structure defect corrected the computation value of elastic critical buckling pressure (this is different from the correction for buckling pressure by precious rules). in the newest (ABS 2010) rules, the computation of shell ultimate bearing capacity will be divided into two stages according to shell thickness radius ratio (ffR): when ffR is really small (namely thin shell), the shell will encounter buckling break first, and then the central computation formula is elastic critical buckling pressure; when ffR is bigger than some value, the shell will be broken by yield stress and then the computation formula is based on buckling pressure formula. It could be noted some development track of mechanics: strength problems are focused in precious period; then buckling are found as important as strength; after further understanding the relation of buckling and strength, clarifying when break happens due to strength and when due to buckling.

When verify the manned cabin design of serving manned deep-ocean submersible by these rules, it is found that most serving manned cabins cannot pass the verification of these rules, see Table 7.5. This reflects that currently used computation formula of manned submersible rules cannot accurately compute manned cabin's bearing capacity, when these rules is used for RBDO, it means that the state function itself has obvious deviation and this will significantly impact result of RBDO, so it is necessary to establish more accurate manned bearing capacity computation formula.

According to comparison of available formulas of rules, it is apparent that the computation of manned cabin bearing capacity is divided into two stages as per shell thickness radius ration (ffR): when ffR is small (thin shell), the shell will first encounter buckling break, then the major computation formula is computation

Table 7.5 Verification of rules for serving titanium alloy manned cabin (Pan and Cui 2011b, c)

Name of submersible	"Lingshi" ("Ross")	"Pearly nautilus"	Deep ocean 6500	"Alvin"	"New alvin"	"Jiaolong"
Depth/m	6000	6000	6500	4500	6500	7000
Work pressure equated as per formula of GL (2009) /MPa	60.60	60.60	65.65	45.45	65.65	70.70
Internal radius of pressure shell/m	2.1	2.1	2.0	2.0	2.1	2.1
Actual thickness of pressure shell/mm	71	62–73	75	49	71–72	76
Safety coefficient of pressure shell	1.50	1.50	1.55	1.50	1.50	1.50
Rules	Min thickness as per rules (keep one decimal place)					
DNV (1988)	78.2	78.2	79.5	59.7	83.5	88.7
BV (1989)	86.9	86.9	90.4	62.6	94.9	103.0
LR (1989)	99.0	99.0	101.0	75.6	106.1	113.4
RS (2004)	68.9	68.9	72.7	51.1	74.0	79.2
CCS (1996)	72.7	72.7	77.8	51.5	78.9	85.2
ABS (2010)	97.6	97.6	103.3	72.0	105.1	112.9
GL (2009)	83.0	83.0	85.9	59.2	90.2	97.5

formula of elastic critical buckling pressure; when ffR is bigger than some value, the shell will be broken due to stress reaching material's yield stress, and then the computation formula shell be based on theoretical yield stress. So the new formula shall include buckling break and yield break, in order to establish comparatively precise computation formula, there shall be enough shells to be researched, while cost of implementing pressure cyclinder model experiment is too high, too long time, the number of experiment sample is limited. In addition, the published data of model spheres break experiment is very few, so the method of totally adopting model spheres break experiment is not feasible in cost and time. As the nonlinear finite element is generally accepted as effective method of computing shell's bearing capacity, including out-of-roundness (Lu et al. 2004; Wang et al. 2007; Depei and Changchun 1991), this book will adopt many nonlinear finite element computation to research the bearing capacity of shell with different thickness radius and out-of-roundness. This book also carried pressure cylinder break experiment of four model spheres of internal radius 500 mm to verify new formula and manufacturing level of domestic titanium alloy shell and obtain relevant data.

More detail about this section could be obtained in documents (Pan and Cui 2010; 2011b, c).

7.1.1.2 Nonlinear Finite Element Computation and Establishing New Formula

After investigation and research of available finite element analytic documents about manned cabin, there are two kinds of modeling shell out-of-roundness, and the relevant analysis procedure could be divided into two kinds (Lu et al. 2004; Wang et al. 2007):

(1) The first out-of-roundness modeling method is based on first-order modal of linear buckling analysis. So linear buckling analysis shall be done prior to nonlinear analysis. After extraction of first-order buckling modal, update node location according to deformation of modal and appointed out-of-roundness deviation, the shell finite element model including out-of-roundness could be obtained. The later nonlinear element analysis will compute this model, and the analysis procedure of this type in this book is called analysis procedure I, see Fig. 7.10. It is clear that this type procedure is distributed according to most hazardous first-order modal given out-of-roundness waveform, so the shell's actual ultimate strength is usually higher than the ultimate strength of this type nonlinear finite element computation.
(2) The second method of out-of-roundness modeling mainly researches single local out-of-roundness and ultimate arc length. In the beginning of modeling, the local shell radius is equated according to ultimate arc length and appointed out-of-roundness, and the spherical surface of ultimate arc length is established local shell radius, then the nonlinear computation will be carried in the finite element model with single local out-of-roundness, this type analysis in this book is call analysis procedure II, see Fig. 7.11.

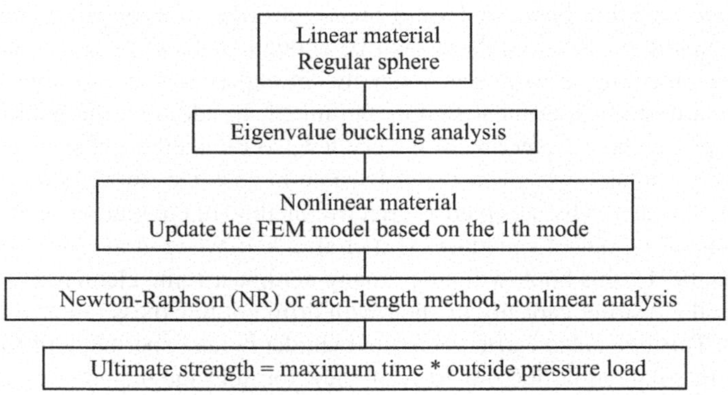

Fig. 7.10 Analysis of manned cabin nonlinear element

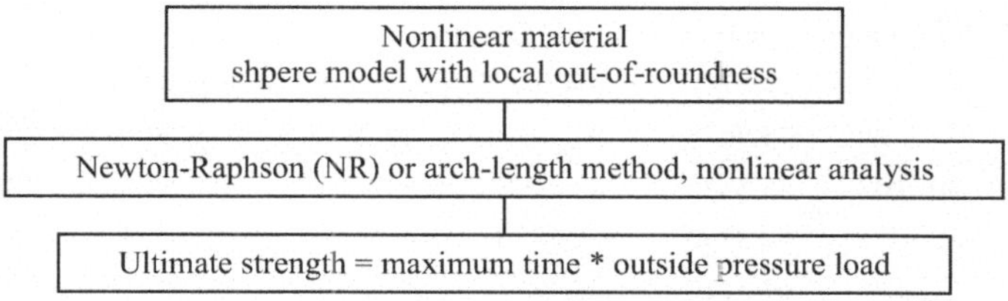

Fig. 7.11 Manned cabin nonlinear finite element analysis procedure II

This book adopts the APDL of Ansys to write the parameterized analysis documents of the two procedures. After analyzing bearing capacity nonlinear finite π of manned cabin with internal radius 1050 mm in documents (Lu et al. 2004; Wang et al. 2007), the difference of result is not obvious, see Table 7.6. The main parameters of finite element model and nonlinear analysis procedure in this book see Table 7.7.

After compiling the parametarized analytic documents of APDL, orthogonal method is adopted to make shell experiment design (DOE) with different thickness and sphericity (out-of-roundness), and the bearing capacity of different shells is computed in two nonlinear analysis procedures, the result of procedure I and II see Tables 7.8 and 7.9. In addition, in the computation of procedure II, the different ultimate arc length experience formulas is computed and compared, it is concluded that the bearing capacity nearing ultimate arc length is not sensitive to arc length variation, and the bearing capacity is basically same after modeling out-of-roundness by available ultimate arc length experience formula and computing.

After analyze the data of computed result in two procedures, the relation of bearing capacity and thickness radius and out-of-roundness is close to linear relation, as

Table 7.6 Comparison of nonlinear finite element result (Pan and Cui 2010)

Type I		Type II	
Lu et al. (2004) result	Result of this book	Document (Wang et al. 2007) result	Result of this book
105.84 MPa	105.61 MPa	114.8 MPa	113.87 MPa

Table 7.7 Model description of finite element analysis in this book (Pan and Cui 2010)

Selected elements	Solid186 3D 20 node hexahedron high order elements
Model of nonlinear material	Multilinear kinetic hardening mises plastic
Mesh generation	At least two elements in direction of thickness, 36–56 elements in direction of circumference, for different t/R, the number circumferential mesh shall be adjusted to make the nonlinear solving converged
Nonlinear equation server	Adopt Full Newton-Raphson method, because after test it is found that Ansys' arc length robustness is not enough. Adopt the load of single payload with 500 substeps, and parallel computation is used

Table 7.8 DOE result of procedure I (Pan and Cui 2010)

DOE	$\Delta = 0$	$\Delta = 0.002$	$\Delta = 0.004$	$\Delta = 0.006$	$\Delta = 0.008$	$\Delta = 0.0010$
$t = 0.025$	42.2264	35.5645	31.7912	28.2557	26.2309	24.2418
$t = 0.030$	50.2044	44.5520	41.3998	37.9980	35.2206	32.7937
$t = 0.035$	58.5220	52.9750	50.1870	47.3267	44.6434	42.1828
$t = 0.040$	67.6643	61.3210	59.1814	57.0389	54.8175	52.5882
$t = 0.045$	75.2425	69.6655	67.5164	65.3997	63.2712	61.1309
$t = 0.050$	83.6028	78.0414	75.8233	73.7998	71.7054	69.6306
$t = 0.055$	91.9631	86.3533	84.3484	82.2885	80.2270	78.1672
$t = 0.060$	100.3234	94.6132	92.7007	90.6407	88.7124	86.7011
$t = 0.065$	108.6837	102.9706	100.9250	99.0990	97.1634	95.1613
$t = 0.070$	117.0439	110.3607	108.8252	106.2833	105.8379	103.4340
$t = 0.075$	125.4042	117.6252	116.1127	114.1949	112.6258	110.1402
$t = 0.080$	133.7645	125.6373	123.9327	121.5143	120.5034	117.8048

Table 7.9 DOE result of procedure II (Pan and Cui 2010)

DOE	$\Delta = 0$	$\Delta = 0.002$	$\Delta = 0.004$	$\Delta = 0.006$	$\Delta = 0.008$
$t = 0.025$	34.7410	30.3286	26.7306	23.8279	21.4552
$t = 0.030$	43.6015	39.5735	35.9156	32.7443	29.9900
$t = 0.035$	52.0498	48.5915	45.0736	41.8223	38.8767
$t = 0.040$	60.3975	57.2364	54.0507	50.8763	47.8709
$t = 0.045$	68.7244	65.7761	62.7771	59.7515	56.8159
$t = 0.050$	77.0194	74.1874	71.3098	68.4680	65.6002
$t = 0.055$	85.2787	82.5778	79.7956	77.0422	74.2601
$t = 0.060$	93.3880	90.8576	88.1938	85.5086	82.8479
$t = 0.065$	101.6130	99.1246	96.5957	93.9674	91.3558
$t = 0.070$	100.6996	107.3586	104.8778	102.3370	99.7932
$t = 0.075$	117.9411	115.4806	113.1037	110.6780	108.2162
$t = 0.080$	126.0550	123.5726	121.2987	118.9746	116.5056

shown in FigS. 7.12 and 7.13, this book proposes new bearing capacity computation formula for manned cabin with formulas in available rules as guide (Pan and Cui 2010):

$$P_u = \left(1 - k\frac{\Delta}{R}\right)P_{up} = \left(1 - k\frac{\Delta}{R}\right)\left(\frac{\sigma_b t}{R} + \frac{\sigma_b t}{R + t/2}\right) \tag{7.48}$$

Note:

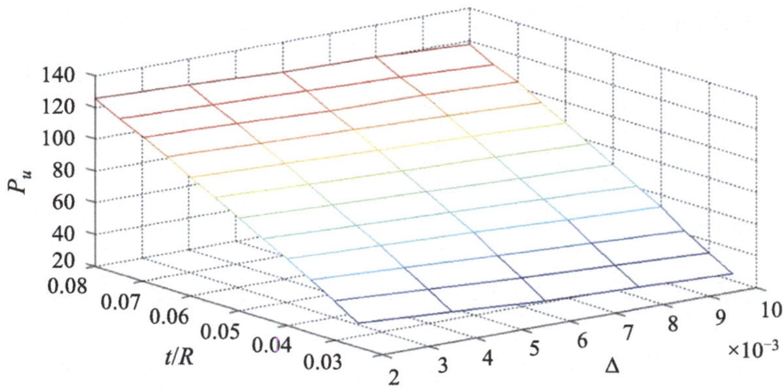

Fig. 7.12 Procedure I bearing capacity and relation t/R and Δ

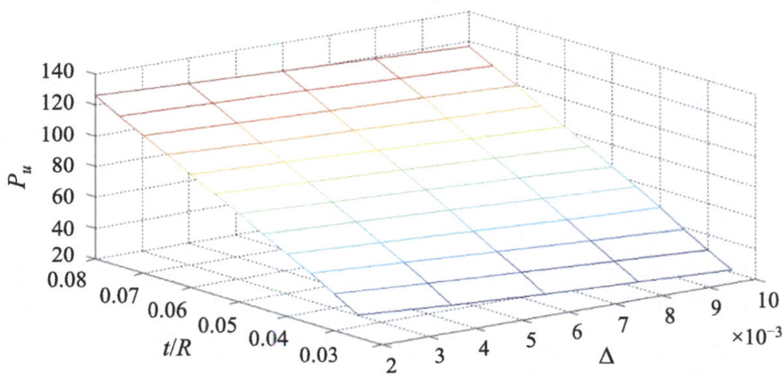

Fig. 7.13 Procedure II bearing capacity and relation t/R and Δ

$$k = a + b \times \exp\left(- c\left(\frac{\Delta}{R}\right) - d\left(\frac{t}{R}\right) + j\left(\frac{\Delta}{R}\right)^2 + f\left(\frac{t}{R}\right)^2 - g\left(\frac{\Delta}{R}\right)^3 - h\left(\frac{t}{R}\right)^3\right)$$

R is internal radius of manned cabin, t is shell thickness of manned cabin, Δ is sphericity of manned cabin (manufacturing deviation) (see Fig. 7.14), σ_b is material's ultimate tensile strength, the coefficient of k see Table 7.10.

To verify new formula, first adopt new formula to compute the model sphere in published documents (Yokota and Murate 1987), and compare the result with result of pressure cylinder break experiment, see Table 7.11, it is clear that the biggest deviation between computation result of new formula and break experiment result in documents (Yokota and Murata 1987).

In order to further verify new formula and clarify material performance of domestic titanium alloy shell and complete manufacturing technique, this book witnesses the pressure cylinder break experiment of four domestic model spheres

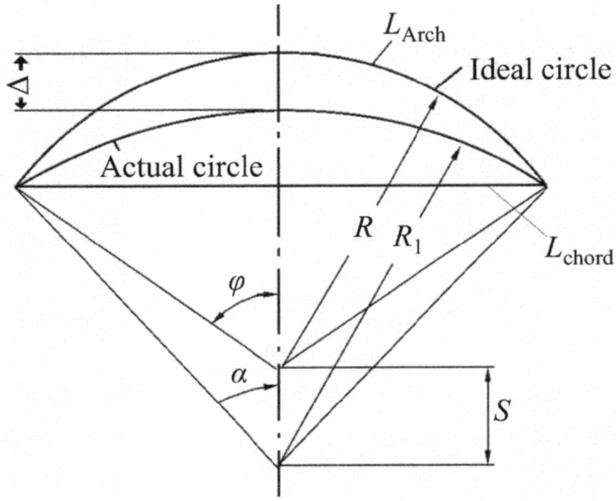

Fig. 7.14 Verification of model sphere

Table 7.10 Coefficient of bearing capacity formula of manned cabin

a	b	c	d	e	f	g	h
15.63	606.6	264.6	72.72	3×10^4		1.2×10^6	3969

Table 7.11 Comparison of collapse pressure of model spheress of article (Yokota and Murata 1987)

No. of model spheres of article (Yokota and Murata1987)	MT-1	MT-2	MT-3
Test collapse pressure (MPa)	120.62	123.56	124.93
Caculated collapse pressure of the proposed new formula (MPa)	121.77	122.76	123.1

(see Fig. 7.15) with internal radius 500 mm including one strengthened enclosure bulkhead.

(1) Data collection prior to experiment.

Four model spheress are coded according to manufacturing date: model spheress 1 and 2 are domestic ball TC4 manufactured by manufacturer A when researching manned submersible "Jiaolong" a few years ago; model spheres 3 is domestic model spheres Ti80 by manufacturer A when researching the project of the first stage of 4500 m manned submersible titanium alloy manned cabin; model spheres 4 is named TC4ELI manufactured by another domestic manufacturer B.

The parameters should be input when adopt new formula to compute shell's bearing capacity, such as material performance, shell's geometry parameters, and manufacturing deviation. So the data from ① to ③ shall be obtained before computing and Pressure test.

Fig. 7.15 Model spheres

① Material property

The model spheres 1 and 2 were not manufactured for break experiment, so there is no the detailed material property about welded joint in report, but welding joint factor, WJF, with its value as 1, (namely the ratio between property of welded joint and parent material is 1); model 3 and 4 were manufactured with consideration of needed parameters by break experiment, and the weld of manned cabin of 4500 m manned submersible adopts narrow gap welding technique. The weld technique gets the equal strength of large thickness weld joint by the meshing effect of week weld stick and strong parent material. Model 3 and 4 adopts weld sticks similar with actual manned cabin to weld, but the weld technique is not narrow gap weld due to model spheres's thin thickness, so the weld joint factor cannot reach 1. The material strength property of four model spheress provided by vendor see Table 7.12.

② Thickness measurement

The project has required vendors to provide shell thickness measurement report of four model spheres, but in order to ensure accuracy, this book adopts

Table 7.12 Material strength property of model spheress

Model spheres No.	1#	2#	3#	4#
Base material yield strength $\sigma_s(MPa)$	925	925	890	888.33
Base material tensile strength $\sigma_b(MPa)$	990	990	958.33	925
Weld joint yield strength factor	$WJF = 1$	$WJF = 1$	$WJF = 0.8858$	$WJF = 0.8668$
Weld joint tensile strength factor	$WJF = 1$	$WJF = 1$	$WJF = 0.8991$	$WJF = 0.9189$

PANAMETRICS-NDT ultrasonic thickness meter 26MG (accuracy 0.01 mm) to measure and verify shell's thickness, the end socket is used as standard test block to mark wave velocity of ultrasonic thickness meter during measurement, the statistic information of four model spheres thickness see Table 7.13.

③ Sphericity measurement

Model 1, 2 and 3 adopts HEXAGON stationary trilinear coordinates measuring instrument to measure manufacturing deviation, see Fig. 7.16; while in the report of model 4 the manufacturer B adopts mobile trilinear coordinates measuring instrument to adjust the sphericity measurement of whole ball, see Fig. 7.17, and the exterior surface of model 4 is polished after welding, so this book will apply the measurement result of manufacturer B and will not measure any more. The sphericity of all model spheress see Table 7.14.

(2) Pressure test procedure

When relative data is collected, four model spheress are delivered to pressure laboratory, which is responsible for pasting strain gauge and preparing experiment. To avoid the constraint effect by supporting mechanism for the deformation and break of model spheres, the model spheres is put into the special frame freely without any fixation, as Fig. 7.18.

In order to obtain as much information as possible from Pressure test, the experiment procedure of each model spheres is arranged carefully. Each model spheres

Table 7.13 Model spheres thickness measurement

Model spheres No.	1#	2#	3#	4#
Scope of northern hemisphere/mm	8.23–8.68	9.42–9.71	9.56–9.87	9.28–9.47
Average thickness of northern hemisphere /mm	8.3671	9.5916	9.7373	9.3763
Scope of southern hemisphere/mm	7.96–8.73	9.41–9.79	9.52–9.76	9.06–10.1
Average thickness of southern hemispherer/mm	8.4849	9.5824	9.5827	9.2436

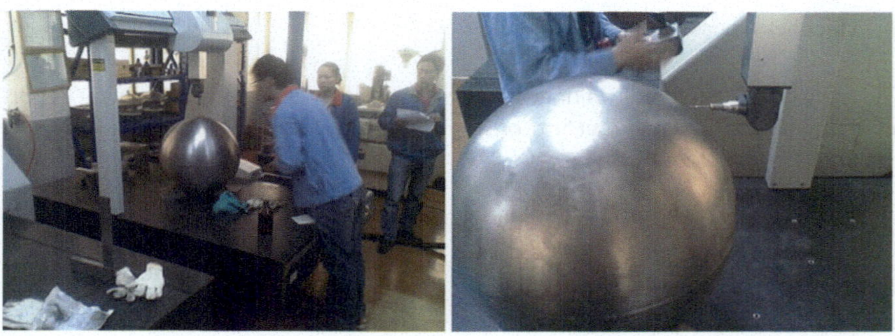

Fig. 7.16 Stationary trilinear coordinate measurement instrument

Fig. 7.17 Mobile trilinear coordinate measurement instrument

Table 7.14 Sphericity of model spheres

Code of model spheres	1#	2#	3#	4#
Hemisphere/mm	0.2868	0.3292	0.2597	–
Complete ball/mm	0.6132	1.8124①	1.0625	0.6②

Note ① the sphericity of the complete ball of 2# surpasses the allowed value in design rules (250*0.5% = 1.25, unit mm), and it's caused by the visible assembly deviation in the equator weld of two hemispheres, so actually this value includes both sphericity and assembly deviation. ② this data is obtained from manufacturer's report

will experience three cyclic pressure loading in pressure cylinder (Fig. 7.19): the first cyclic pressure loading is pre-loading cycle, which is intended to test the sealing of the set of experiment devices and their reading, meanwhile to eliminate stress of weld joint, during this cycle 5 MPa shall be increased in each loading step, namely the pressure step is 5 MPa, 3 min later the pressure shall be increased 5 MPa, the same shall be done until the pressure reaches work pressure and then the pressure shall be decreased to atmospheric pressure according to same step and pressure keeping period; the second cycle is strain measurement cycle, the pressure increasing and decreasing shall be same as the first cycle; the third cycle is implementing break experiment, the pressure keeping period is still 3 min, initial pressure step is 5 MPa, but when strain gauge shows that plastic deformation starts to appear the pressure step shall changed to 3 MPa, 1 MPa, even 0.5 MPa, in order to catch the ultimate bearing capacity of model spheres as accurately as possible.

(3) Experiment result and comparison

The broken model spheres is shown as Fig. 7.20, the comparison of experiment break pressure and break pressure computed by new formula is shown in Table 7.15. In

Fig. 7.18 Lifting model spheres into pressure cylinder

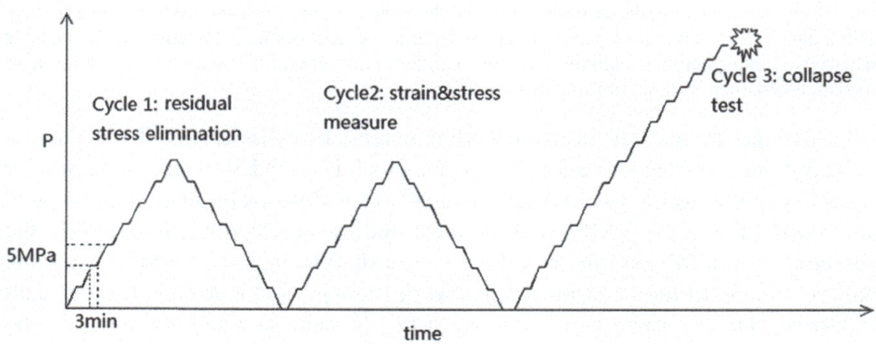

Fig. 7.19 Pressure test pressure increasing procedure (Pan et al. 2012)

Table 7.15, the pressure low limit is computed according to min thickness, sphericity of whole ball and strength of weld joint; and the computation is according to smaller average thickness of southern and northern hemisphere, hemispheric sphericity and parent material strength.

It is shown that both computation result of new formula and experiment break pressure deviation are within the scope of engineering accuracy.

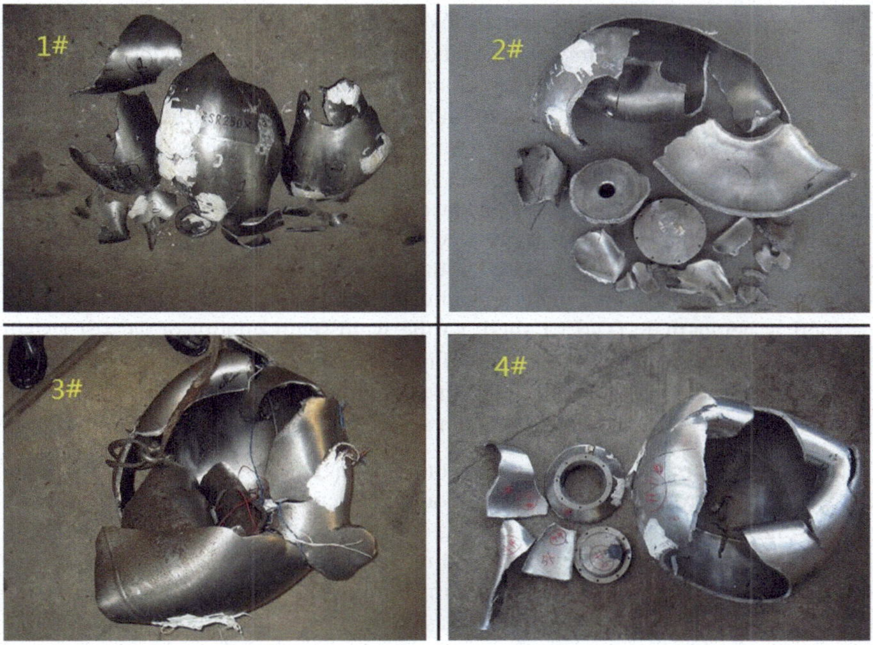

Fig. 7.20 Condition of model spheres after break

Table 7.15 Comparison of break pressure computed by new formula and experiment break pressure

No. of model spheres	1#	2#	3# 4#	
Break pressure computed by new formula/MPa	52.8–56.4	55.5–56.9[a]	53.8–60.4	52.9–59
Experiment break pressure/MPa	56	58.29	57.8	55

Note [a]due to the docking deviation at equator weld seam between southern and northern hemisphere (see the specification of Table 7.14), this value is computed according to whole ball sphericity, if computed according to southern hemispheric sphericity $\Delta = 0.3292$, the upper limit value of estimated pressure will be increased to 68.7 MPa

It is shown in Fig. 7.20 that the model 1 and 2 are broken into many fragments, while the major part of model spheres 3 and 4 is not separated with shell, which reflects that the toughness of new domestic titanium alloy material is better than old TC4. It can be concluded after entrusting units concerned to make detailed fracture analysis for shell:

① Experiment titanium alloy shell is mainly broken by shearing stress.
② The titanium alloy pressure shell is broken into many fragments after buckling, the main reason is because of titanium alloy's high yield strength and relevant low toughness, which makes the shearing stress very high under buckling pressure,

even close to shearing yield strength. When model spheres deforms due to buckling, the crack starts from weld seam because stress condition changes suddenly and the change happens with initial sunken area as center and high toughness weld seam as edge. And stress continuously exists on the broken tank because the break transmits faster than tank's pressure relief. The stress on sunken area makes the break expand outward with sunken area as its center, meanwhile other part starts to sink inward under outer pressure and breaks the tank with more clefts. Due to the sudden stress change on tank and different expanding paths of clefts, fragments will be separated with tank once the paths meet.

③ In conclusion, the parent material and weld seam's mechanics property of Ti80 shell is better than TC4ELI shell, so the weld seam quality of TC4ELI shell shall be strengthen; while one pretty big layer defect parallel with shell surface is detected in Ti80 shell, thus the inspection shall be strengthened in actual manufacturing to avoid layer defect in actual manned cabin.

This book is going to pay attention to comparison of new formula and experiment result, will not introduce more content about this section.

Up to now, new manned cabin bearing capacity computation formula has been established after systematic research, and its accuracy has been verified by aboard and domestic experiment result, with verification of china classification society the formula has been listed into principle of classification review of manned submersible "Jiaolong", also has been accepted as manned cabin computation formula of underwater system and manned submersible classification and manufacturing rules 2013 edition. The establishment of new formula provides high accuracy state function for manned cabin's reliability design.

7.1.2 Statistic Information of Titanium Material Strength

Parallel with the project at the first stage of 4500 m manned submersible domestic titanium alloy manned cabin, we conducted more than 15 mechanical experiments under technological conditions of thick plate rolling, ball clack pressing and welding ball clack into quarter ball of domestic Ti80 and TC4ELI material, including stretching, compressing, shock, fracture toughness (KIC), stress corrosion fracture toughness (KISCC), cold bend, creep, dynamic tear, high-cycle fatigue, fatigue crack growth rate, fatigue crack growth threshold, load spectrum fatigue crack growth rate and compression fatigue. This book obtains many property data of domestic titanium alloy under the above three technique conditions, such as strength, toughness, fatigue and creep, provides many material property information for manned cabin RBDO, the detailed process of sampling, processing, experiment and data statistics, will not be introduced here, nor the detailed experiment data. And only material's ultimate tensile toughness property is used in computing bearing capacity in new formula, so this book will just provide statistic and analytic result of material's ultimate tensile property. Besides there are 36 efficient samples for Ti80 plate, TC4ELI plate and

Ti80 ball clack; 29 efficient samples for TC4ELI ball clack; 13 and 12 for Ti80 and TC4ELI weld joints respectively.

In statistic process, this book adopts normal distribution, log-normal distribution, Weibull distribution, extreme value distribution and logistic distribution to conduct experiment data statistics and analysis, and selects suitable distribution function according to average deviation of cumulative distribution value (CDF) and experience cumulative distribution value (ECDF), finally obtains the most suitable distribution type and its statistic parameters as in Table 7.16.

Given that SLRBDO algorithm only supports normal distribution, this book adopts normal distribution analyze the distribution of ultimate tensile toughness of domestic titanium alloy under three technique conditions, the result is listed in Table 7.17 for the utilization of SLRBDO algorithm.

It is shown in Tables 7.16 and 7.17 that the ultimate tensile toughness of two domestic different titanium alloy have the following characters under different technique conditions:

Table 7.16 Statistic information of domestic titanium alloy material ultimate tensile toughness

Material property parameter	Statistic character	Plate		Ball clave		Weld joint	
Ultimate tensile toughness	Material	Ti80	TC4ELI	Ti80	TC4ELI	Ti80	TC4ELI
	Distribution type	Weibull	Weibull	Logistic	Logistic	Logistic	Logistic
	Average value	867.9867	919.5666	839.8414	925.3538	857.3934	859.9860
	Standard deviation	20.3397	22.8569	14.5366	18.9302	41.4467	35.1294
	Density function parameter[a]	$a =$ 867.9867 $b =$ 11.2139	$a =$ 929.7650 $b =$ 50.8827	$\mu =$ 839.8414 $\sigma =$ 8.0145	$\mu =$ 925.3538 $\sigma =$ 10.4367	$\mu =$ 857.3934 $\sigma =$ 22.8507	$\mu =$ 859.9860 $\sigma =$ 19.3678

Note [a]This row of data could also be computed according to distribution type, average value and standard deviation

Table 7.17 Ultimate tensile toughness normal distribution statistic parameters of domestic titanium alloy material

Material property parameter	Statistic character	Plate		Ball clave		Weld seam	
Ultimate tensile toughness	Material	Ti80	TC4ELI	Ti80	TC4ELI	Ti80	TC4ELI
	Average value	866.7194	919.7806	839.5278	924.1724	859.3846	861
	Average deviation	20.4587	21.1646	14.4963	20.9286	39.7524	35.3913

Table 7.18 Material property table of min statistic deviation distribution of manned cabin RBDO

Material		Ti80	TC4ELI
	Distribution type	Logistic	Logistic
Ultimate tensile toughness	Average value	839.8414	925.3538
	Standard deviation	14.5366	18.9302
	WJF	1	0.93

Table 7.19 Material property manned RBDO normal distribution

Material		Ti80	TC4ELI
	Distribution type	Normal	Normal
Ultimate tensile toughness	Average value	839.5278	924.1724
	Standard deviation	14.4963	20.9286
	WJF	1	0.93

① Under the technique condition of plate and ball clave (namely manned cabin titanium alloy parent material toughness), the ultimate tensile toughness of TC4ELI is higher than that of Ti80;
② Under the technique condition of narrow gap weld joint (namely manned cabin titanium alloy weld seam toughness), the ultimate tensile toughness of TC4ELI and Ti80 is close to each other;
③ The toughness of Ti80 decreases after pressed into ball clave;
④ The toughness of TC4ELI changes a little after pressed into ball clave;
⑤ The toughness of Ti80 weld joint almost equals that of parent material;
⑥ The coefficient of TC4ELI weld joint is around 0.93.

In summary, the material property of Ti80 and TC4ELI in manned cabin RBDO shall be the ball clave property closest to actual manned cabin and the weld coefficient of ball clave shall be computed. Material ultimate tensile toughness is shown in Tables 7.18 and 7.19 when manned cabin's RBDO is done under conditions of min deviation distribution type and normal distribution.

7.1.3 Uncertainty of Other Parameters

The input parameters of new formula (formula 7.48) includes not only material ultimate tensile toughness, but also thickness, internal radius and sphericity, the statistic information of these parameters cannot be obtained through large amount of experiment, so this book will determine their random character by referring to available standards and project experience combined with some subjective judgment.

7.1.3.1 Manned Cabin Internal Radius and Sphericity

This chapter has pointed out that there is some relation between manned cabin internal radius R and sphericity (manufacturing deviation) Δ, while in design stage, domestic manufacturer's capacity of accurately controlling manned cabin internal radius and sphericity has not been verified, namely there is not historic data for reference, so the requirement f or processing parameters of these two manned cabins' dimension is generally as:

① $(1-0.5\%)R_{req} \leq R_{act} \leq (1+0.5\%)R_{req}$, R_{act} is the actual internal radius, R_{req} is designed internal radius (the designed internal radius of 4500 m submersible manned cabin is 1050 mm); R_{req}

② $-0.5\% \times R_{req} \leq \Delta \leq 0.5\% \times R_{req}$, Δ is actual sphericity, R_{req} is defined as ①.

Thus this book will be based on these two requirements to establish random characters of R and Δ.

Firstly, R is one mechanical processed dimension which usually subjects to normal distribution, so this book assumes that R subjects to normal distribution and that actual internal radius will meet the requirement of ① with confidence of 95%. Finally $P\left((1-0.5\%)R_{req} \leq R_{act} \leq (1+0.5\%)R_{req}\right) = 0.95$ will reflect R's standard deviation as 2.6786, and R$\sim$N (1050,2.6786).

Due to regulations in available rules, sphericity A is computed as max allowed value ($\Delta = 0.5\% \times R_{req}$) in design stage,, so this book will regard sphericity as determined value $\Delta = 0.5\% \times R_{req}$ in actual computation.

7.1.3.2 Thickness

After negotiating with domestic manufacturers about processing ability, the thickness processing accuracy is confirmed to vary with 1 mm, with a view to actual thickness of "Jiaolong" manned cabin, namely:

$$t_{design} - 1mm \leq t_{act} \leq t_{design} + 1mm \tag{7.49}$$

Note: t_{act} is actual thickness, t_{design} is design thickness.

$t \sim N(\mu_t, 0.5102)$. Similarly, according to the process of confirming internal radius R statistic parameters of manned cabin, it is reasonable to think that thickness subjects to normal distribution:

7.1.3.3 Bore Pressure of Manned Cabin

Safety scope of manned cabin bearing capacity (namely the probability constrain conditions of RBDO) is:

$$P_u - P \geq 0 \tag{7.50}$$

Note: P is the exterior pressure bore by manned cabin during its lifetime.

During manufacturing verification, ocean test and re-inspection, manned cabin will experience Pressure test, 1.15 times as big as work pressure (Pan and Cui 2011b, c), manned submersible will experience the sea water pressure not more than pressure of 4800 m in its lifetime, (namely 300 m surpassed).

However, the density of seawater differs at different sea area, and varies in different climates and ocean current. In addition, the acceleration of gravity changes other dimensionalities and altitudes, so the ocean pressure at depth of 4500 m is not fixed, so is P.

The measured pressure in depth 4500 m during sea trial of "Jiaolong" manned submersible is 45.7635 MPa, the max load of manned cabin in its lifetime is 1.15 * 45.7635. This book assumes that P subjects to normal distribution, 45.7635 MPa is read as average value of P, $(1 + 0.15) \times 45.7635$ as confidence upper limit, (1–0.15) $\times$ 45.7635 as lower confidence limit, degree of confidence is 95%, so this is similar with solution procedure of other parameters, and result is P–N(45.7635,3.5024).

7.1.4 Reliability Analysis of Traditional Safety Factor Method

The objective function of manned cabin design is lightest in weight, constrain condition is that manned cabin's bearing capacity shall be better than the required design pressure (design pressure = safety factor x submersible work pressure P), design variable is thickness, safety factor in available manned submersible rules is 1.5, this book will also adopts sf = 1.5. when design manned cabin through traditional safety factor method, its optimization model is:

$$\begin{aligned} \min f &= 4450 \times \frac{4\pi}{3}((R+t)^3 - R^3) \\ s.t.\quad & P_u - sf \times P \geq 0 \\ & P_u = WJF \times \left(1 - k\frac{\Delta}{R}\right)\left(\frac{\sigma_b t}{R} + \frac{\sigma_b t}{R + t/2}\right) \\ & sf = 1.5 \\ & P = 45.7635 \\ & R = 1050 \\ & \Delta = 0.5\% R \\ & 1 \leq \mathrm{t} \leq 70 \end{aligned} \tag{7.51}$$

Note: other parameters of P_u refers to formula (7.48), σ_b, WJF is read as relevant values in Table 7.18.

This problem has only on design variable, it is easy to get its optimal solution, see Table 7.20.

Table 7.20 Design manned cabin optimal thickness according to traditional safety factor method

	Material	Ti80	TC4ELI
Ultimate tensile toughness	Ultimate tensile toughness	839.8414	925.3538
	WJF	1	0.93
Optimization result of safety factor method	*t* (mm)	50.6168	49.5826
	Objective function value(t)	3.2735	3.2035

7.1.5 Manned Cabin RBDO

In the former three sections of this chapter, the necessary probability constrain conditions and random variables statistic characters of manned cabin RBDO design has been already prepared, thus the RBDO model of manned cabin could be presented as:

$$
\begin{aligned}
& \min f = 4450 \times \frac{4\pi}{3}\left((\mu_R + \mu_t)^3 - \mu_R^3\right) \\
& s.t.\quad P(P_u - P \geq 0) \geq R_{req} \\
& \qquad P_u = WJF \times \left(1 - k\frac{\Delta}{R}\right)\left(\frac{\sigma_b t}{R} + \frac{\sigma_b t}{R + t/2}\right) \\
& \qquad P \sim N(45.7635, 3.5024) \\
& \qquad \Delta = 0.5\% \times \mu_R \\
& \qquad \sigma_b, WJF\text{见表29或表30} \\
& \qquad R \sim N(1050, 2.6786) \\
& \qquad t \sim N(\mu_t, 0.5102) \\
& \qquad 1 \leq \mu_t \leq 70
\end{aligned}
\tag{7.52}
$$

Note: Rreq is the required reliability, which is generally read as 0.9987 (around milli failure probability), while for important parts, reliability shall be read as 0.999999 (namely one millionth failure probability), even as 0.999999999 (one billionth failure probability).

Algorithm SFSORA and SLRBDO will be adopted in manned cabin thickness reliability design.

7.1.6 Manned Cabin Design by Algorithm ISFSORA

It is easy to modify formula (7.52) into optimal model conforming to algorithm SFSORA, as shown in formula (7.53), no extra parameter is involved, so the sf has actual physic sense of safety factor.

$$
\begin{aligned}
\min f &= 4450 \times \frac{4\pi}{3}\left((\mu_R + \mu_t)^3 - \mu_R^3\right) \\
s.t.\ & \frac{\mu_P}{P_u} - \frac{1}{sf} \le 0 \\
& P_u = WJF \times (1 - k\frac{\Delta}{\mu_R})(\frac{\mu_{\sigma_b}\mu_t}{\mu_R} + \frac{\mu_{\sigma_b}\mu_t}{\mu_R + \mu_t/2}) \\
& \Delta = 0.5\% \times \mu_R \\
& \mu_{\sigma_b}, WJF\text{见表29或表30} \\
& \mu_R = 1050 \\
& 1 \le \mu_t \le 70
\end{aligned}
\tag{7.53}
$$

According to algorithm SFSORA, when Rreq is very high, the calculated quantity of CMC method, after analyzed by variance, will be so large that regular PC could not distribute enough memory space to complete computation. Thus this book adopts importance sampling algorithm compiled in fourth section, third chapter to conduct reliability analysis, to finally figure out solution for this question, see Table 7.12.

It is shown in algorithm SFSORA:

As a matter of fact that model spheres TC4ELI has high-toughness parent material and low-property weld joint and model Ti80 has low-toughness parent material and high-property weld joint, the final property of the two material is almost same. Whether the thickness is designed according to 3 σ (R = 0.9987) or 6 σ (R = 0.999999999) with high requirement, the design thickness of these two marks does not change much.

(1) As requirement for reliability arises, the safety factor of two materials arises relevantly, which conforms to direct understanding.
(2) Compared with traditional safety factor method: when design according to 6 σ with high requirement of reliability, the safety factor of Ti80 and TC4ELI shall be read as 1.5064 and 1.5360 to meet requirement, whose manned cabin thickness shall be 50.8001 mm and 50.5877 mm; while in traditional safety factor method, the safety factor is read as 1.5, the thickness of two kinds of material is 50.6168 mm and 49.5826 mm respectively. It is clear that the 1.5 times safety factor in traditional safety factor method of manned cabin design is required according to failure probability close to one billionth, namely 6 σ design. If adopt other algorithm to conduct RBDO, it is not possible to find out the relation between safety factor and reliability of manned cabin.
(3) For nonlinear systemic design of high requirement on reliability, algorithm JC has big deviation and algorithm CMC cannot accept the calculated quantity, so importance sampling algorithm becomes the most suitable reliability analysis algorithm.

7.1.6.1 Algorithm SLRBDO in Manned Cabin Design

When adopt algorithm SLRBDO to solve RBDO of formula (7.52), the algorithm SLRBDO only supports normal distribution, so material property parameters will be read as statistic value of normal distribution list in Table 7.19. Optimal result is easy to get through the above mentioned programme SLRBDO, see Table 7.22.

It is shown in Tables 7.21 and 7.22:

(1) Very little deviation between algorithm SLRBDO result and algorithm SFSORA result. For example, Rreq = 0.999999999, thickness deviation is only 0.28 mm, thickness deviation for TC4ELI is only 0.556 mm.
(2) For same uncertain parameter, when adopt different distribution type to conduct statistics and analysis, very little deviation between results.

Table 7.21 Result of manned cabin computed by algorithm SFSORA

Rreq	Titanium alloy mark	Ti80	TC4ELI
	Safety factor (sf)	1.2482	1.2518
0.9987	μ_t/mm	43.4701	42.7238
	Objective function value	2.7925	2.7426
	IS computing reliability (keep ten decimal places)	0.9987049099	0.9987027062
	Safety factor sf	1.3948	1.4072
0.999999	μ_t/mm	47.6147	47.0038
	Objective function value(t)	3.0707	3.0295
	IS computing reliability (keep ten decimal places)	0.9999990019	0.9999990026
	Safety factorsf	1.5064	1.5360
0.999999999	μ_t/mm	50.8001	50.5877
	Objective function value(t)	3.2859	3.2715
	IS computing reliability (keep ten decimal places)	0.9999999990	0.9999999990

Table 7.22 Algorithm SLRBDO result for manned cabin

Rreq	Titanium alloy marks	Ti80	TC4ELI
0.9987	μ_t/mm	43.4107	42.7737
	Objective function value(t)	2.7885	2.7460
0.999999	$\mu_t/mm.$	47.5126	46.9414
	Objective function value(t)	3.0638	3.0253
	μ_t/mm	50.5201	50.0321
0.999999999	Objective function value(t)	3.2669	3.2339

(3) Algorithm SLRBDO transforms reliability analysis into nonlinear formula of solving coordinate about MMP, so reliability requirement has little effect on computation efficiency, even if it is question with high reliability requirement.

This chapter conducts complete RBDO research, from establishing state function, confirming statistic parameters of random variables, establishing RBDO model to conducting RBDO. It is found in algorithm SFSORA that the safety factor in traditional safety factor method is to promise manned cabin design could meet requirement 6 σ, the conclusion could not be found by other SFSORA algorithms, that is why there is not relevant documents about manned submersible research. It means SFSORA algorithm could not only be used to conduct RBDO, but also to confirm safety factor through one time solution by SFSLRA algorithm for similar problems, this safety factor could also be used in other similar problems and finally used to substitute complex RBDO by certainty design optimization with safety factor. After safety factor obtained by SFSORA algorithm is widely used, it could even update safety factor of industry standard, and make safety factor of industry standard not as experience parameters any more, but trackable and reasonable. It is found by SLRBDO algorithm that result of RBDO varies a little when adopt different distribution types to describe one same uncertain variable, so the deviation caused by frequently describing uncertain variables by normal distribution in projects is not acceptable in most situations.

Besides manned cabin thickness, the manned cabin design needs trepannings to strengthen the dimension of trunk bulkhead and according to available manned submersible rules, design for strengthening trunk bulkhead shall be strong enough, it is not allowed to affect shell's bearing capacity due to trepanning. Thus the design of manned cabin thickness and strengthening trunk bulkhead could be conducted separately, and the stress state of strengthening trunk bulkhead is so complicated that the design shall be confirmed by some value computation methods, like finite element. The document (Pan and Cui 2011b, c) conducts manned cabin design optimization based on finite element, it is shown that when units are almost full of hexahedral elements, the calculated quantity is still pretty huge, and the weight-reducing effect by strengthening trunk bulkhead design is smaller than thickness optimization, thus this book will not conduct RBDO of strengthening trunk bulkhead based on finite element, but adopt stress constrain method based on rules as shown in document (Pan and Cui 2011b, c).

When review this project, we could find out that the process of establishing new manned cabin ultimate bearing capacity could be regarded as process of establishing high-accuracy approximation model, this model will simplify RBDO solution by stopping using nonlinear finite element computation model with huge calculated quantity, and greatly lower calculated quantity of RBDO (each nonlinear analysis needs several hours and even several days, while approximation model will only needs several seconds), so the computation that is hard to complete on PC will become very easy. The reduction of calculated quantity makes us able to adopt CMC and other reliability analysis methods, and we can also adopt two-cycle SFSORA and other RBDO methods to solve.

7.2 Manned Submersible General Design Optimization

Manned submersible is one complex engineering system composed of many parts and devices, see "Jiaolong" in Fig. 7.21). The manned submersible design will be not only involved with structural mechanics introduced informer chapter, which is necessary for manned cabin design and design of other structural parts, but also many other disciplines, such as hydromechanics, electronic engineering, optics, machinery and hydraulic engineering. The design of manned submersible "Jiaolong" is divided into many subsystems according to different disciplines, when conduct general design, the input and output of each subsystem shall be coordinated to achieve a balance in whole system, so it is necessary to adopt multi-dicipline design optimization to modeling general design.

Combined with the key technology of manned submersible design introduced in former part, this book will first introduce each subsystem of manned submersible general design and modeling of general system.

7.2.1 Manned Submersible General Design Model

In the process of "Jiaolong" design, some people, like Cao Anxi, Liuwei and Goupeng and others (Cao and Cui 2008; Liu 2007) have already done fruitful research on

Fig. 7.21 Retrieve of "Jiaolong" in 7000 m sea trial

manned submersible multi-discipline design optimization, the established multi-discipline design optimization model is involved with six main subsystems of manned submersible design. Based on the "Jiaolong" unpowered snorkeling sea trail data and new design principle of manned cabin and other pressure container, this book will focus on the key techniques of each subsystem on manned submersible design to improve this model and establishes multi-discipline design optimization model about manned submersible concept including 11 subsystems, see Fig. 7.22. Five subsystems in MDO model, including loading, structure, outfit, adjustable ballast and slope adjustment, load discharging structure and devices, only output weight (M), volume of displacement (V), and their moment; while other five subsystems, including propulsion, observation and communication navigation, acoustics, hydraulic and life support, will not only output weight, volume of displacement and their moment but also power consumption P. P will be regarded as one input parameter of power distribution system, the necessary battery capacity, weight and volume shall determined by power consumption and capacity density of battery of whole

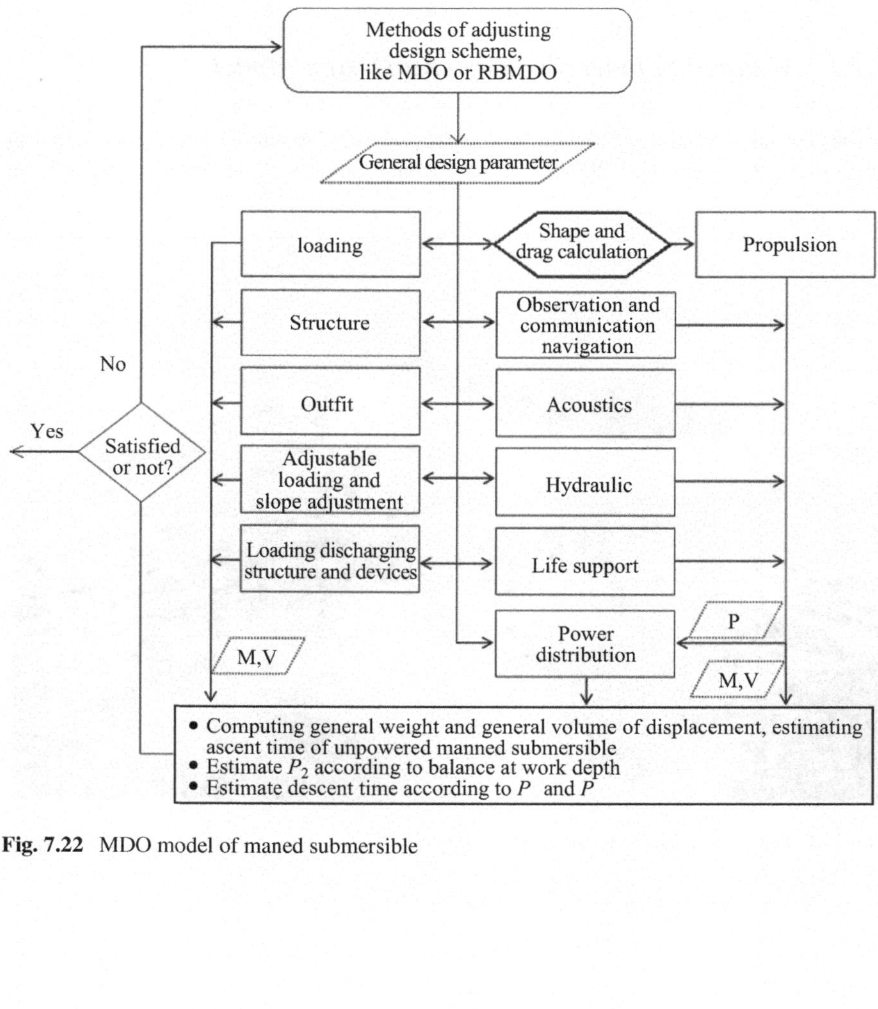

Fig. 7.22 MDO model of maned submersible

submersible, and the weight and volume of battery box and other attachments shall be estimated according to experience data of "Jiaolong". To the end, standard weight and volume of manned submersible could be obtained by weight of eleven subsystems.

MDO或RBMDO等调整设计方案的方法	Methods of adjusting design scheme, like MDO or RBMDO
总体设计参数	General design parameter
是	Yes
否	No
是否满足	Satisfied or not
载荷	loading
结构	Structure
舾装	Outfit
可调压载和纵倾调节	Adjustable loading and slope adjustment
抛载机构和装置	Loading discharging structure and devices
外形和阻力计算	Shape and drag calculation
观通导航	Observation and communication navigation
声学	Acoustics
液压	Hydraulic
生命支持	Life support
推进	Propulsion
电力配电	Power distribution
计算总重和总排水体积, 由此可估算潜器无动力上浮时间	Computing general weight and general volume of displacement, estimating ascent time of unpowered manned submersible
根据工作深度处平衡可估算 P_2	Estimate P2 according to balance at work depth
由P_1 和P_2 可估算无动力下潜时间	Estimate descent time according to P1 and P2

7.2.1.1 Shape and Drag Calculation Module

This book will adopt simplified hypothesis for manned submersible shape in documents (Cao and Cui 2008; Liu 2007), thinks that shape of submersible could be determined by parameters in Fig. 7.23. Submersible is mainly divided into three parts: stern part (La), parallel middle part (Lm) and ship bow part, which could be simplified as cone, cylinder missing angle and ball-lacking type.. There is introduction of computation formulas of surface area and volume in geometry manual and regular mechanical design booklet, they won't be introduced here. In the MDO model, there is one computation sub-module of submersible shape to compute wetted surface, volume of displacement, and surface when moving forward, laternal movement, descent and ascent, and other geometry parameters in drag computation.

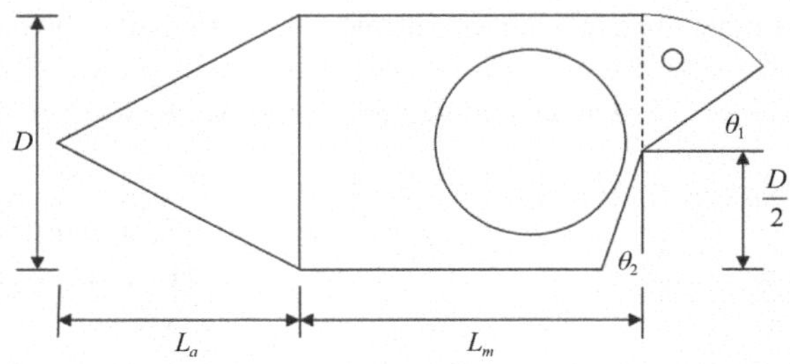

Fig. 7.23 Simplified shape of manned submersible

After computing submersible shape parameters, the drag from four directions of moving forward, lateral movement, descent and ascent shall be computed. At the beginning of establishing drag computation module, Fluent is adopted to compute drag on naked boat of "Jiaolong" to find out the possibility of applying CFD. Unstructured grid and body-fitted grid are combined to control height of first layer grid, and the area arising along ship body y + is around 60, after adopting different turbulence models to compute, the paralleled CFD computation speed is found able to basically meet the requirement of engineering, but deviation between numerical convergence solution and model experiment data is still big. Also grid is re-distributed by structured grid, it is found that there is also big deviation between results of regular turbulence model and laminar flow. At the next stage of designing submersible shaped lines, model experiment data could be used to correct computation result of CFD model and CFD model if necessary, and the corrected CFD model could be used to estimate drag of submersibles with similar shape. Further, it is possible to conclude formula which are suitable for more submersible shapes based on large amount of CFD computation result and experiment data, to update available experience formula and provide more accurate drag computation module for submersible general design. The drag computation module in this book will correct coefficient of experience formula by "Jiaolong" model experiment data, to make experience formula result close to tank test result, this module will adopt different experience computation formula according to shape character of manned submersible in different directions.

(1) Drag in moving forward

Drag in moving forward is composed of two parts: frictional drag and shape drag.

After comparing regular computation formula of frictional drag coefficient, it is found that result of formula ITTC1957 is close to data in property design report of "Jiaolong", so this book will adopt formula ITTC1957 to compute frictional drag coefficient of manned submersible.

$$C_f = \frac{0.075}{(\lg(R_e) - 2)^2} \tag{7.54}$$

Note: R_e is Reynolds number.

Frictional drag is:

$$R_f = \frac{1}{2}(C_f + \Delta C_f)\rho S V^2 \tag{7.55}$$

Note: ΔC_f is coefficient of subsidies; ρ is seawater density in work depth of submersible, it could be computed by depth-density curve measured by "Jiaolong" sea trial; S is submersible's wetted surface; V is speed in direction of moving forward.

There are two main computation methods of computing shape drag: the first method proposes shape drag is in direct proportion to wetted surface of submersible [formula (7.56)]; the second proposes shape drag is in direct proportion to surface of submersible facing water.

$$R_{pv} = \frac{1}{2} C_{pv} \rho S V^2 \tag{7.56}$$

$$R_{pv} = \frac{1}{2} C_{pv} \rho A V^2 \tag{7.57}$$

Note: C_{pv} is shape drag coefficient, when compute shape drag by formula (7.57), Bapmuir's formula is always adopted $C_{pv} = \frac{0.09A\sqrt{\sqrt{A}/(2L_a)}}{S}$; A is front water facing area in direction of moving forward (Fig. 7.23 shows: $A = \frac{\pi D^2}{4}$); other parameters are same as formula (7.55) and Fig. 7.23.

It is verified that large deviation exists in computation of formula (7.57), so this book will adopt formula (7.56) to compute shape drag, C_{pV} is reckoned according to tank test result of model "Jiao Comparison of general drag computation result in direction of moving forward and tank model test result see Fig. 7.24.

(2) Drag in other direction

Drag in direction of lateral movement, descent and ascent is divided into frictional drag and shape drag, the general drag is directly computed by below formula:

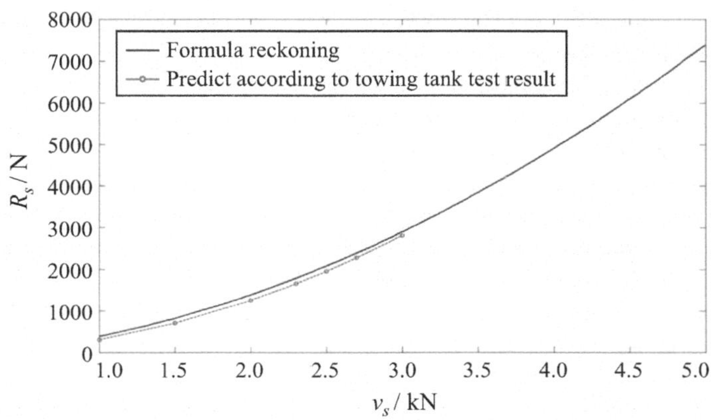

Fig. 7.24 Comparison of general drag computation result in direction of moving forward and tank model test result

$$R_s = \frac{1}{2} C_s \rho S_f V^2 \tag{7.58}$$

Note: C_s is drag coefficient, computed by model test result; S_f is submersible's water facing area when moving in all directions; V is moving speed in all directions.

(3) Appendage drag.

It is clear in Fig. 7.21 that there are many appendages on submersible, such as sampling basket, sampling device and work tools, mechanical hand, lights, camera, holder, sensors and antenna, bottom bracket, propeller, stabilizer fin, and possible channel propulsion. These appendages make flow field around obtuse submersible more complex, it is hard to accurately compute or simulate the drag effect of appendages on submersible, although model boat in tank model test does not have all of the appendages. Besides, according to experience of "Jiaolong", the layout of submersible's light and camera in sea trail and actual application will be adjusted, and sampling device and work tools are also allocated according to task, so the drag in design stage is different from drag during submersible descending.

Three computation methods for drag of submersible appendages:

① Directly compute drag of boat with all appendages. It can only be computed by software CFD, but modeling and grid distribution is very hard, the calculated quantity is also large and there is always big deviation.
② Compute respectively drag of naked boat and appendages, composite matrix could be used to reflect mutual effect between boat and appendages and effect among appendages. The calculated quantity is smaller than first method, and the drag of boat and each appendage could be computed at the same time, the computation period is shortened. But it is hard to confirm composite coefficient matrix, in general it is confirmed by experience.
③ Reckon according to proportion of naked boat drag. This method is totally decided by designer's experience or similar data of available boats with similar shape, this method is usually used at the stage of concept design.

The documents (Cao and Cui 2008; Liu 2007) in this book regards appendage's drag as 25% of naked boat.

Shape and drag computation module is the "pre-procession" of propulsion module, the computation of drag in all directions is prepare for choosing suitable thrust in this module:

Parameters input geometry shape sub-module: ① general length of submersible; ② general length of parallel middle boat; ③ diameter of submersible; ④flare angle of submersible's upper bow θ_1 (Fig. 7.23); ⑤sweepback angle of submersible's upper bow θ_2; output parameters: ①wetted surface of submersible shape; ②water facing area in direction of lateral movement; ③water facing area in direction of descent; ④ water facing area in direction of ascent; ⑤ envelope volume; ⑥ length of boat stern.

Input parameters of drag sub-module in direction of moving forward: ① submersible's general length; ② wetted surface of submersible; ③ speed in direction of moving forward; ④ seawater density in sailing depth of submersible; ⑤ seawater's coefficient of viscosity. Output parameter is direction drag.

Input parameters in drag sub-modules of lateral movement, ascent and descent: ① water facing area in the direction; ② speed in the direction; ③seawater density in sailing depth of submersible. Output parameters are drag in direction of lateral movement, ascent and descent.

7.2.1.2 Propulsion Module

In order to control manned submersible's movement, propeller is installed in the three direction of submersible, the required moving speed of submersible in each direction is different, drag is also different, the number of propeller shall be determined by propulsion requirement. At the propulsion module of this book, two propellers is arranged in the direction of moving forward, which are rotary type and could supply propulsion in the direction of moving forward, ascent and descent; one for boat bow and boat stern respectively, the propeller at boat stern is rotary type, could supply propulsion in the direction of lateral movement and moving forward; two propellers in the direction of ascent and descent.

The propeller of manned submersible is composed of three parts in general, namely motor, variable speed drive and screw propeller.

(1) Motors are divided into hydraulic motor and electronic motor. Hydraulic motor is easy to control and maintain, small, could share one hydraulic source, it is generally used in large submersible which needs great propulsion and more than one propellers; electronic motor is more efficient than hydraulic motor, noise is lower than hydraulic motor, is the most regular propelling motor. Generally to lighten electronic motor, the seal of rotation axis of propeller is achieved by electronic insulting oil pressure compensator or magnetic coupling drive. Brushless direct current motor or alternating current motor is applied in modern electronic motor to avoid efficiently arc spark and its seal oil pollute of brush direct current motor. But most submersibles adopt battery as their power source, if adopt alternating current motor current transformer is needed to transform direct current into alternating current. Brush direct current motor is free of maintenance, long lifetime, high rotational speed, small, light. As development of high performance permanent magnet, rotary optical encoder, Hall element and pulse width modulation technology, the efficiency of brushless motor is constantly rising, its noise and electromagnetic interference are also constantly lowered, and it gradually becomes main motor form of underwater electronic motor.
(2) Variable speed drive. Electronic motor usually has high rotary speed but low torque, the efficiency of low speed motor is lower than high speed electronic motor. Generally in order to make screw propeller get enough torque and improve its efficiency, moderator shall be installed between electronic motor and transmission shaft, which will cause noise and lower efficiency. So adopting

moderator or low speed electronic motor shall be determined by actual need. Transmission shaft is connected directly with screw propeller, the seal between transmission and motor shell is where the biggest difference between underwater motor and land motor. Pressure compensator or magnetic coupling drive is generally used to solve seal problem of transmission shaft. Currently the underwater propeller researched by our country adopts magnetic coupling drive, without moderator, so its noise performance is better than imported propeller.

(3) Screw propeller. Underwater sailing speed of manned submersible is regularly very low, so accelerating ducted screw propeller is applied. Besides, to keep the same propulsion of forward and reverse, protect screw propeller from damage when absorbing other items, duct and screw propeller shall be verified and re-designed.

When design "Jiaolong", Tecnadyne company provides performance parameters of underwater propeller products of Model series, see Table 7.23. Domestic underwater propeller has been used in 7000 m sea trial of "Jiaolong", their performance has basically caught up with level of imported devices, but domestic underwater propellers is lack of serial model, and their performance data is not complete, so this book will adopt parameters of Model propeller by Tecnadyne company to select submersible's propeller. Besides power coefficient of integrated underwater propeller is unknown, but their relation of propulsion and input power is almost confirmed, so this book will reckon propulsion by drag, and then select the process of propeller power by propulsion.

Firstly, thrust deduction fraction shall be same as submarine ($t = 0.18$), propulsion of propeller could be determined by drag in all directions:

Table 7.23 Performance parameter of Model series integrated propeller

Model	Input power/kW	Propulsion/kN	Weight/kg	Volume/m^3	Diameter of screw propeller/m
Model 260	0.350	0.0529	0.9	0.2	7.6
Model 280	0.350	0.0529	1.0	0.2	11.7
Model 300	0.500	0.0804	1.0	0.3	8.9
Model 520	0.500	0.1019	1.8	0.4	11.5
Model 540	0.975	0.0931	1.9	0.4	14.9
Model 560	0.975	0.1695	1.9	0.5	15.8
Model 1020	1.050	0.2450	2.7	0.7	15.3
Model 1040	1.500	0.2548	2.8	0.7	20.3
Model 1080	2.100	0.5586	2.9	1.8	18.1
Model 2010	4.900	0.9800	10.5	2.6	20.4
Model 2020	5.500	1.1368	10.2	2.5	22.9
Model 8020	12.900	2.2540	25.5	10.0	30.5

$$T = \frac{R}{1 - t} \tag{7.59}$$

The approximation relation of propulsion and input power of propeller shall be established according to Table 7.23, as in Fig. 7.25, once the propulsion is confirmed, the input power of propeller could be computed; similarly, propeller's weight and volume could also be computed by Figs. 7.26 and 7.27. Then the weight and volume of appendages of submersible could be recknote that computation process of propulsion module in this book is different from documents (Cao and Cui 2008; Liu 2007), this

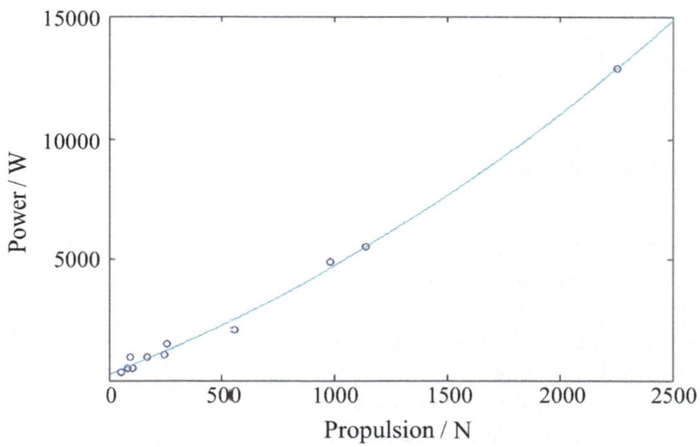

Fig. 7.25 Propeller's propulsion-power curve

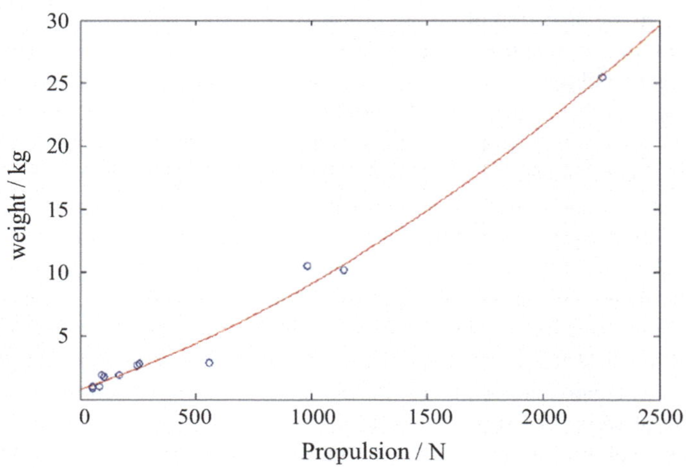

Fig. 7.26 Propeller's propulsion-weight curve

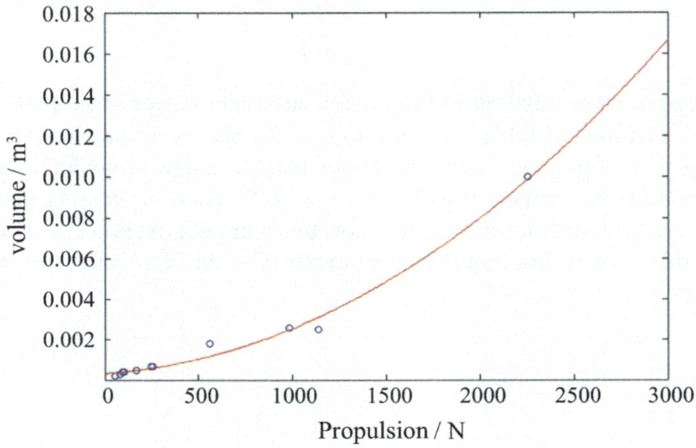

Fig. 7.27 Propeller's propulsion-volume curve

book did not conduct model selection from the perspective of power transmission, but conduct model selection directly by propulsion matching of propulsion module:

Input parameters of sub-module in the direction of moving forward: ① general length of submersible; ② submersible's width; ③ submersible's height (assuming height equals width); ④ drag of submersible in the direction of moving forward; ⑤ number of propeller in the direction of moving forward. Output parameters: ①general input power of submersible in the direction of moving forward; ② general weight of submersible in the direction of moving forward; ③ general displacement volume of submersible in the direction of moving forward; ④ general weight torque of submersible in the direction of moving forward; ⑤ general discharge volume of submersible in the direction of moving forward.

Input parameters of sub-module in the direction of lateral movement: ① general length of submersible; ② submersible's width; ③ submersible's height (assuming height equals width); ④ drag of submersible in the direction of lateral movement. Output parameters: ① general input power of submersible in the direction of lateral movement; ② general weight of submersible in the direction of lateral movement; ③ general displacement volume of submersible in the direction of lateral movement; ④general weight torque of submersible in the direction of lateral movement; ⑤ general discharge volume of submersible in the direction of lateral movement.

Input parameters of sub-module in the direction of perpendicular: ① general length of submersible; ② submersible's width; ③ submersible's height (assuming height equals width); ④ drag of submersible in the direction of perpendicular. Output parameters: ① general input power of submersible in the direction of perpendicular; ②general weight of submersible in the direction of perpendicular; ③ general displacement volume of submersible in the direction of perpendicular; ④ general weight torque of submersible in the direction of perpendicular; ⑤ general discharge volume of submersible in the direction of perpendicular.

7.2.1.3 Loading Module

Loading of submersible is mainly divided into two parts: the first is submersible passenger's weight, and the second is weight of work tools and sampler, etc. According to design specification, three passengers are allowed, each person is weighed 80 kg, center of gravity shall be reckoned according to height of people's gravity center and manned submersible diameter; loading target is 220 kg, this book will design according to 2 × 220 kg, namely, fixed ballast of 220 kg is installed at regular period, when necessary it could be discharged. Please note that work tools and sampler may need power supply when they work, so this module will reserve power interface with max power 2000 W and voltage 24 V for work tools.

Input parameters in this module: ① general length of submersible; ② submersible's width; ③submersible's height (assuming height equals width); ④ passenger weight; ⑤ work loading weight; ⑥ manned cabin shell internal radius. Output parameters: ① effective loading weight; ② volume of displacement under work loading; ③effective loading weight torque; ④ effective loading displacement torque.

7.2.1.4 Control, Observation and Communication and Navigation Module

The devices in this module are video recording device for observation, communication and navigation, light, holder, VHF wireless communication, display devices and video recording devices in cabin, computer tank devices, CTD, but underwater acoustic communication and sonar is not included, sonar will be designed as single module. The devices are similar with "Jiaolong".

(1) Light

Professional underwater light is needed for light module. Attenuation of light in water is faster than in air, which is caused by absorption function of water itself, items dissolved in water, underwater plankton and suspension of cuttings. Light absorption may be weak and strong according to wave length. Scattering effect is actually not related with wave length, because granular size is larger than wave length of visible light spectrum. The wave length of light with most strong penetrating power underwater is wave length 5000A of green light, so the light near that wave length will be selected as light of manned submersible, regular ones are quartz-halogen lamp, electric arc light and thallium iodide lamp.

The following aspects shall be considered as selecting light as video recording light (the above three kinds of light is of power 250 W):

① Optical output efficiency: thallium iodide lamp is twice the optical output efficiency of electric arc light, six times the optical output efficiency of quartz-halogen lamp.
② Attenuation in water. Attenuation with two meters: 90% for quartz-halogen lamp, 80% for electric arc light, 70% for thallium iodide lamp.

③ Contrast: thallium iodide lamp is better than other two lights on contrast, it is more obvious when faster. Electric arc light and quartz-halogen lamp has similar contrast.

④ Color reproduction: after comparing spectral color diagram lighting of each light with power of 250 W at the distance of 1 m, it is found that green and blue attenuation of quartz-halogen lamp is pretty big, its purple is almost changed into red. Color reproduction of electric arc light is also bad except for blue and green area. Thallium iodide lamp could reproduce some red, blue and purple, mainly green. Advantage and disadvantage and characters of three kinds of light see Table 7.24.

As requirement for underwater video recording become higher, manned submersibles are developing towards high definition video recording. "Jiaolong" submersible has been installed with high definition video recording system in sea trail, it is found that color temperature around 5000 K has big effect on video recording color rendition of HID electric arc light, see Fig. 7.28. In order to improve color rendi-

Table 7.24 Comparison of character of regular light source

Light type	Character	Advantage	Disadvantage
Quartz-halogen lamp	Able to output in the scope of whole visible spectrum, from blue to red	Short time for launching, good color reproduction	Low efficiency
Electric arc light	Light is presented as purple, blue, green and yellow	Pretty high efficiency	Long time for launching and re-launching
Thallium iodide lamp	Mainly for blue light	Extreme high efficiency	Pretty long time for launching

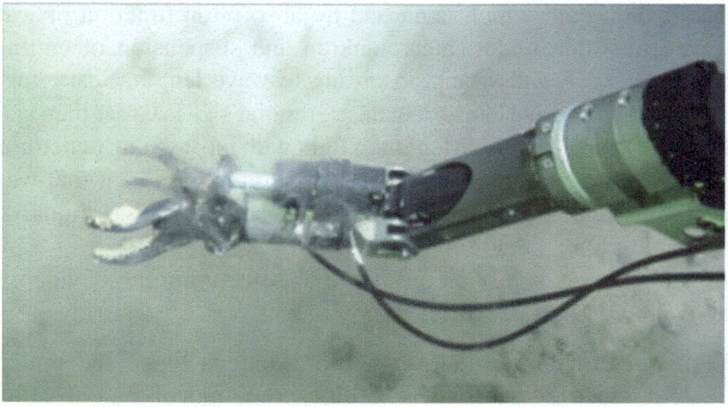

Fig. 7.28 Light of green HID affects color of video recording

tion of high definition video recording, HID electric arc light is changed from color temperature around 5000 K into 5500 K.

In addition, LED light source develops very quickly, underwater LED light also develops quickly, its energy conservation and lifetime is beyond comparison, so "Jiaolong" replaced Quartz-halogen lamp with LED light in sea trail. Light module in this book will adopt "Jiaolong" new configuration:

① Two 400 W HMI lights. HMI is Halogen mercury-iodine lamp, it is a new underwater light source improved from electronic arc light, its color temperature is 5600 K, close to day light, suitable for color and monochrome video recording. The above mentioned thallium iodide lamp belongs to HMI light, it could be used as thallium iodide lamp when replace its bulb with thallium iodide bulb. HMI light will not produce high temperature, and it could be restarted promptly after turned off.

② Two 400 W HID electric arc light. HID belongs to high strength gas discharge lamp, it is 3 to 4 times the efficiency of tungsten halogen lamp, without heater, resistant to vibration and shock, color temperature at 5500 K, it is suitable for video recording, lower price than tungsten halogen lamp, long lifetime.

③ Ten 60 W underwater LED lamps. It is of long lifetime, energy conservation, new type underwater lamp, its color reproduction needs verification. With that configuration, operator could make reasonable arrangement according to site condition, then it could be used to provide enough illumination intensity for submersible observation, video recording, work, keep pretty enough underwater visibility, and provide reasonable underwater illumination spectrum, operator of submersible could launch the relevant lamps or combine several kind of lamps according to work.

(2) Video recording.

Manned cabin of submersible is set with observation window, its field of view is limited due to size of window and direction, only part of outside environment could be viewed through one window, so it is hard to finish sailing and work by simply relying on observation window. Thus camera is needed to expand observation scope to observe outside environment and observation target from multiple directions. And the video could also be recorded and saved, provide site documents for subsequent research and proving.

Camera on submersible is usually selected according to application. For example, camera on holder and mechanical hand is responsible for observing researched items, so high definition color camera is installed; before submersible descending to the bottom of sea, operator shall observe conditions of sea bottom below submersible to judge land or not, while the light from below is frequently weak, low-light level camera is usually used.

The basic theory of video recording system on submersible is similar with land video recording system, parallel light is still parallel after refraction by prismatic borrow light, so it is okay to simply install land video recording system inside pressure seal structure but in big depth, the thickness of prismatic borrow light will be big, the window seat of pressure structure will also becomes big, to lower the size of borrow light and strengthened trunk bulkhead, it is usually changed into hemispherical borrow light. While hemispherical borrow light is strong in concave lens function, so the focusing system of land video recording devices should be revised. Recently, definition of video recording system is improved quickly, high definition video recording devices, like 720P、1080P、4Kx2K, are gradually installed on submersible. Besides 3D video recording and display technique is developing towards practical application. In the future, submersible will be installed with more advanced video recording system, to make scientists feel the real environment from 3D video.

(3) Communication and others.

This part is involved with VHF communication, CTD and display control, etc.

Antenna of wireless communication equipment of VHF is usually installed on the top of manned submersible. When submersible is underwater, antenna of VHF is put into submersible's light shell, or it works outside. Main performance of wireless communication equipment is sent and received frequency, available signal channel, work voltage, power consumption, work temperature, weight, etc.

Salinity sensor and temperature sensor are maturer products. Depth sensor is divided into acoustic depth sensor and hydraulic depth sensor. The theory of acoustic depth sensor is sending signal towards sea surface, counting time difference between sending and receiving, multiplying sound speed by the time difference, finally obtaining the distance. The theory of hydraulic depth sensor, measuring seawater pressure according to seawater density, relation of depth and pressure, computing the depth reversely by combining it with density sensor data. The regular hydraulic depth sensor is divided into Burdon tube pressure sensor, strain pressure sensor and Crystal oscillator manometer. CTD of 4500 m manned submersible will basically adopt configuration of "Jiaolong".

Monitor includes displays, video hard disk recorder, video controller, etc.

The underwater communication is involved with acoustics, domestic professional research institute has conducted relative works. The acoustic devices of 4500 m manned submersible is entrusted to relative units to design, thus acoustic system is separated from communication system in this book, as an independent subsystem.

Input parameters in this module: ① general length of submersible; ② submersible's width; ③ submersible's height; output parameters: ① input power of light system; ②subsystem weight; ③ volume of displacement of subsystem; ④ subsystem weight torque; ⑤ subsystem displacement volume torque.

7.2.1.5 Structure Module

Submersible structure is divided into pressure and non-pressure structure.

(1) Pressure structure.

It has been found in the former chapter that after considering weld performance, manned cabin thickness of two kinds of domestic titanium alloy design has only a little difference. Thus one domestic titanium alloy material will be applied in the structure module of general design.

Great depth submersible's pressure structure can be divided into two parts according to function: manned cabin and other small pressure tank.

Manned cabin as core part of manned submersible is usually designed independently, the manned cabin design has been introduced in detail in the former chapter, it needs no description any more.

Other pressure tanks usually have small diameter, two structure forms of ball-shape and cylinder.

① Sphere-shape

Manned cabin and pressure adjustable water warehouse and high pressure tank of manned submersible are going to bear high pressure and it is better to have big volume, so ball-shape is generally applied.

According to China classification society's recent design principle (China Classification Society 2011), ultimate pressure bearing capacity shall be 1.5 times of work pressure at least [namely ultimate bearing capacity formula established in Chap. 7, see formula (7.48)].

Beside ultimate bearing capacity, average film stress of ball-shape pressure structure [see formula (7.60)] shall not surpass two thirds of material yield strength.

$$\sigma_m = \frac{PR_o^3}{2(R_o^3 - R_i^3)}[2 + (\frac{R_i}{R_m})^3] \tag{7.60}$$

Note: R_i is internal radius; R_m is radius of middle surface; R_o exterior radius; P is submersible's designed work pressure.

Ball-shape pressure structure thickness could be determined by these two principles, and weight of pressure structure's major part could also be reckoned, and weight of appendages, like trunk bulkhead after opening hole, end socket, fixing lug, could also be reckoned according to experience data of "Jiaolong".

② Cylindrical end socket with hemispheres on both sides.

Many tanks of submersible, like computer tank, acoustic communication tank, deep side scan sonar measuring tank, power distribution tank, are regularly designed as cylindrical pressure container, they are sealed by hemispherical end cap. This kind of pressure structure shall be confirmed about its cylindrical wall thickness and wall thickness of hemispherical end cap on the two ends. Computation method of hemispherical end caps on the two end is similar with ball-shape pressure structure as mentioned before. Design of cylinder shall meet the following two principles:

a. Ultimate bearing capacity [see formula (7.61)] shall be at least 1.5 times of work pressure (China classification society, 2011).

$$P_u = 0.85 C_s P_e \tag{7.61}$$

$$P_e = \frac{0.6E(t/R_m)^2}{u - 0.37} \tag{7.62}$$

$$u = \frac{0.643 L_c}{\sqrt{R_m t}} \tag{7.63}$$

$$\sigma_e = \frac{P_e R_m}{t} \tag{7.64}$$

Note: C_s is ratio of σ_e and σ_s (material's yield buckling strength), read as in Fig. 7.29; L_c is read as result of cylinder length plus 40% of each hemispheric end cap depth on the two ends (namely 40% of hemispherical end cap's internal radius), other parameters are defined as before.

b. Average file stress [see formula (7.65)] shall not exceed two thirds of material's buckling strength.

$$\sigma_m = \frac{P R_o^2}{R_o^2 - R_i^2}[1 + (\frac{R_i}{R_m})^2] \tag{7.65}$$

Its parameters are defined as formula (7.60).

Computation of relevant formula for ultimate bearing capacity and average film stress's design principle could be used to confirm wall thickness of ball-shape and cylindrical pressure structure and its hemispherical end caps. There is only

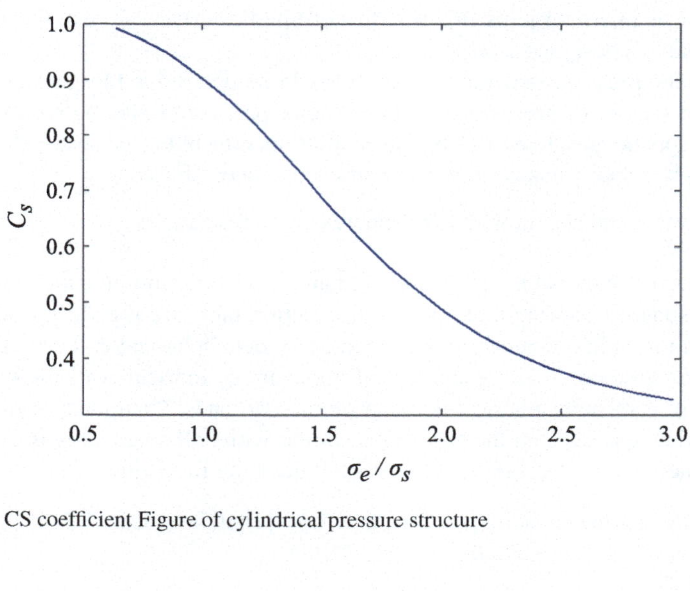

Fig. 7.29 CS coefficient Figure of cylindrical pressure structure

one unknown number in formula, so one-parameter nonlinear formula numerical method fzero could be used to compute wall thickness.

(2) Non-pressure structure.
Non-pressure structure includes main body frame, light shell, stabilizer fin and buoyant material of submersible.

① Main body frame

Frame design shall be optimized frequently according to layout of devices, stress from work conditions of hiking up, and shelving. The nodes of multiply beam cross and gathering together shall be optimized locally to lower stress concentration and make weld convenient.

② Light shell

Light shell is generally made of fiberglass to keep submersible's molded lines unchanged and protect internal devices. It is generally designed and processed according to layout of devices and buoyant materials. When design the thickness of light shell, it is necessary to consider the impact force when submersible entering water under work sea conditions and attached water pressure when getting out of water.

③ Stabilizer fin

The stabilizer fin of "Jiaolong" adopts fiberglass hollow thin wall structure, it is stuffed with prefabricated low-density, hyper high pressure resistance buoyant material. Its exterior high strength fiberglass shell could improve its ability of bearing wave impact, could also keep exterior molded lines of stabilizer fin; internal buoyant material is composed of pressure hollow glass beads, which is distributed evenly in epoxy resin, it cannot only prevent buckling and local deformation of stabilizer fin, but bear stress of hyper high hydrostatic pressure and lower weight of stabilizer fin. The most terrible work condition of stabilizer fin is the impact when submersible enters into water, herein the stress of stabilizer fin could be simplified as the working arm. The root segment of stabilizer fin bears biggest stress under work condition, so in order to lower stabilizer fin's weight and improve its lengthways strength and stiffness, the thickness of exterior glass fin shell is changing unequally, namely the thickness from top to root increases gradually.

④ Buoyant material.

Four major parameters of buoyant material have to be satisfied: a, density, the space to install buoyant material in submersible is limited, it is necessary to choose suitable buoyant material density and volume in limited installation space, the buoyant force shall be enough and charge of buoyant material shall be as low as possible; B. amount of shrinkage, water will shrink under high pressure, and its density will increase, also the buoyant material will shrink under high pressure,if amount of shrinkage is too big, namely displacement volume of buoyant material is reduced more than increment of seawater density, the buoyant force will become smaller when descending deeper; c. pressure strength, buoyant material shall not be broken under submersible's design pressure (and have enough safe storage); d. water absorption rate, when buoyant material is steeped in water for long period, water will penetrate into material, and penetrating rate will increase as water pressure becomes bigger, namely material's water absorption rate increases as water pressure becomes bigger. Water penetrating into material will increase buoyant material's weight, so it is necessary to keep absorbed water will not cause too much weight increment, concrete water absorption rate requirement is regulated in submersible rules, it can also be confirmed according to submersible's gravity and buoyant force adjustment ability.

Regular buoyant material:

Gas. It is the earliest buoyant material, with density around 0.75–0.8, due to its possible environment, it is not used any more.

Metal lithium. With density of 0.54, it shall be put into container and sealed after filling oil. However in case of affair and container broken, violent chemical reaction will happen once lithium and seawater meets together, explosion is also possible, no practical application.

Wood. Wood is cheap and easy to process. Wood's low density is caused by the air in wood's fiber, so its water absorption rate under high pressure is very big. When wood is chosen as buoyant material, high pressure resistance waterproof coating is needed to paint on its surface. The amount shrinkage of wood shall also be researched. It is necessary to check whether there is damage on time during application.

Single size glass beads syntactic foam. This material is produced by fixing hollow glass beads with resin. There is not precise description about its birth. Once it is thought that in 1960s scientists discovered one technique to improve product's percent of pass when researching defective products about the bubble appearing in process of producing reflective material for road sign, the scientist also discovered how to control bubble and get even hollow glass beads.

Multiply glass beads syntactic foam. To increase the proportion of hollow glass beads among buoyant material, small size hollow glass beads is filled into big size hollow glass beads. Typical example for this material is: scattering sand into the bottle full of stone, sand will stuff all of the gap of stone. Multiply size glass beads syntactic foam has more complex technique than single size glass beads, has more cost, but lower density could be got.

Ceramic ball syntactic foam. Hollow ceramic ball has high pressure resistance ability, but the joint surface of resin and ceramic ball will be damaged if directly fix hollow ceramic ball by resin as buoyant material due to their different amount of shrinkage. Thus the ceramic ball shall be coated by rubber before producing buoyant material composed of ceramic ball, rubber and resin. Besides ceramic ball has terrible shock performance, when one hollow ceramic ball in buoyant material breaks under high pressure, the shock wave of implosion may arise chain reaction of other hollow ceramic ball. So this material still needs more research and demonstration.

At the initial design stage, the non-pressure structure in this book basically is reckoned according to similar configuration of "Jiaolong".

Input parameters of whole structure module: ① general length of submersible; ② submersible's width; ③ submersible's height (assuming height equals width); ④ work depth; ⑤ manned cabin shell's internal radius; ⑥ submersible's hoisting weight index; ⑦ adjustable ballast water weight; ⑧ volume of buoyant; ⑨ density of buoyant material. Output parameters: ① general weight of structure system; ② general displacement volume of structure system; ③ general weight torque of structure system; ④ general discharge volume torque of structure system; ⑤ parameters of all pressure structure.

7.2.1.6 Acoustic Module

The devices in acoustic module are: equipment in acoustic communication device tank, transducer of acoustic communication device, devices in tank of deep side scan sonar measuring, transducer of deep side scan sonar, transducer of remote ultra short base lines sonar, obstacle avoidance sonar, Doppler velocity sonar, imaging sonar, movement sensor, manned cabin acoustic computer and other appendages.

According to function, acoustic module is divided into navigation and communication.

(1) Navigation.

When manned submersible is in the depth of 100 m, the deep sea is dark, the pilot can only see distance from several meters to more than ten meters illuminated by lamp on submersible, it is impossible for pilot to judge direction. To make manned submersible sail in the correct preset route, acoustic system has to be used to help navigate. Electromagnetic signal of GPS does not work underwater, so the regular underwater navigation and position techniques of submersible includes acoustic position technique [including long base ling, LBL and ultra short base line, USBL], automatic navigation technique as well as integrated navigation technique of these techniques.

① Long base line acoustic position system (Fig. 7.30) shall be equipped with several acoustic transponders at the bottom of sea, the distance between transponders is very long, usually kilometer level, so it is called "Long Base Line". Transponder is distributed from submersible at the bottom of sea in form of submerged buoy, their 3D coordinates could be measured by some calibration programme. Transponder

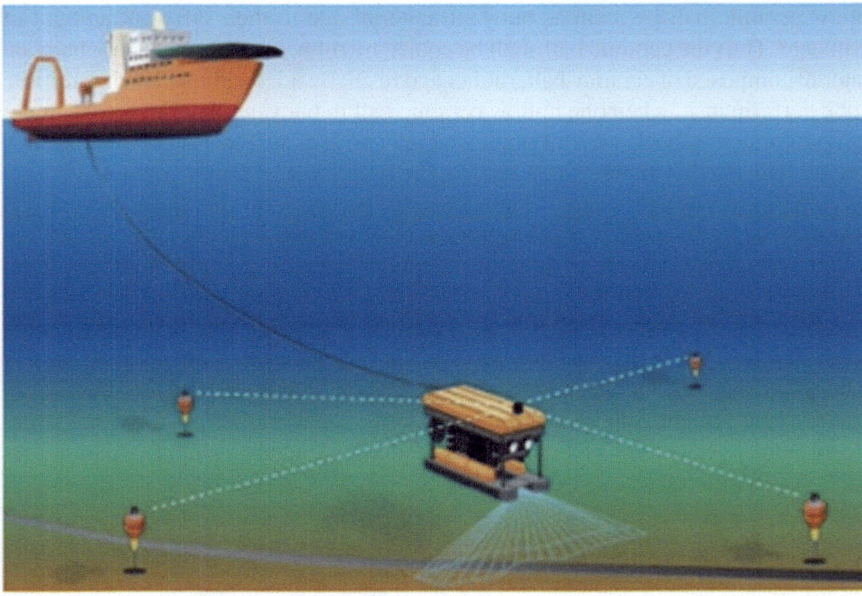

Fig. 7.30 Long base ling acoustic position system

has internal battery, could work underwater for from several days to more than ten days, after work, they can be retrieved after submersible sending instructions, transponder will release heavy items and emerge from water. When used for submersible positioning, acoustic beacons of submersible will send out one enquiry signal, each transponder reply one signal after receiving. The acoustic beacon could measure time period between sending and receiving signal, the direct distance from transponder to acoustic beacon could be computed with sound speed profile of current condition. The 3D coordinates of transponder has been known, then acoustic beacon's 3D coordinates could be computed according to solid geometry formula, namely the 3D coordinate of underwater submersible.

The advantage of long base line acoustic position system is high accuracy position, measurement accuracy is not affected much by work depth, submersible could get its position by computation. Its disadvantages are work area limited by transponders, troublesome in changing work area; too much time is wasted for setting, calibrating and retrieving transponder; there is possibility of failing to retrieve them.

② Ultra short base line acoustic position system (Fig. 7.31) is positioned by acoustic transducer of surface boat, without setting acoustic transponders in the bottom of sea. The distance between parts of acoustic transducer is very short, only more than 10 m, that is why it is called ultra short base line. Two work patterns of ultra short base line acoustic position system: a. transducer on mother ship will send out a enquiry signal, signal beacon will send a answering signal after receiving

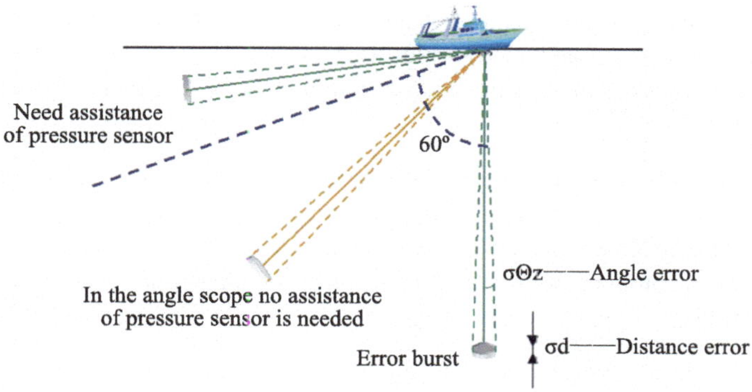

Fig. 7.31 Ultra short base line acoustic position system

it; b. signal beacon will send a answering signal after triggered by synchronizing signal, synchronizing pulse can be produced by cable or synchronizing clock installed on mother ship and submersible respectively. A pattern is two-way sound wave transmission, work cycle is once longer than pattern b, but it needs not synchronizing pulse. Pattern b does not require beacon to receive enquiry signal, is suitable for manned submersibles with high self noise and too many acoustic devices.

Comparison of USBL and LBL measurement theory. LBL gets beacon's coordinate by measuring distance between beacon and transponder, while USBL gets beacon's coordinate by measuring distance and horizontal and vertical angle of transducer compared with beacon, and then converting polar coordinate into land coordinate. No transponder is needed for USBL, so its operation is simple, movement is good. Because USBL gets the coordinate by converting measured angle and distance, absolute error will be bigger with longer distance under the same angle error condition. And the error is closely related with angle, bigger vertical angle will cause bigger error, when angle exceeds 60 degree (120 degree cone angle), pressure sensor is needed to assist to get enough position angle.

需要压力传感器辅助	Need assistance of pressure sensor
在此角度范围内不需要压力传感器辅助	In the angle scope no assistance of pressure sensor is needed
误差区间	Error burst
角度误差	Angle error
距离误差	Distance error

③ Acoustic Doppler Log (ADL) applies acoustic Doppler theory, it could measure submersible's moving speed compared with sea bottom. Combined with navigation sensor, submersible's underwater moving track could be got through integral operation. It could achieve underwater navigation and position by its own sensor, so this method is called "autonomous navigation". Another form of autonomous navigation is adopt inertial navigation system, it is not applied frequently due to its high cost.

Compared with position system, autonomous navigation does need underwater subsurface buoy, surface system's support, it is simple and economic; with high definition of positioned time and space, it is suitable for short and small scope application. Its disadvantages are drifting, especially for long time work, accumulative error will be big.

④ There are disadvantages and advantages of LBL and USBL position system to achieve navigation position and autonomous navigation, they shall be combined together to form combined navigation by applying Doppler velocity log, course sensor, pressure sensor, position system to get measurement data submersible's own movement character. Update rate of LBL and USBL is low, once per seconds, the combine navigation could lower effect of position system error, and provide position of submersible during two position cycles, it indeed improve relative position accuracy. Pressure data and position as absolute position information could eliminate accumulative error of autonomous navigation computation. Thus combined navigation algorithm improves submersible's underwater navigation performance obviously, it could help get precise position, depth and speed of underwater submersible, it is the development direction of underwater navigation technique.

(2) Communication

The communication among submersible, its mother ship, and other underwater and surface devices. Three reasons: first is dark deep sea and small space will make passengers terrified and nervous, and cause wrong judgment and operation; second is when passenger cannot make a decision, it is necessary to contact with surface command system and transmit measurement data and site pictures to surface command system and get support and decision; third is surface scientists or command system needs to deliver instruction to passengers.

Signal of wireless communication has high dissipation speed, it cannot be used for great depth submersible. Underwater acoustic technology is the only communication method between submersible and mother ship, including underwater acoustic telephone and digital underwater acoustic communication, underwater communication in short.

① Underwater acoustic telephone

The function of regular underwater acoustic telephone is single, it is used to achieve underwater voice conversation. The theory of underwater acoustic telephone is after unilateral modulation sending voice from transmitting terminal, displaying at receiving terminal after demodulation. The analog signal transmitting underwater is

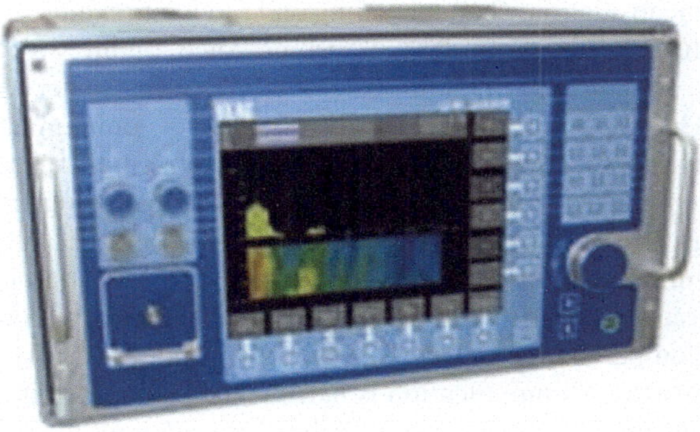

Fig. 7.32 The AIO machine UT300 of underwater acoustic telephone and underwater acoustic communication by Germany company ELAC

easy to be affected by environmental noise, interference signal, signal channel distortion and other factors, the distortion is so big that the underwater acoustic telephone cannot transmit data or only has simple data transmitting function. The recent products integrates functions of underwater acoustic telephone and underwater acoustic communication equipment (Fig. 7.32).

② Underwater acoustic communication

The communication process of underwater acoustic communication equipment sees Fig. 7.33. The digitized analog signal of voice and Figure, words and observation data as information source will be processed by source coding, data organization

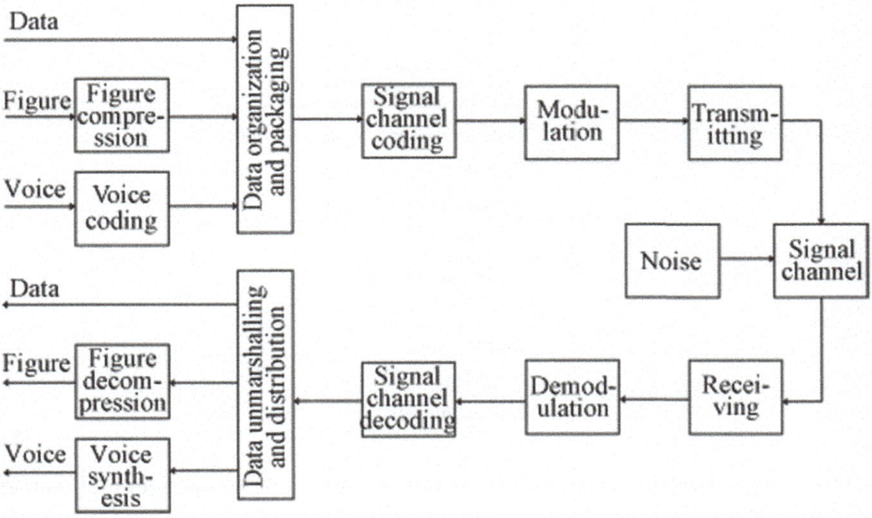

Fig. 7.33 Procedure of underwater acoustic communication

encapsulation, channel coding and modulation and sent out at transmitting terminal, the receiving terminal will resume voice, Figure, words and data reversely.

According to communication distance, underwater acoustic communication technology is divided into short-range communication (<1 km), middle-range communication (1–10 km), distant-range communication (10–200 km) and hyper-remote-range communication (>200 km) technologies. The communication between great depth manned submersible and mother ship belongs to middle-range communication with communication distance around 5 to 10 km. The work frequency for this distance is around 10 kHz, band width could be around 5 kHz. Compared with wireless wave, the band width that can be used for underwater acoustic communication is narrow, the data transmitting rate is pretty low, so when transmitting large amount of data information of voice, Figure and video, they shall be processed by high ratio compression. The transmitting speed of acoustic wave is also low, around 1500 mfs, 0.2 times slower than wireless wave. More than 4 s is needed for acoustic wave transmitting from 6000 m depth to mother ship, which will cause bad effect on timeliness, its application shall be considered carefully.

数据	Data
图像	Figure
图像压缩	Figure compression
语音	Voice
语音编码	Voice coding
数据	Data
图像	Figure
图像解压	Figure decompression
语音	Voice
语音合成	Voice synthesis
数据组织封袋	Data organization sealing bag
数据解包分发	Data unmarshalling and distribution
信道编码	Signal channel coding
信道解码	Signal channel decoding
调制	Modulation
解调	Demodulation
噪音	Noise
发射	Transmitting
接收	Receiving
信道	Signal channel

Underwater acoustic signal channel is a time-varying delay and Doppler double diffusion channel, it can be directly and apparently affected by human activities, such as temperature, salinity, depth, wind, wave, stream, shipping and ocean engineering, so underwater acoustic communication is hard than wireless communication. With the fast development of electronic technology and digital signal procession technology, high performance underwater acoustic communication technology is finally be achieved. According to work theory, underwater acoustic communication technology is divided into incoherent underwater acoustic communication technology and coherent underwater acoustic communication technology and spread spectrum communication technology, and spread spectrum communication technology has low transmitting speed, has not been seen any application on manned submersible.

Incoherent underwater acoustic communication technology applies multiple frequency shift key (MFSK) and coding technology to overcome the interference by multiple path, it had been developed pretty well in 1990s. It has low utilization ratio of band width,, with transmitting speed around hundreds of bit per second Due to its good robustness, it has been applied widely on manned submersible.

Coherent underwater acoustic communication technology applies many techniques, such as multiphase shift keying signal (MPSK), space diversity, adaptive equalizer, error correction of coding and Doppler compensation. Its band width utilization ratio has higher orders of magnitude than incoherent underwater acoustic communication technology, with regular transmission speed at thousands of bit per second.

Foreign manned submersibles apply mainly underwater acoustic telephone to communicate with mother ship when working underwater. American "Alvin" manned submersible is installed with underwater acoustic communication equipment, but its utilization is very few. Underwater acoustic communication equipment developed by Japan is installed on Japanese "Deep ocean 6500" and French "Nautilus", used specially to upload Figure. Its work frequency is 16 to 24 kHz, it adopts modulation DPSK, with max communication speed as 16 kbs, max communication distance 6.5 km, its sonar wavefront beam angle width is 35°, it adopts vertical up-and-down communication.

"Jiaolong" deep ocean manned submersible is installed with middle distance high speed underwater acoustic communication system researched by our country. Many advanced underwater acoustic communication technology and signal procession algorithm is applied, it could transmit Figure, voice, data and word in horizontal and vertical direction within distance from 8 to 10 km. Its underwater acoustic communication function is the strongest among all deep ocean manned submersible, its performance index see Table 7.25.

This book will adopt basically the acoustic communication scheme of "Jiaolong". For general design integration, only power consumption, weight and volume is needed by this system.

Input parameters of this module: ① general length of submersible; ② submersible's width; ③ submersible's height (assuming height equals width); Output parameters: ① weight of subsystem; ② displacement volume of subsystem; ③ weight torque of subsystem; ④ displacement volume torque of subsystem.

Table 7.25 performance of "Jiaolong" manned submersible underwater acoustic communication equipment

Max work distance	8–10 km	
Max work depth	7000 m	
Work frequency	8–13 kHz	
Transmitting Figure	Transmitting rate	10 kb/s
	Probability of bit error	10^{-2}–10^{-3}
Transmitting voice	Transmitting rate	1 kb/s
	Probability of bit error	10^{-2}
Transmitting command	Transmitting rate	256 bit/s
	Probability of bit error	10^{-4}–10^{-5}

7.2.1.7 Outfit Module

Outfit module needs to compute weight of exterior coating, internal outfit, propeller protect frame, hand rail and device of fixation, volume of displacement, weight torque, displacement volume torque.

Input parameters of this module: ① general length of submersible; ② submersible's width; ③ submersible's height (assuming height equals width); ④ internal radius of manned shell. Output parameters: ① general weight of outfit system; ② general displacement volume of outfit system; ③ general weight torque of outfit system; ④ general displacement volume torque of outfit system.

7.2.1.8 Hydraulic Module

The devices that needs to output pretty much force or torque is generally driven by hydraulic pressure, such as mechanical hand of manned submersible, propeller rotation device, buoyant force and slope adjustment devices, mechanical ballast release devices. Hydraulic drive is good at small volume, big output, cheap, durable, easy for achieving automatic control, pressure limiting valve protecting hydraulic system from overburden operation, but is not good at low efficiency, too many pipelines, system flow change caused by hydraulic oil, which is inflammable with high temperature and viscous with low temperature.

Hydraulic system of manned submersible is composed of hydraulic source, pipeline, valve, hydraulic drive, pressure-limiting valve, environment pressure compensation and electromagnetic control devices.

(1) Hydraulic source

Hydraulic source is composed of oil tank, hydraulic pump, electric motor, pressure-limiting valve, pressure indicator. The core parts of hydraulic source are hydraulic pump. Regular pump are plunger pumps, gear pumps and vane pumps. Axial plunger pumps are widely used due to its high efficiency, small volume, but it has big noise. In order to lower noise, some submersible will also choose low efficient but low noisy pumps, such as screw rod hydraulic pump. Vane pump could also be used when hydraulic system has low pressure.

(2) Pipeline.

Hydraulic pipeline of submersible usually adopts flexible pipe. Compared with rigid pipe, flexible pipe has many advantages, such as easy to install, able to absorb deformation and shock from other parts, convenient for dismantle and maintenance. Generally heat dissipation device is installed in pipeline to keep temperature of hydraulic oil varying within allowed scope.

(3) Control valve

Submersible works in deep ocean, so it is not possible to change the control valve of hydraulic system, only remote control is possible. Electromagnetic control will be used in general, which needs suitable electromagnetic control system and software. According to its function, control valve of hydraulic system can be divided into Stop valve, direction control valve, flow control valve. One special valve in valve group of hydraulic system, it is pressure-limiting valve. Pressure-limiting valve is generally set at the output end of hydraulic source, when pressure in oil supplying pipeline of hydraulic system surpasses set value (the value is a little higher than work pressure of hydraulic system, decided by pressure capacity of whole system, rated pressure of hydraulic drive and rate load of drive motor), the valve will open automatically, release pressure to drain line of hydraulic system. This is used to protect the whole hydraulic system.

(4) Hydraulic drive

There are two kinds of hydraulic drive: cylinder and hydraulic motor. For hydraulic oil motor, if noise could meet requirement, small and high efficient axial piston motor is ideal choice; if rotation angle is not whole circle, oscillating vane motor is also workable. For cylinder, it should be noted that if seawater is used as low pressure end of cylinder, the volume of whole hydraulic system will change greatly when cylinder moves within design route. It is required that compensator of hydraulic system shall have enough compensation volume.

(5) Pressure compensation device

For great depth manned submersible, in order to lower weight and volume of hydraulic system, the low pressure pipeline (namely drain pipeline) of whole hydraulic system shall be kept a little higher than seawater pressure or equal seawater pressure by pressure compensation device.

When design hydraulic system, the key point could be concluded as:

① Consider corrosion resistance.
② Note pressure compensation.
③ Drive will become bigger when hydraulic oil become viscous under low temperature (0–2 °C).

④ If whether hydraulic system can still work with seawater in hydraulic circuit.

The hydraulic module of general module in this book includes hydraulic source, propeller turning mechanism, mechanical hand and its control, sampling basket, pipeline and its oil. This module needs to provide general design with power consumption requirement, weight and weight torque, volume and volume torque.

The input parameters of this module include: ① overall length of submersible; ② width of submersible; and ③ height of submersible (the height is equal to the width here). The output parameters include: ① total weight of hydraulic system; ② total displacement volume of hydraulic system; ③ total gravity moment of hydraulic system; and ④ total moment of displacement volume in hydraulic system.

7.2.1.9 Module for Shifting Ballast and Trim Adjustment

This module is composed of three subsystems, i.e. main ballast tank, shifting ballast water system and trim adjustment system.

(1) Main ballast tank.

There is a vent valve on the main ballast tank and an exhaust hole at the bottom of it. The vent valve is opened when the submersible dives, then the submersible dives under the action of gravity. The seawater pours in from the exhaust hole and the air flows out from the vent valve till the main ballast tank is fully filled with water and the vent valve is closed. When the submersible nearly floats to the sea surface after completing the work, the high pressure air is poured into the main ballast tank and the seawater is discharged from the exhaust hole, which will accelerate the floating speed of submersible and ensure the sufficient freeboard of submersible on sea surface. In order to ensure the functional reliability of main ballast tank, the pressure of 15 MPa is set for high pressure gasholder so that the air storage capacity of high pressure gasholder under high pressure can blow the seawater out of the main ballast tank completely at the water depth of 50 m.

(2) Shifting ballast water system.

The gravity and buoyancy of submersible are adjusted by discarding the solid ballast at a greater depth. But the fine adjustment of gravity and buoyancy is usually achieved through shifting ballast water system. The shifting ballast water system is composed of variable ballast tank, high pressure seawater pump, driving motor, valve bank and pipeline, etc. Thereinto, the variable ballast tank is generally of a spherical pressure structure that shall be designed according to the pressure vessel subjected to external pressure. The high pressure seawater pump is the core component of shifting ballast water system. Now, the axial piston is generally adopted for high pressure seawater pump. Due to such characteristics as strong corrosivity, poor lubricity and multiple impurities of seawater, the titanium alloy is usually used for the cylinder and piston of piston pump. Besides, the ceramic lining is used for the internal surface of cylinder and the tungsten carbide is sprayed onto the surface of piston. In order to ensure the flow under high pressure, the fit tolerance of cylinder and piston shall be controlled within micron order so as to avoid abrasion and heating in the case of tight fit and avoid the reduction in efficiency and output flow in the case of loose fit. Due to the small fit clearance between cylinder and piston, the precision of piston in linear motion shall be guaranteed to prevent the piston from off-centering due to the tilt power of tilting frame and thus avoid abrasion and even the damage to ceramic lining of cylinder due to collision. Hence, the guide mechanism of pump shall meet a higher precision requirement. It is thus obvious that the high pressure seawater pump poses a higher challenge to the manufacture of precision instrument in China. The sub-topic on the high pressure seawater pump with large flow has been established for 4500 m manned submersible and the domestic forces have been gathered to overcome this technical difficulty. The deep-sea submersible shall bear a high pressure of seawater, but the frequently-used solenoid electric valve cannot be used under such high pressure, so the hydraulic drive valve is usually used for the valve bank of shifting ballast water system.

With the technical progress of high pressure seawater pump with large flow, the adjustment methods of gravity and buoyancy focusing on the solid ballast in manned submersible will be substituted by the large shifting ballast water system and the trim adjustment system may even be substituted by the shifting ballast water system with the bow and the stern equipped with variable ballast tank, such as Russia's "MIR-I" and "MIR-II".

(3) Trim adjustment system.

Mercury and oil are often used as control agents in trim adjustment system of submersible. Besides, the pressure compensating device is installed to reduce the structural weight of system. The environment will be polluted seriously in case of any leakage of mercury, so with the technical progress of shifting ballast water system, the trim adjustment system based on mercury is gradually substituted by shifting ballast water system.

The input parameters of this module include: ① overall length of submersible; ② width of submersible; ③ height of submersible (the height is equal to the width here); ④ weight of shifting ballast water; and ⑤ average weight of mercury in trim adjustment tank. The output parameters include: ① total weight of ballast and trim adjustment system; ② total displacement volume of ballast and trim adjustment system; ③ total gravity moment of ballast and trim adjustment system; and ④ total moment of displacement volume in ballast and trim adjustment system.

7.2.1.10 Life Support Module

The life support system of manned submersible can control the oxygen concentration in manned compressive cabin, remove the carbon dioxide, moisture and odor from manned compressive cabin and monitor the pressure, temperature and humidity of manned compressive cabin so as to provide a suitable living environment for crew. In order to ensure the safety of crew, the life support system must be equipped with emergency devices to deal with emergency situations. In addition, the emergency devices are also used as redundant backups of devices under normal operation to guarantee the reliability of life support.

In this module, the capacity of high pressure oxygen and the amount of carbon dioxide absorbent are calculated according to the average oxygen consumption and carbon dioxide emission under normal working and emergency conditions and the life support time under normal diving and emergency conditions.

On this basis, the power consumption and weight of oxygen storage device, carbon dioxide absorption device (including the cabin gas-circulating system, gas composition detection system and control system), display panel and control panel, pipelines and valves and other corollary equipment are calculated based on existing experience.

The input parameters of this module include: ① overall length of submersible; ② width of submersible; ③ height of submersible (the height is equal to the width here); ④ time for unpowered diving and floating; ⑤ submarine cruising time; ⑥ operating time; and ⑦maximum survival time of submersible. The output parameters include: ① weight of life support system; ② gravity moment of life support system; ③ capacity of oxygen tank needed for normal mask, emergency open mask and emergency oral-nasal mask; and ④ amount of carbon dioxide absorbent needed for normal mask, emergency open mask and emergency oral-nasal mask.

7.2.1.11 Load Rejection Module

This module includes the release mechanism for diving and floating kentledge under normal diving conditions and the devices to discard the main battery box and manipulators under emergency conditions. The load rejection mechanism and devices independently designed for "Jiaolong" have reliable performance and the mature high pressure seawater pump with large flow has not been manufactured in the sub-topic on high pressure seawater pump, so this module will tentatively use the load rejection devices with mechanical kentledge of "Jiaolong" and other emergency load rejection devices.

The input parameters of this module include: ① overall length of submersible; ② width of submersible; and ③ height of submersible (the height is equal to the width here). The output parameters include: ① weight of load rejection system; ② displacement volume of load rejection system; ③ gravity moment of load rejection system; and ④ moment of displacement volume in load rejection system.

7.2.1.12 Power Distribution Module

According to the literature (Cao and Cui 2008; Wei et al. 2008), the typical voyage of submersible is divided into 11 stages, including preparation, layout, diving, adjustment, cruising, operating, emergency escape, adjustment, floating, recycle and reservation. The power consumption table of typical voyage is made by analyzing the power consumption of propulsion, hydraulic pressure, observation and communication, navigation, acoustics, life support and other power consumption subsystems in each stage. In this book, the power consumption table is preliminarily modified based on the sea trial data of "Jiaolong". And the capacity of main battery and auxiliary battery can be determined by combining the power consumption of each subsystem and the time consumption of each stage. Similarly, the capacity of standby battery and cabin emergency battery can be determined based on the power consumption under emergency conditions.

Most systems of submersible are drove by power. The power of self-propelled submersible is mostly supplied by rechargeable batteries. The lead-acid battery and nickel-cadmium battery are firstly used for submersible. These batteries are cheap and reliable, but the energy density cannot meet the requirements of submersible. Later, the silver-zinc battery with high energy density is used for submersible, but this battery is expensive, with short lifespan and high cost. Now, all the countries are attempting to adopt the lithium battery with high energy density and moderate cost to substitute the silver-zinc battery. The lithium battery has been used for Japan's "Deep Sea 6500" Manned Submersible since 2004 and kept in a good service condition. But the lithium battery with high energy has had accidents like spontaneous combustion, so its safety and reliability shall be further examined. The performance comparison of batteries that are commonly used for submersible is shown in Table 7.26.

In addition to these batteries, the fuel battery is also used for underwater vehicle, like Germany's Submarine U1. The proton exchange membrane fuel battery that is used underwater is characterized by superhigh energy density, stable high power output and high efficiency, etc., but the volume is huge and the safety is to be examined. However, it is important to realize that most fuel batteries cannot achieve the

Table 7.26 Performance comparison of popular submersible batteries

Type of battery	Gravimetric energy density/(Wh/kg)	Volumetric energy density/(Wh/m^3)	Gravimetric power density/(W/kg)	Times of charge and discharge	Efficiency (%)
Lead-acid battery	30–50	200–300	200–300	>700	>80
Silver-zinc battery	75–140	150–250	30–400	>1000	>80
Nickel-cadmium battery	35–45	150	95–105	>800	>70
Lithium battery	95–120	280–300	280–300	>1200	>80

balance of internal and external pressures by means of pressure compensation and consequently they shall be placed in a special pressure structure, which will bring about a sudden increase in weight and volume of energy system.

America launched the "Curiosity" Mars Rover in 2012 and even used the miniature nuclear reactor on the modest body of "Curiosity". The application of similar miniature nuclear reactor technologies to manned submersible will greatly reform the energy of underwater vehicle, such as manned submersible, AUV and submarine space station, etc. Thus, the great progress will be made in respect of operating time and operating range.

The sub-topic on domestic lithium battery has been established following the sub-topic on 4500 m manned submersible. The lithium battery will be preliminarily used in this module and the weight and volume needed for the battery will be calculated according to battery capacity and energy density of the lithium battery. On this basis, the data provided by the supplier and the empirical data of "Jiaolong" are used to estimate the weight of pressure equalizing battery box, control circuit for battery charge and discharge, cables and connectors, power distribution equipment and junction box.

The input parameters of this module include: ① overall length of submersible; ② width of submersible; ③ height of submersible (the height is equal to the width here); ④ energy density of main power supply, secondary power supply, reserve power supply and emergency power supply; ⑤ mass density of main battery, auxiliary battery and standby battery; ⑥ cruising in working direction, highest speed of working direction, and input power of lateral and vertical propulsion systems; ⑦ power of hydraulic system; ⑧ power of lighting system; ⑨ time for unpowered diving and floating; ⑩ submarine cruising time; ⑪operating time; ⑫maximum survival time of submersible; ⑬vertical position of main battery box; and ⑭vertical position of auxiliary battery box. The output parameters include: ① capacity of main battery; ② capacity of auxiliary battery; ③ capacity of standby battery; ④ capacity of emergency battery; ⑤ weight of energy system; ⑥ displacement volume of energy system; ⑦ gravity moment of energy system; and ⑧ moment of displacement volume in energy system.

7.2.1.13 Module for Unpowered Diving and Floating Motion

Through the summary about the calculation of above-mentioned modules are the weight and displacement volume of submersible obtained. This refers to the standard weight and standard displacement volume of submersible because the kentledge is not mounted.

The sea trial of "Jiaolong" aims to obtain the actual data of unpowered diving and floating motion; analyze the changes of seawater density with depth and the buoyancy variation of submersible caused by the amount of volume compression under the action of seawater pressure; and consider the water absorption and amount of volume compression of buoyancy materials in the cruising process of submersible at working depth and those in the pressure maintaining process. On this basis, a more practical loading model for unpowered diving and floating motion of submersible has been established and the program has been written (Binbin et al. 2012), which achieves a better application effect in sea trial (with the loading error controlled within dozens of kilograms). In the sea trial of 7,000 m, it approximately takes 3 h to dive to the depth of 7,000 m according to the normal loading of "Jiaolong" and the diving and floating time approximately accounts for half the time of the whole diving motion. The effective working time of submersible can be increased by reducing the diving and floating time. It can be seen that the diving and floating speed is an important index for manned deep-sea submersible and the problem faced by 4500 m manned submersible is how to accelerate the diving speed. So in this book, MDO model has been added into the unpowered diving and floating motion program and the diving and floating speed has been included into the overall design since the conceptual design.

In this module, the floating speed and time can be estimated based on the difference between standard weight and standard displacement volume of submersible and the weight of floating load rejection P_2 can be determined based on the balance of gravity and buoyancy at working depth. Then, the diving speed and time can be estimated merely by mounting the diving load rejection P_1. The detailed unpowered motion model and related algorithms are shown in references (Binbin et al. 2012), which will not be repeated in this book. In this module:

The input parameters of submodule used to calculate the weight of floating load rejection include: ① equilibrium location (i.e. diving depth); ② standard weight of submersible (excluding the weight of kentledge); ③ standard displacement volume of submersible; ④ bottom-sitting force; and ⑤ design parameters of all pressure structures. The output parameter is the weight of floating load rejection.

The input parameters of submodule used to calculate the diving time include: ① diving depth; ② standard weight of submersible; ③ standard displacement volume of submersible; ④ weight of diving load rejection; ⑤ weight of floating load rejection; and ⑥ design parameters of all pressure structures. The output parameter is diving time.

The input parameters of submodule used to calculate the floating time include: ① diving depth; ② standard weight of submersible; ③ standard displacement volume of submersible; and ④ design parameters of all pressure structures. The output parameter is floating time.

It can be seen that the overall design of manned submersible involves various disciplines, such as structural mechanics, fluid mechanics, acoustics, optics, electricity, machinery, control and material, etc. After integrating these subsystems into the multidisciplinary design optimization model of submersible, the optimization algorithm will be used to solve the overall MDO model and the local optimization of some subsystems. For example, as for the structural system, the optimization algorithm shall be used to determine the wall thickness and other parameters of all pressure structures based on the parameters transmitted by the overall model; and as for the diving and floating module, the optimization algorithm shall be used to determine the suitable floating load rejection P_2 based on the difference of gravity and buoyancy obtained from the overall model. According to MDO theory, the local variables of these subsystems can be returned to the overall system so as to combine with design variables of the overall model and thus constitute the design variables with higher dimensionality. Then, the optimization algorithm is used for the general solution in the layer of overall model, but this will increase the complexity and dimensionality of overall model and thus greatly increase the calculated amount. In this book, the local design variables of subsystems are determined by local optimization, which means that the optimization of subsystems shall be solved in each operation process of the overall model. There are only a few of subsystems that shall be optimized locally in MDO model of manned submersible and the design variables of local optimization have a lower dimensionality, so this practice can simultaneously reduce the complexity and calculated amount of the overall system model.

There is no feedback but only feedforward in the multidisciplinary design optimization model of manned submersible after taking these treatment measures, as shown in Fig. 7.22. The multidisciplinary feasible method (MDF) can be used directly to establish the multidisciplinary design optimization model.

Here, taking the simplest estimation of conceptual design for example, the total diving and floating time is taken as objective function. Furthermore, in the premise of ensuring the balance of gravity and buoyancy and meeting the constraint conditions of structural strength, power demand and other subsystems, the constraint condition of the overall MDO model is that the total hoisting weight shall not exceed the rated working weight of hoisting equipment on master ship (The constraints among other subsystems are automatically satisfied through the optimization within subsystems. For example, the power distribution is matched with battery capacity; the strength of pressure structure meets the requirements; and the floating load rejection is matched automatically.), as shown in Formula (7.66).

$$\begin{aligned}
&\min \quad f(X) = t_a + t_d \\
&s.t. \quad g(X) = M - 20 \times 10^3 \leq 0 \\
&X = [D, L_m, \rho_b, V_b, M_{P_1}] \\
&\underline{X} \leq X \leq \overline{X} \\
&\underline{X} = [2.5, 4.25, 520, 5, 100] \\
&\overline{X} = [3.2, 5.5, 550, 15, 1500]
\end{aligned} \tag{7.66}$$

Thereinto, f refers to objective function; t_a refers to floating time; t_d refers to diving time; g(X) refers to constraint function; and M refers to the hoisting weight after the submersible is mounted with all load rejections, where the weight shall not exceed 20t as required (i.e. the allowable hoisting weight of hoisting equipment on master ship). In the design variable $X = [D, L_m, \rho_b, V_b, M_{P_1}]$, the components respectively refer to shape diameter, parallel middle body length, density and volume of buoyancy materials and weight of diving load rejection M_{P_1}.

7.2.2 Uncertain Parameter Modeling

There are a lot of parameters for the overall model of submersible, but many parameters and constraint conditions are regulated in *Design Specification for 4500 m Manned Submersible*. For instance, the internal diameter of manned cabin shall be 2.1 m; there shall be 5 viewing windows; the overall length of submersible shall be 8.2 m; the battery shall be lithium battery; the portable load shall be 200 kg; the number of crew shall be 3; the submerged cruising speed shall be 1 knot; the highest submerged speed shall be 2.5 knots; and the submerged life support time shall be 72 h, etc. In addition to the research on previous key technical topics, the performance and specifications of products manufactured by domestic manufacturers have been understood basically. Many parameters and indexes become named parameters that cannot be designed in this model, so there are only 5 design variables left.

As known from Chap. 5 of this book, the security coefficient of 1.5 is equivalent to the design of spherical pressure structure with a reliability of 6ˆ, so the structural module of MDO model in this section still uses the security coefficient of 1.5 for deterministic structural design so as to ensure the reliability of structural system and avoid the RBDO with huge calculated amount in structural subsystems. Besides, the approximate treatment and fitting treatment have been performed for all subsystems when MDO model of manned submersible is established above, so there is no feedback but only feedforward in the data transmission for MDO model of manned submersible. Hence, in the process of RBMDO for MDO model of manned submersible, it can be regarded as a traditional RBDO problem, which can effectively control the calculated amount and calculation difficulty.

As for the MDO model of manned submersible established above, there are no complete statistical data about these 5 design variables, so in this book, the statistical information of design variables is assumed according to the empirical data and engineering experience of "Jiaolong", as shown in Table 7.27.

Compared with the tight RBDO for manned cabin, the RBMDO model is relatively rough in conceptual design stage. Firstly, MDO model contains a lot of empirical data and estimation methods, which are complied based on the design reports and final reports about all subsystems of "Jiaolong" and thus basically integrate the design levels of all subsystems into the overall model. But it can be seen that the designs and researches of some subsystems are not intensive enough and the theories and methods of calculation reports are still rough, which leads to the large deviation between overall model and actual design. However, the calculation methods for several core subsystems of submersible in this book have been verified through actual data. It can be said that the overall model of manned submersible in this book is the one that is closest to engineering practice. Secondly, it can be seen that most design variables of MDO model in the overall design of submersible are overall parameters, such as external dimension, volume of buoyancy materials and loading, etc. But there are no reports or papers about the uncertain statistics of these parameters, which shall be determined through the consultation with manufacturers in the design process. The uncertain variable statistics that are determined in this

Table 7.27 Statistical parameters for RBMDO design variables of manned submersible

Uncertain variable	Distribution pattern	Standard deviation	Source
D	Normal distribution	0.05	$3\sigma = 3.0 \times 5\%$ when the engineering precision is 5%
Lm	Normal distribution	0.075	$3\sigma = 4.5 \times 5\%$ when the engineering precision is 5%
pb	Normal distribution	3.6	Based on the density determination of buoyancy materials used for "Jiaolong", the density of buoyancy materials can be controlled within 2%, taking $3\sigma = 540 \times 2\%$
Vb	Normal distribution	0.15	$3\sigma = 9 \times 5\%$ when the engineering precision is 5%
Mp	Normal distribution	4	The kentledge is made of sheet irons that are piled up and the weight of each square sheet iron is about 12 kg, taking $3\sigma = 12$

way will be subject to the subjective influence caused by designers of overall design and related subsystems. Thirdly, the lifting capacity of lifting equipment on master ship generally contains a large allowance, so 20t is not the real limit state surface of submersible and thus the surface support system shall be used to provide a more accurate lifting capacity for lifting equipment. These problems are those that shall be solved when RBMDO theory is applied to practical engineering problems. Besides, the solutions are not limited to RBMDO and all subdisciplines shall be enhanced. For example, the precision of RBMDO model will not be improved until the designers of all subsystems consistently enhance the calculation and analysis methods of their own disciplines.

7.2.3 Determination of Design Indexes

Various design indexes of submersible have been proposed in *Design Specification for 4500 m Manned Submersible* and many parameters and constraint conditions have also been regulated here. For instance, the internal diameter of manned cabin shall be 2.1 m; there shall be 5 viewing windows; the overall length of submersible shall be 8.2 m; the battery shall be lithium battery; the portable load shall be 200 kg; the number of crew shall be 3; the submerged cruising speed shall be 1 knot; the highest submerged speed shall be 2.5 knots; and the submerged life support time shall be 72 h, etc. Therefore, these indexes have been regarded as named parameters (that cannot be designed) in the process of establishing MDO model and the indexes related to this subsystem are taken as constraint conditions for the internal optimization of all subsystems. So these indexes are not taken as constraint conditions in system layer.

In addition to the rigid indexes stipulated in *Design Specification*, the energy, weight, static moment and other constraints are also converted in this model. According to the overall design ideas of submersible, the power consumption is determined by equipment, the energy is determined by power consumption and the weight of power supply and accessories is determined by energy, so the energy will be converted into weight finally. This means that the constraint conditions, like energy, are transmitted among all subsystems in the form of coupling parameters and gotten in balance automatically. Similar parameters that can automatically achieve the balance in subsystems also include floating load rejection and kentledge P_2, etc. In addition, there is no working condition of leaving cabin on sea surface for the current design of submersible, like "Jiaolong", so the metacentric height is not strictly regulated in submersible specifications. Besides, the general arrangement is not accurate in schematic design stage, so the static moment is not constrained in this book. However, the gravity moment exists in the output module programs of all subsystems and the static moment can be output for constraint merely by updating the positions of all equipment, so only one constraint function is needed in this model.

7.2.4 Calculation Results and Discussion

7.2.4.1 Results of Certainty MDO for Manned Submersible

The algorithm of Latin Hypercube Search (LHS) is taken as the PS optimization algorithm of pre-search (LHSPS) to solve the overall MDO model, with the optimization results shown in Table 7.28.

It can be seen from above calculation results that in order to control the diving and floating time within 3 h, the density of buoyancy materials shall be 520, the weight of diving kentledge P_1 shall be 885 kg or so and the volume of submersible

Table 7.28 Results for certainty multidisciplinary design optimization of manned submersible

Parameter	Initial point	Lower limit	Upper limit	Optimization point
Design variable				
D	3.0	2.5	3.2	2.5
L_m	4.45	4.25	5.5	4.25
Pb	520	520	550	520
V	9.2	5	15	8.9282
Mp_x	500	100	1500	883.56
Constraint function				
	48.5			0
Objective function				
f	3.4			2.969

shall be reduced in the premise that the hoisting weight does not exceed 20t. Of course, the results of this model are based on the fact that the shape of 4500 m manned submersible is similar to that of "Jiaolong". In order to ensure that the internal space of submersible can accommodate the equipment of all subsystems and there is enough space for the layout of buoyancy materials, the hydrodynamic performance optimization can be made to the shape of submersible in diving or floating direction, such as changing the round transverse section of submersible into an oval one, increasing the trim angle of submersible in diving and floating processes, optimizing the molded lines of the bow and installing the transparent openable fairwater, etc. (The fairwater will be closed in the process of diving, floating and cruising to wrap up such equipment as manipulators, sampling baskets, lights and video cameras and thus reduce the resistance; and the fairwater is opened and folded under submersible in the operating process so that the submersible can operate normally. This is like the opening and closing of the car roof of convertibles.) Or the battery with a higher power density is used and the vertical propeller is opened in diving and floating processes to achieve acceleration.

7.2.4.2 RBMDO Results of Manned Submersible

According to the specified value of hoisting weight in existing design levels and *Design Specification*, the objective reliability of constraint function is set as 0.9987 in this RBMDO model. Then, the SLRBDO algorithm is used to solve this RBMDO model, with the optimized solution shown in Table 7.29.

Compared with the results of certainty MDO, it can be seen that the shape of submersible shall be as small as possible both in certain and uncertain optimization results, which means that the resistance of submersible and the density of buoyancy materials shall be as small as possible. But in order to ensure the reliability of hoisting weight, the volume of buoyancy materials and the weight of kentledge for diving load

Table 7.29 Overall RBMDO results of manned submersible

Variables	Initial point	Lower Bound	Upper bound	Optimized point
Design variables				
D	3.0	2.5	3.2	2.5
L_m	4.45	4.25	5.5	4.25
ρ_b	520	520	550	520
V_b	9.2	5	15	8.9282
M_{P_1}	500	100	1500	883.56
Constraint function				
g	48.5	–		0
Objective function				
f	3.4	–		2.969

rejection in uncertain optimization shall be lower than those in certain optimization. And the corresponding diving and floating time is also increased by $\frac{3.433-2.969}{2.969} = 15.6\%$.

It can be seen from this chapter that there are a lot of difficulties arising in the application of RBMDO method to practical engineering problems, including the precision of MDO model, acquisition of parameter statistics and determination of limit state surface in addition to the huge calculated amount, which shall be solved through corresponding measures. The basic handling methods are displayed in this book by examples.

References

Agarwal H, Renaud JE, Mack JD. A decomposition approach for reliability-based multidisciplinary design optimization. In: 44th AIAA/ASME/ASCE/AHS structures, structural dynamics, and materials conference. Norfolk, Virginia, AIAA 2003-1778

Agarwal H, Renaud JE, LeeJ C, Watson LT (2004) A unilevel method for reliability based design optimization. In: Proceedings of the 45th AIAA/ASME/ASCE/AHS/ASC structures, structural dynamics, and materials conference, AIAA2004–2029. Palm Springs, CA

Anxi C (2008) Multidisciplinary design optimization methods and application research for HOV. PhD thesis of Shanghai Jiaotong University. (操安喜. 载人潜水器多学科设计优化方法及其应用研究. 上海交通大学博士学位论文, 2008)

Binbin P, Weicheng C, Cong Y, Zhengyuan L (2012) Development of the unpowered descent and ascent system for jiaolong HOV. J Ship Mech 16(1–2):58–71. (潘彬彬, 崔维成, 叶聪, 刘正元. 蛟龙号载人潜水器无动力潜浮运动分析系统开发. 船舶力学, 2012,Vol. 16, No.1-2, pp. 58-71.)

Binbin P, Weicheng C (2010) Multidisciplinary design optimization of whole ship structure based on finite element method. China Ship Build 51(1):47–54. (潘彬彬,崔维成. 基于有限元的整船结构多学科设计优化. 中国造船, 2010, Vol. 51, No.1, pp. 47-54.)

CCS (2011) Principles of drawing review and inspection of deep-sea manned submersible. (中国船级社. 深海载人潜水器审图和检验原则. 2011.)

Cao AX, Cui WC (2008) Multidisciplinary design optimization method applied to a HOV design. In: International conference of INMARTECH2008, All Seasons Hotel. France

Chen XG, Hasselman TK, Neill DJ, Reliability based structural design optimization for practical applications. AIAA-97–1403, pp 2724–2732

Choi KK, Youn BD, Du L (2005) Integration of reliability- and possibility-based design optimization using performance measure approach. In: Keynote paper 2005 SAE world congress. Detroit, MI

Choi KK, Youn BD, Yang R (2001) Moving least square method for raliability-based design optimization. In: Proceedings of the 4th world congress of structural and multidisciplinary optimization. Dalian, China

Committee on Future Needs in Deep Submergence Science (CFNDSC). Future Needs in Deep Submergence Science: Occupied and Unoccupied Vehicles in Basic Ocean Research, National Research Council, ISBN: 0-309-52917-4, 152 pages, The National Academies Press, 2004. This PDF is available from the National Academies Press at: http://www.nap.edu/catalog/10854.html

Cui W, Hu Y, Guo W et al (2014) A preliminary design of a movable laboratory for hadal trenches. Methods Oceanogr 9:1–16

Depei S, Changchun L (1991) Structural strength of submersible. Shanghai jiao tong university press. (施德培,李长春. 潜水器结构强度[M]. 上海交通大学出版社,1991.)

Deremaux Y, Pietremont N, Négrier J, Herbin E, Ravachol M (2008) Environmental MDO and uncertainty hybrid approach applied to a supersonic business jet. In: 12th AIAA/ISSMO multidisciplinary analysis and optimization conference. Victoria, British Columbia Canada, AIAA 2008-5832

Dong WM, Shah HC, Vertex method for computing functions of fuzzy variables. FSS24, pp 65–78

Fan H, Li WJ (2008) An efficient method for reliability-based multidisciplinary design optimization. Chin J Aeronaut 21:335–340

Geethaikrishnan C, Mujumdar PM, Sudhakar K, Adimurthy V (2010) A Hybrid MDO architecture for launch vehicle conceptual design. In: 51st AIAA/ASME/ASCE/AHS/ASC structures, structural dynamics, and materials conference < BR > 18th. Orlando, Florida. AIAA 2010-2757

Guo J (2008) Uncertainty analysis and sentivity analysis for multidisciplinary systems design. PhD thesis of Missouri University of Science and Technology

Gündüz ME, Software integration for automated stability analysis and design optimization of a bearingless rotor blade. PhD thesis of Daniel Guggenheim School of Aerospace Engineering in Georgia Institute of Technology

Hawkes G (2009) The old arguments of manned versus unmanned systems are about to become irrelevant: new technologies are game changers. Marine Technol Soc J 43(5):164–168

Helton J, Johnson J, Oberkampf W (2007) An exploration of alternative approaches to the representation of uncertainty in model predictions. Reliab Eng Syst Saf 85(1–3):39–71

Jianguo Z, Duo S, Yingwei L (2008) Reliability analysis and optimization of mechanical products. Electronic industry press. (张建国,苏多,刘英卫(编著). 机械产品可靠性分析与优化[M]. 北京:电子工业出版社,2008.)

Jimao Z (1992) Submersible design. Shanghai jiao tong University Press. (朱继懋. 潜水器设计[M]. 上海交通大学出版社, 1992.)

Kim HM (2001) Target cascading in optimal system design. PhD Dissertation, Department of Mechanical Engineering, University of Michigan

Kohnen W (2009) Human exploration of the deep seas: fifty years and the inspiration continues. Marine Technol Soc J 43(5):42–62

Li M, Azarm S (2008) Multiobjective collaborative robust optimization with interval uncertainty and interdisciplinary uncertainty propagation. J Mech Des 130

Liu B (1999) Uncertain Programming. Wiley, New York

Liu W (2007) Application of multidisciplinary design optimization method in the overall design of 7000m Manned submersible [D]. Shanghai Jiao Tong University, 2007. (刘蔚. 多学科设计优化方法在7000 米载人潜水器总体设计中的应用[D]. 上海: 上海交通大学,2007.)

Liu B, Iwamura K (1998b) Chance constrainted programming with fuzzy parameters. Fuzzy Sets Syst 94:227–237

Liu B, Liu YK (2002) Expected value of fuzzy variable and fuzzy expected value models. IEEE Trans Fuzzy Syst 10:445–450

Liu B, Iwamura K (1998) A note on chance constrained programming with fuzzy coeffieients. Fuzzy Sets Syst 100:229–233

Liu B (2006) A survey of credibility theory. Fuzzy Optim Decis Making 5(4)

Liu B (2002) Toward fuzzy optimization without mathematical ambiguity. Fluzzy Optim Decis Making l:43–63

Lu B, Liu T, Cui WC (2004) Ultimate strength of pressure spherical hull in deep-sea manned submersibles. J Ship Mech 8(1):51–58 (in Chinese)

Lévêque JP, Drogou JF (2006) Operational overview of NAUTILE deep submergence vehicle since 2001. New Orleans, LA, Underwater Intervention 2006. In: Proceedings of Underwater Intervention Conference. New Orleans, LA. Published by Marine Technology Society

Momma H (1999) Deep ocean technology at JAMSTEC. Mar Technol Soc J 33:49–64

Nanba N, Morihana H, Nakamura E et al (1990) Development of deep submergence research vehicle "SHINKAI 6500". Tech Rev Mitsubish Heavy Ind Ltd 27:157–168

Neufeld D, Chung J, Behdinan K (2009) Aircraft conceptual design optimization with uncertain contributing analyses. In: AIAA modeling and simulation technologies conference. Chicago, Illinois, AIAA 2009-6237

Ohno D, Shibata Y, Tezuka H, Morihana H, Seki R (2004) A design study of manned deep submergence research vehicles in Japan. MTS J 38(1):40–51

Padula SL, Gillian RE (2006) Multidisciplinary environments: a history of engineering framework development. In 11th AIAA/ISSMO multidisciplinary analysis and optimization conference. Portsmouth, Virginia. AIAA 2006-7083

Paliy OM (1991) Weight characteristics, reliability and operational safety of deep-sea submersible hulls. In: International symposium on marine structures (ISM'91). Shanghai, China, pp 197–199

Pan BB, Cui WC (2010) An overview of buckling and ultimate strength of spherical pressure hull under external pressure. Marine Struct 23:227–240

Pan BB, Cui WC, Leng WH (2009) Multidisciplinary design optimization of surface vessels. J Ship Mech 13(3):378–387

Pan BB, Cui WC, Shen YS (2012) Experimental verification of the new ultimate strength equation of spherical pressure hulls. Marine Struct. https://doi.org/10.1016/j.marstruc.2012.05.007

Pan BB, Cui WC (2011a) On an appropriate design and test standard for spherical pressure hull in a deep manned submersible. In: Presented at advanced underwater technology for the ocean symposium on underwater technology 2011 and workshop on scientific use of submarine cables & related technologies 2011, IIS conference Hall, "Haricot". Tokyo, Japan

Pan BB, Cui WC (2011b) A comparison of different rules for the spherical pressure hull of manned submersible. J Ship Mech 15(3):276–285

Pan BB, Cui WC (2011c) Structural optimization of the manned pressure hull of a deep manned submersible based on a new safety standard. In: International conference on computation, and experimental engineering and science (ICCES). Nanjing, China. Session: Theory, Analysis and Design of Shell Structures

Peng G (2009) Application of sequential collaborative optimization method in the design of deep-sea space station structure system. PhD thesis of Shanghai Jiaotong University. (苟鹏. 序列协同优化方法在深海空间站结构系统设计中的应用. 上海交通大学博士学位论文, 2009.)

Qu X, Venkataraman S, Haftka RT, Johnson TF (2001) Reliability, weight, and cost tradeoffs in the design of composite laminates for cryogenic environments. In: AIAA Paper 2001-1327, Proceedings of the AIAA/ASME/ASCE/AHS/ASC 42nd structures, structural dynamics and materials conference. Seattle, Washington

Rules for certification/classification of submersibles, published by Det Norske Veritas (DNV) in 1988

Rules for building and classing underwater vehicles, systems and hyperbaric facilities 2010, published by American Bureau of Shipping (ABS) in 2010

Rules and Regulations for the classification of submersibles, published by Bureau Veritas (BV) in 1989

Rules and Regulations for the construction and classification of submersibles and underwater systems, published by Lloyd's Register (LR) in 1989

Rules for the survey and construction of steel ships contents, published by Nippon Kaiji Kyokai (NK) (publishing year unknown)

Rules for the classification and construction of manned submersibles, ship's diving systemsand passenger submersibles, published by Russian Maritime Register of Shipping (RS) in 2004

Rules for the classification and construction of diving systems and submersibles, publish by China Classification Society (CCS) in 1996

Rules for Classification and Construction,1-Ship Technology,5-Underwater Technology,2-Manned Submersibles, published Germanischer Lloyd Aktiengesellschaft (GL) in 2009

Sagalevitch AM (2009) From the bathyscaph trieste to the submersibles mir. Marine Technol Soc J 43(5):79–86

Sagalevitch AM, Experience of the use of manned submersibles in P. P. Shirshov Institute of Oceanology of the Russian Academy of Sciences. In: Proceeding of the 1998 international symposium on under-water technology, p 403

Sakalkar V, Hajela P (2011) Multilevel decomposition based nondeterministic design optimization for structural systems. Adv Eng Softw 42:1–11

Savoia M (2002) Structural reliability analysis through fuzzy number approach, with application to stability. Comput Struct 80:1087–1102

Taylor L, Lawson T (2009) Project deepsearch: an innovative solution for accessing the oceans. Marine Technol Soc J 43(5):169–177

Tu J, Choi KK (1999) A new Study on reliability-based design optimization. J Mech Des 121(4):557–564

Vadim VK, Jury KM (1999) About designing experience of submarines and deep submersibles from titanium alloys in Bureau “MALACHITE”. In: Titanium’99, p 1125

Van Dover C (eds) (2009) 6500 m HOV Project: A-4500 HOV project execution plan

Walsh D (2009) In the beginning… a personal view. Marine Technol Soc J 43(5):9–14

Wang ZL, Wang RH, Yu MH, Li LB (2007) The influence of the initial imperfections on the ultimate strength of manned deep-sea submersible pressure sphere hull. Shipbuild China 48(2):45–50

Weck OD, Agte J, Sobieski JS, Arendsen P, Morris A, Spieck M (2007) State-of-the-Art and future trends in multidisciplinary design optimization. In: 48th AIAA/ASME/ASCE/AHS/ASC structures, structural dynamics, and materials conference < br > 15th. Honolulu, Hawaii. AIAA 2007-1905

Wei L, Anxi C, Peng G, Weicheng C (2008) BLH based MDO frame for deepsea HOV general design. Ship Mech 12(1):110–117. (刘蔚, 操安喜,苟鹏,崔维成. 基于BLH框架的大深度载人潜水器总体性能的多学科设计优化. 船舶力学,2008, 第12卷第1期, pp.110-117.)

Wei L (2007) Application of MDO method in the design of 7000 m rating HOV design. PhD thesis of Shanghai Jiaotong University. (刘蔚. 多学科设计优化方法在7000米载人潜水器总体设计中的应用. 上海交通大学博士学位论文, 2007.)

Weicheng C, Zhen H, Cong Y, Binbin P (2011) Current status and future trends for deep manned submersibles. J Central South Univer (Science and Technology) 42(Suppl. 2):13–20. (崔维成, 胡震, 叶聪, 潘彬彬 (2011) 深海载人潜水器技术的发展现状与趋势, 中南大学学报 (自然科学版), 第42卷增刊 2, 13-20)

Yao W, Chen XQ, Luo WC, Tooren MV, Guo J (2011) Review of uncertainty-based multidisciplinary design optimization methods for aerospace vehicles. Prog Aerosp Sci 47:450–479

Yokota K, Murata T (1986–1987) Research on pressure hull for deep submergence research vehicle made of titanium alloy. Tech Rev Mitsubish Heavy Ind Ltd 23(4):440–2

Youn BD, Choi KK (2001) Hybrid analysis method for reliability-based design optimization. J Mech Des ASME 125(2):221–232 and Proceedings of 2001 ASME design engineering technical conferences: 27th design automation conference, Number DETC/DAC-21044. Pittsburgh, PA

Zhengyuan L, Jianmin G (1992) Four elements method. J Nanjing Univer Sci Technol (Nature Science) 6(016). (刘正元, 顾建民. 系统四要素[J]. 南京理工大学学报 (自然科学版), 1992, 6: 016.)

iSIGHT9.0 help—Quality Engineering Methods. Engineous Software

The manufacturer's authorised representative in the EU is Springer Nature Customer Service Centre GmbH, Europaplatz 3, 69115 Heidelberg, Germany. If you have any concerns regarding our products, please contact ProductSafety@springernature.com

Printed and bound by CPI Group (UK) Ltd, Croydon, CR0 4YY
30/09/2026
02225559-0012